# Word Processing in Groups

This book was written under the auspices of the University of Minnesota Geometry Center (NSF Science and Technology Research Center for the Computation and Visualization of Geometric Structures).

# Word Processing in Groups

David B. A. Epstein, *University of Warwick*

*with*

James W. Cannon, *Brigham Young University*
Derek F. Holt, *University of Warwick*
Silvio V. F. Levy, *University of Minnesota*
Michael S. Paterson, *University of Warwick*
William P. Thurston, *University of California, Berkeley*

CRC Press is an imprint of the
Taylor & Francis Group, an **informa** business

CRC Press
Taylor & Francis Group
6000 Broken Sound Parkway NW, Suite 300
Boca Raton, FL 33487-2742



**Visit the Taylor & Francis Web site at**
**http://www.taylorandfrancis.com**

**and the CRC Press Web site at**
**http://www.crcpress.com**

# Contents

Preface ix

I An Introduction to Automatic Groups 1

1 Finite State Automata, Regular Languages and Predicate Calculus 3
1.1 Languages and Regular Languages . . . . . . . . . . . . . . 4
1.2 Finite State Automata . . . . . . . . . . . . . . . . . . . . 7
1.3 Simply Starred Languages . . . . . . . . . . . . . . . . . . 18
1.4 Predicate Calculus and Regular Languages . . . . . . . . . 22

2 Automatic Groups 27
2.1 Groups As Languages . . . . . . . . . . . . . . . . . . . . . 28
2.2 Cayley Graphs and Isoperimetric Inequalities . . . . . . . . 33
2.3 Automatic Groups: Definition . . . . . . . . . . . . . . . . . 45
2.4 Invariance under Change of Generators . . . . . . . . . . . 52
2.5 Improving the Automatic Structure . . . . . . . . . . . . . 56

3 Quasigeodesics, Pseudoisometries and Combings 63
3.1 Metric Spaces, Path Metrics and Geodesics . . . . . . . . . 64
3.2 Shortest Strings . . . . . . . . . . . . . . . . . . . . . . . . 66
3.3 Pseudoisometries . . . . . . . . . . . . . . . . . . . . . . . 71
3.4 Applications to Fundamental Groups . . . . . . . . . . . . 76
3.5 Counterexample to the Use of Shortest Strings . . . . . . . 80
3.6 Combable Groups . . . . . . . . . . . . . . . . . . . . . . . 83

4 Abelian and Euclidean Groups 87
4.1 General Results . . . . . . . . . . . . . . . . . . . . . . . . 87
4.2 Euclidean Groups are Biautomatic . . . . . . . . . . . . . . 91
4.3 Abelian Groups and ShortLex . . . . . . . . . . . . . . . . . 96
4.4 A Euclidean Counterexample . . . . . . . . . . . . . . . . . 97

5 Finding the Automatic Structure: Theory 101
5.1 Axiom Checking . . . . . . . . . . . . . . . . . . . . . . . . . 102
5.2 A Naive Algorithm . . . . . . . . . . . . . . . . . . . . . . . . 108

6 Finding the Automatic Structure: Practical Methods 113
6.1 Semigroups and Specialized Axioms . . . . . . . . . . . . . . . 114
6.2 The Knuth–Bendix Procedure . . . . . . . . . . . . . . . . . . 116
6.3 Knuth–Bendix and Word Differences . . . . . . . . . . . . . . . 126

7 Asynchronous Automatic Groups 135
7.1 Asynchronous Automata . . . . . . . . . . . . . . . . . . . . . 136
7.2 Asynchronous Automatic Groups: Definition . . . . . . . . . . . 139
7.3 Properties of Asynchronous Automatic Groups . . . . . . . . . . 149
7.4 Asynchronous but not synchronous . . . . . . . . . . . . . . . . 154

8 Nilpotent Groups 161
8.1 The Heisenberg Group . . . . . . . . . . . . . . . . . . . . . . 161
8.2 Nilpotent Groups Are Not Automatic . . . . . . . . . . . . . . . 167
8.3 Regular Subgroups and Nilpotency . . . . . . . . . . . . . . . . 173

II Topics in the Theory of Automatic Groups 179

9 Braid Groups 181
9.1 The Braid Group and the Symmetric Group . . . . . . . . . . . 182
9.2 Canonical Forms . . . . . . . . . . . . . . . . . . . . . . . . . 190
9.3 The Braid Group Is Automatic . . . . . . . . . . . . . . . . . . 195
9.4 The Conjugacy Problem . . . . . . . . . . . . . . . . . . . . . 201
9.5 Complexity Issues . . . . . . . . . . . . . . . . . . . . . . . . . 204

10 Higher-Dimensional Isoperimetric Inequalities 211
10.1 Cell Complexes and Lipschitz Maps . . . . . . . . . . . . . . . 211
10.2 Estimates for Cell Complexes . . . . . . . . . . . . . . . . . . 214
10.3 Combable Groups and Riemannian Manifolds . . . . . . . . . . 221
10.4 The Special Linear Groups . . . . . . . . . . . . . . . . . . . . 230

11 Geometrically Finite Groups 243
11.1 Groupoids . . . . . . . . . . . . . . . . . . . . . . . . . . . . . 243
11.2 Generators of Differing Lengths . . . . . . . . . . . . . . . . . 249
11.3 Geodesics and Horoballs . . . . . . . . . . . . . . . . . . . . . 253
11.4 Geometrically Finite Groups of Hyperbolic Isometries . . . . . 266

**12 Three-Dimensional Manifolds** **273**
12.1 Taking the Problem to Pieces . . . . . . . . . . . . . . . . . . . . 273
12.2 The Basic Seifert Fibre Space . . . . . . . . . . . . . . . . . . . . 282
12.3 Fitting Pieces Together along the Boundaries . . . . . . . . . . 297
12.4 The Automatic Structure on a Three-Manifold . . . . . . . . . . 302

**Bibliography** **315**

**Index** **321**

# Preface

Connections between the theory of hyperbolic manifolds and the theory of automata are deeply interwoven in the history of mathematics of this century. The use of symbol sequences to study dynamical systems originates in the work of Koebe [Koe27, Koe29] and Morse [Mor87], who both used symbol sequences to code geodesics on a surface of constant negative curvature. Ergodic theorists have been motivated by the consideration of geodesic flows on hyperbolic manifolds, amongst other things, to consider shifts of finite type. The concept of a "sophic system" is simply the ergodic theorist's specialized name for what is known in other branches of science as a finite state automaton. In [BS79], Bowen and Series describe Markov partitions related to the action of certain groups of hyperbolic isometries on the boundary space.

Another fundamental contribution was made by Max Dehn [Deh87]. He was the first person to point out the importance of the word problem in group theory. His solution in the case of fundamental group of a surface is very much in the spirit of what we are doing, namely a geometric approach to group theory. He was aware that geodesics in the Cayley graph of such a group follow certain rules (formalized in this book by the use of a finite state automaton), and that rules also characterize pairs of geodesics ending at the same point. His work was confined to the use of generators with geometric significance.

In 1978 Thurston and Cannon had a conversation at the International Congress of Mathematicians in Helsinki about growth functions of groups, in which Thurston conjectured that the growth function of a hyperbolic group is rational. Cannon had already worked out cone types (page 66) for a number of examples of group presentations, and he later discovered that his computations of cone types could easily be used to find growth functions and show that they were rational. Further investigations by Cannon [Can84] showed once again the connections between recursive processes and hyperbolic phenomena. Thurston observed that Cannon's results had an appropriate expression in terms of finite state automata in two variables.

In 1985, Epstein was developing a computer program for making group invariant drawings in the hyperbolic plane, and battling with the complexities of floating point inaccuracy and the gross inefficiencies of a naive approach to enumerating group elements up to a certain length. Around the same time, Al Marden was beginning to formulate his plans for a research project involving geometry and computing, centred at the University of Minnesota in Minneapolis. During the discussion of these plans and the preparation of an associated grant proposal, Epstein learned from Thurston and Cannon an outline of what was to become known as the theory of automatic groups.

On returning to Warwick, Epstein explained these ideas to Holt and Paterson, and the three of them started to work out in detail many basic aspects of the theory. In 1986 Holt started to produce computer programs to find the automata associated to an automatic group. The original programming was based on the theory underlying Cannon's proofs. However, this method seems to lead to an exponential explosion in complexity and very few examples could be handled successfully in this way.

Another important strand was the work of Knuth and Bendix. Several authors had considered "confluent groups", in which the word problem could be solved by using the Knuth–Bendix procedure. The experimental work of Bob Gilman [Gil84b] was specially important to us in this regard; it led Holt to make the Knuth–Bendix procedure the basis for our computer programs. Other significant contributions by Gilman are [Gil87], [Gil84a] and [Gil79]. Reading Gilman's papers, one clearly discerns the outlines of the theory of automatic groups in the experimental results he obtained, although his prophetic insight lacked theoretical explanation at the time he wrote.

The Knuth–Bendix approach turned out to be quite effective, and it is a central element of our current suite of programs. These programs were written by Epstein, Holt and Sarah Rees.

In 1987 Epstein started to write a paper describing the results so far achieved. Cannon, Epstein, Holt, Thurston and Paterson were originally to be the joint authors of this research paper, as they had each made vital contributions to the theory. The paper expanded steadily as additional results were discovered, and was widely distributed in preprint form, in many different versions. Comments have been received from many readers, and have been incorporated into the work. They have not been recorded systematically, and, in any case, are too numerous to mention, but we are grateful for the improvements suggested.

Eventually the paper expanded to such an extent that the only reasonable way to publish the work was as a book. Most of the original labour of writing the book was undertaken by Epstein, though substantial pieces were also contributed by Holt and by Thurston. Since then the work has been thoroughly edited, revised and rewritten by Levy, who also made original contributions. The LaTeX macros we use are also due to Levy. We hope that their use has made our work easier and more pleasant to read and understand—in particular, they have made it easy to include many figures and diagrams.

We are glad to make the following acknowledgements. Uri Zwick has made extensive comments and his suggestions have improved the exposition and removed many typos. John Sullivan has made even more extensive comments and has rewritten and recast several of the proofs in better form. Most of the pictures were produced by Epstein using Adobe's *Illustrator* program. They were postprocessed by a program written by Nathaniel Thurston which, together with Levy's macros, causes the labels from the *Illustrator* file to be typeset by LaTeX. In particular this enables subscripts and superscripts to be placed correctly and uniformly. Other pictures were produced using Mathematica. Figures 2.9 and 2.10 were produced using *gm*, a graphics hyperbolic geometry program written by Epstein and Steve Rumsby.

Part I of this book is suitable for use as a graduate level introduction to the theory of automatic groups; we have gone to some trouble to keep the amount of prerequisite material in this part of the book down to a minimum. Part II gives an account of research by Epstein and Thurston where a more general background in mathematics is assumed; this is suitable for more advanced graduate students.

As mentioned above, the whole project had its seeds in the preparation of the grant application for the Geometry Supercomputer Project, directed by Al Marden. Subsequently this project and its successor, the NSF Geometry Center, provided Epstein with the conditions in Minneapolis which enabled the daunting task of writing the book to be completed. Financial support provided by the NSF and the SERC have been of crucial importance. The SERC's Computational Science Initiative has been particularly helpful, and the SERC's Mathematics Committee has also lent support. The SERC and NSF have provided the computers on which the book was written and the programming done, and financed visits which allowed the authors to discuss their ideas face to face, resulting in much more rapid progress. Finally, the University of Warwick and its Mathematics Department provided the congenial environment in which most of the book writing was carried out, and a year of leave (1990–91) for Epstein to complete the work.

# Part I

# An Introduction to Automatic Groups

# Chapter 1

# Finite State Automata, Regular Languages and Predicate Calculus

The concept of a finite state automaton has emerged as significant in many branches of human knowledge and understanding, including linguistics, computer science, philosophy, biology and mathematics, particularly in logic. For example, early attempts to make mathematical models of neural nets used the concept. Many of the remarkable capabilities of modern word processors are the result of clever algorithms which manipulate or use finite state automata. Because of their efficiency in certain types of parsing, finite state automata are usually built into compilers of high-level computer languages.

In our work, finite state automata are of fundamental importance: our objective is to use them to understand individual groups. We will end up with fast algorithms for doing computations on words in the generators of certain groups.

A standard theorem (Theorem 1.4.6) states that it is possible to apply the operators of first-order predicate calculus to finite state automata, obtaining other finite state automata. For instance, there are algorithms which take as their input a finite state automaton and give as their output another finite state automaton which is the result of applying the "there exists" operator or the "not" operator. Of course, these statements need some explanation. At this point, we would merely like the reader to note that the possibility of applying first-order predicate calculus in an algorithmic fashion gives us the power to automatically verify whether a given finite state automaton satisfies certain axioms.

Some of the material in this chapter is standard in many undergraduate computer science courses. Section 1.3 (Simply Starred Languages) is new and is due to Paterson.

## 1.1. Languages and Regular Languages

By an *alphabet* $A$ we mean nothing more than a finite set. However, the word alphabet helps to fix the context in which the finite set is being used. An element of $A$ is called a *letter*. A *string* over the alphabet $A$ is a finite sequence of letters, that is, an integer $n \geq 0$ and a mapping $\{1, \ldots, n\} \to A$. If $n = 0$, the domain is the nullset and there is a unique such mapping, called the *nullstring* and generally denoted $\varepsilon$. A string is considered to contain, as an additional piece of information, the alphabet $A$—mention of this is normally suppressed. However, we sometimes denote the nullstring over $A$ by $\varepsilon_A$, which is useful either to distinguish nullstrings over different alphabets or to distinguish the nullstring from the symbol $\varepsilon$, which may already be an element of the alphabet—see, for example, Definition 1.1.3.

The set of all strings over the alphabet $A$ is denoted $A^*$.

It is usual to write a string by simply listing the successive values of the mapping (though one should note the confusion created by this casual approach when $n = 0$). Thus, if $A$ is the alphabet of lowercase letters, "automaton" is a string over $A$ with $n = 9$. If $\omega$ is a string $\{1, \ldots, n\} \to A$, we call $n$ the *length* of $\omega$, and we denote it by $|\omega|$.

Given two strings $\omega : \{1, \ldots, n\} \to A$ and $\tau : \{1, \ldots, m\} \to A$, the *concatenation* $\omega\tau$ of $\omega$ and $\tau$ is defined to be the string $\{1, \ldots, m+n\} \to A$ given by $(\omega\tau)(i) = \omega(i)$ if $1 \leq i \leq n$ and $(\omega\tau)(i) = \tau(i-n)$ if $n+1 \leq i \leq n+m$.

With the operation of concatenation, the set $A^*$ of strings over $A$ forms a monoid, with identity element $\varepsilon$. (A *monoid* is a set with an associative multiplication and an identity. A *semigroup* is a set with an associative multiplication. All semigroups considered in this book will be monoids, so we will use the words semigroup and monoid interchangeably.) In fact, $A^*$ is the free monoid or semigroup on the set of generators $A$.

Suppose that we have $w = puq$, for some (possibly null) strings $w$, $p$, $q$ and $u$ over $A$. We say that $p$ is a *prefix* of $w$, that $q$ is a *suffix* of $w$, and that $u$ is a *substring* of $w$. If $t \geq 0$ is an integer, we denote by $w(t)$ the prefix of $w$ of length $t$, or else $w$ itself if $t$ is greater than the length of $w$.

**Definition 1.1.1 (language).** A *language* over $A$ is a subset of $A^*$, together with the alphabet $A$. Mention of the alphabet $A$ is frequently suppressed. Nevertheless, if we are being rigorous, we must distinguish the null language over the alphabet $\{x\}$ from the null language over the alphabet $\{x, y\}$.

This use of "language" is derived from the usual sense of the word, as applied to "natural languages" such as French, Chinese, etc. Are natural languages languages in the mathematical sense? Not really. Natural languages change with time, which is not part of Definition 1.1.1. One can argue whether a sentence by James Joyce is English or not.

We don't even know how to specify the alphabet for a natural language. The continuum of sounds of a spoken language can be broken down into phonemes, but experts can disagree on where the boundary lies. One must also take into account stress, intonation, pauses... The situation for a written language is not much clearer, even in the case of a highly standardized language such as English. We certainly need all the lowercase and uppercase letters, the space and the standard punctuation marks. But is # a symbol in the alphabet? What about the conventional indentation at the beginning of the paragraph? Moreover, we don't know what strings we should take as the elements of the language. Should we take a sentence, a paragraph, or a whole book?

Despite these difficulties, Definition 1.1.1 provides a reasonable framework for mathematical models of natural languages. As with all mathematical models of natural phenomena, such models cannot be accurate in all respects; nevertheless, there has been considerable cross-fertilization between linguistics and computer science.

Computer scientists and linguists have developed a vocabulary and mathematical techniques and theorems for dealing with a variety of types of languages. Mostly, if a language is to be useful, there has to be some way of understanding it. A fairly weak requirement is that there be some mechanical way of deciding whether a string is in the language or not. For example, there may be a machine capable of examining strings one letter at a time from left to right, and giving the answer Yes if and only if the string is in the language. In this case we say that the machine *recognizes* or *accepts* the language. In general, the machine is not required to give any answer if the string is not in the language, but in this book nearly all the machines considered will also answer No if the string is not in the language.

We can impose restrictions of various types on the machine; in this book we will work mostly with finite state automata (see Section 1.2). A fundamental theorem (Theorem 1.2.7) says that a language is accepted by a finite state automaton if and only if the totality its strings can be expressed using a *regular expression*, a natural concept that we will soon define. Regular expressions are familiar to almost all computer users; most operating systems allow users to refer to files by means of "wildcards," so that, for example, `chap*.tex` might refer to files called `chap1.tex`, `chap2.tex`, `chapABC.tex` and `chap#%^&.tex`. The Unix operating system goes even further: many of its standard utilities, such as the file searcher *egrep*, the stream editor *sed* and the screen editor *vi*, use regular expressions in an essential way.

**Definition 1.1.2 (concatenation of languages).** If $K$ and $L$ are languages over the same alphabet $A$, we define their *concatenation* $KL$ to be the set of strings $w$ for which $w = w_1 w_2$ in $A^*$, where $w_1 \in K$ and $w_2 \in L$. If $K$ or $L$ is empty, so is $KL$. We define the *star closure* of $K$ as

$$K^* = \bigcup_{n \geq 0} K^n,$$

where $K^0 = \{\varepsilon\}$ and $K^n = K^{n-1}K$ for $n > 0$. In particular, if $K = \varnothing$, then $K^* = \{\varepsilon\}$, a language with one element. If $K = A$, we have two definitions of $A^*$, and these give the same result. The star closure is sometimes called the *Kleene closure*.

**Definition 1.1.3 (regular expression).** A regular expression over an alphabet $A$ is a particular type of string (specified below) over the alphabet $E$ formed by adjoining to $A$ the following five symbols, which are assumed not to lie in $A$ already: (, ), $*$, $\vee$ and $\varepsilon$. We pronounce $\vee$ as "or" and $*$ as "star". Informally, parentheses are used for grouping, $*$ denotes repetition, $\vee$ is used to combine alternative patterns, and $\varepsilon$ matches the nullstring $\varepsilon_A$ over $A$. (Notice, however, that in the regular expression itself $\varepsilon$ is not the nullstring, but a letter in the extended alphabet $E$.)

A single regular expression $r$ defines a language $L(r)$ over $A$. Normally, the real object of interest is the language, rather than the regular expression itself. We will give an inductive definition of a regular expression, together with an inductive definition of the language defined. Any language defined by a regular expression is called a *regular language*.

The induction starts with the following "primitive" regular expressions: $\varepsilon$, which gives the language $L(\varepsilon) = \{\varepsilon_A\}$; $a$, for $a \in A$, which gives the language $L(a) = \{a\}$; and $()$, which gives the language $L(()) = \varnothing$. The languages $L(\varepsilon)$ and $L(a)$ have exactly one element each, and $L(())$ has no elements.

If $r$ is a regular expression, $(r)$ is a regular expression, and $L((r)) = L(r)$. If $r$ is a regular expression, $(r^*)$ is a regular expression defining the language $L((r^*)) = (L(r))^*$. If $r = x \in A$, we get $L((x^*))$, which is equal to $\{\varepsilon_A, x, x^2, \ldots\}$. If $r = ()$, we get $L((()^*))$, which is equal to $\{\varepsilon_A\}$; this can also be written $L(\varepsilon)$. If $r_1$ and $r_2$ are regular expressions, $(r_1 \vee r_2)$ is a regular expression defining the language $L((r_1 \vee r_2)) = L(r_1) \cup L(r_2)$, and $(r_1 r_2)$ is a regular expression defining the language $L((r_1 r_2)) = L(r_1)L(r_2)$. All regular expressions are formed in this way. If $r$ is a regular expression and $w \in L(r)$, we say that $w$ *matches* $r$.

As in arithmetic, it is usual to omit parentheses from a regular expression where possible. Since union and concatenation of languages are associative, many parentheses can be omitted, and even more so if we adopt the convention that "$*$" binds more tightly than concatenation, and that concatenation binds more tightly than "$\vee$". Only when it is important to be completely rigorous will we insist on the presence of parentheses.

The following result is obvious, since concatenation of regular expressions is equivalent to concatenation of the corresponding regular languages:

**Proposition 1.1.4 (concatenation semigroup for languages).** *The set of languages over $A$ forms a semigroup under concatenation. The identity element is the language whose only element is the nullstring over $A$. The set of regular languages over $A$ forms a subsemigroup, and the set of finite languages form a subsemigroup of that.* [1.1.4]

## 1.2. Finite State Automata

It is usual to classify languages according to the type of machine capable of recognizing them. A *finite state automaton* is a particularly simple type of machine, and it turns out that a language is regular if and only if it is recognized by some finite state automaton.

**Definition 1.2.1 (finite state automaton).** A *finite state automaton* (or simply *automaton*) is a quintuple $(S, A, \mu, Y, s_0)$, where $S$ is a finite set, called the *state set*, $A$ is a finite set, called the *alphabet*, $\mu : S \times A \to S$ is a function, called the *transition function*, $Y$ is a (possibly empty) subset of $S$ called the subset of *accept states*, and $s_0 \in S$ is called the *start state* or the *initial state*. We sometimes talk of a finite state automaton over $A$. We often suppress $\mu$ and write $sx$ instead of $\mu(s, x)$.

Accept states are also called *success states* or *final states* in the literature. The name $Y$ for the set of accept states stands for Yes. The idea is that the machine starts in the start state $s_0$, and reads a tape on which a string over $A$ is printed. The tape is read one letter at a time. After reading a letter, the state of the machine is changed in accordance with the existing state, with the letter read and with the transition function $\mu$. Then the tape is moved one letter to the left (i.e., the tapehead is moved one letter to the right on the tape). If, after consuming all the input, the state of the machine is in $Y$, then the machine answers Yes; otherwise it answers No. More formally, we have the following definition:

**Definition 1.2.2 (recognized).** Let $M = (S, A, \mu, Y, s_0)$ be a finite state automaton. Let $\mathsf{Map}(S, S)$ be the semigroup consisting of all maps from $S$ to $S$. The map $A \to \mathsf{Map}(S, S)$ given by $\mu$ can be extended in a unique way to a semigroup homomorphism $A^* \to \mathsf{Map}(S, S)$. We say that a string $w$ is *recognized*, or *accepted*, by $M$ if the corresponding element of $\mathsf{Map}(S, S)$ takes $s_0$ to an element of $Y$. The language of strings recognized by $M$ is called the language recognized by $M$, and is denoted by $L(M)$.

We often represent a finite state automaton as a directed graph, with a node for each state and an arrow for each transition (that is, each pair of state and letter). Figure 1.1 shows additional conventions used in drawing finite state automata, and Figures 1.2 and 1.3 show two examples of use.

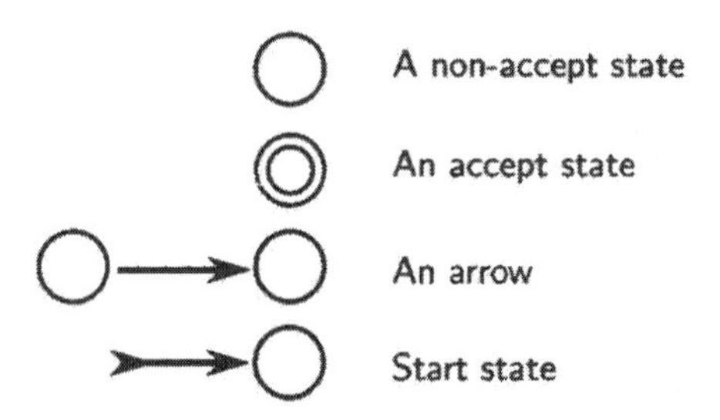


**Figure 1.1. Depicting finite state automata.** Here are the conventions we use for depicting finite state automata. Several different notations are current in the literature.

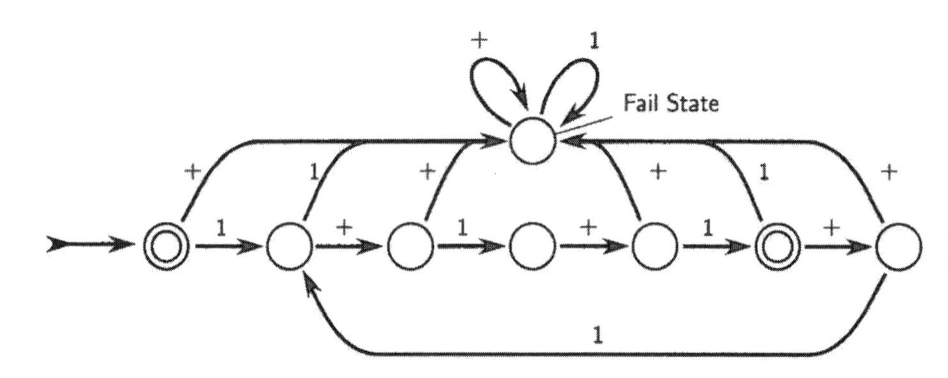


**Figure 1.2. Plus one mod 3.** Here is a deterministic finite state automaton over an alphabet consisting of two symbols, "1" and "+". The empty string is interpreted as representing zero and is accepted. Apart from this, strings of the form $1+1+\cdots+1$ are accepted, provided the string represents 0 modulo 3; all other strings are rejected.

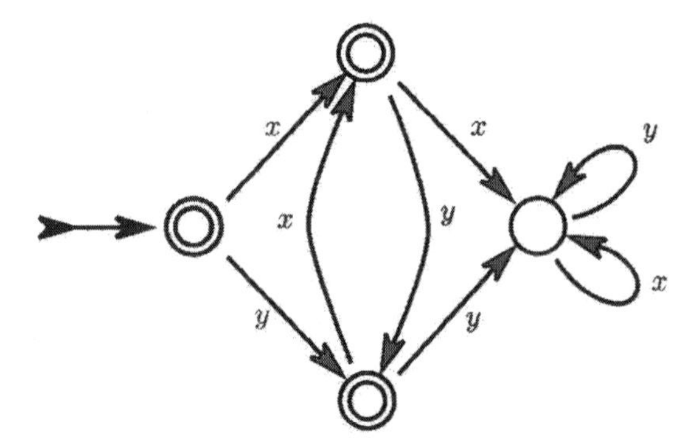


**Figure 1.3. Free product of two groups of order two.** Let $x$ and $y$ be generators of two groups $G_x$ and $G_y$ of order two. Here is a finite state automaton that accepts exactly those strings over $\{x, y\}$ that are reduced in the free product $G_x * G_y$ (that is, that don't have two $x$'s or two $y$'s in a row).

Finite state automata have important applications in word processing and compilers. Their speed as a computational device is primarily due to the fact that the transition function can be realized as an array or lookup table.

Given a finite state automaton, we can sometimes simplify it without changing the language it recognizes. First, we can remove all the automaton's *inaccessible states*, those that cannot be reached from the start state. Next, we say that a state is *live* or *dead* depending on whether some accept state can be reached from it. (Notice that a state may be live, but inaccessible; sometimes an automaton has a simpler description if we do not remove such live states.) We may amalgamate all of an automaton's dead states into a single *failure state*, that is, a state that has transitions to no states other than itself and is not an accept state.

**Convention 1.2.3 (normalized automaton).** Unless we state otherwise, we agree that, when constructing a finite state automaton, we are to remove inaccessible states and combine dead states into a single failure state. An automaton to which this procedure has been applied is called *normalized*.

There are variants of the definition of a finite state automaton, which we will now give. (The type of machine just described is sometimes called a *deterministic* finite state automaton, to distinguish it from the variants; but when we say just "finite state automaton" we will always mean a deterministic one.) In each case we are given a finite set $S$, called the *state set*. An *arrow* is a triple $(s_1, x, s_2)$, where $s_1$ and $s_2$ are elements of $S$ and $x$ is an element of some other set and is called the *label* of the arrow. The *source* of the arrow is $s_1$ and its *target* is $s_2$. An arrow labelled $x$ is sometimes called an *$x$-transition*.

**Definition 1.2.4 (non-deterministic finite state automaton).** A *non-deterministic finite state automaton* is a quintuple $(S, A, \mu, Y, S_0)$, where $A$ is a finite set, called the *alphabet*, $S_0$ is a subset of $S$, called the subset of *initial states*, $Y$ is subset of $S$, called the subset of *accept states*, and $\mu$ is a set of arrows with labels in the enlarged alphabet $A \cup \{\varepsilon\}$ (the symbol $\varepsilon$ is assumed not to lie in $A$, and it is meant to evoke the nullstring).

Figure 1.4 shows an example of a non-deterministic automaton. Clearly, a deterministic finite state automaton can be considered a special case of a non-deterministic finite state automaton, for which there following conditions are satisfied: there are no arrows labelled $\varepsilon$; each state is the source of exactly one arrow with any given label from $A$; and $S_0$ has exactly one element. The language over $A$ assigned to a non-deterministic automaton is the set of elements of $A^*$ obtainable from the labels on a path of arrows from an element of $S_0$ to an element of $Y$. Details, if desired, are given in Definition 1.2.6 (generalized finite state automaton).

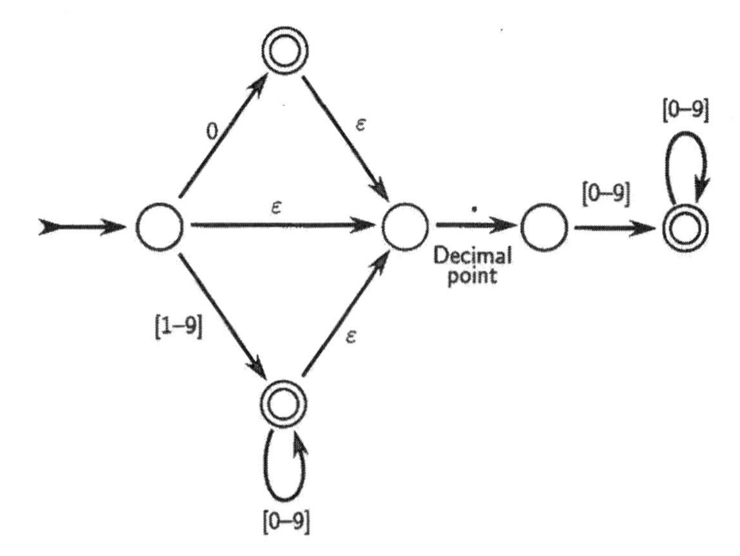


**Figure 1.4. Decimal numbers.** A non-deterministic automaton which recognizes decimal numbers. There are eleven symbols or letters of the alphabet, namely the ten digits plus the decimal point. The only non-determinism is the use of $\varepsilon$ arrows.

The next subtype of automaton is morally equivalent to a deterministic automaton, but it is convenient to be able to pass from one description to the other at will.

**Definition 1.2.5 (partial automaton).** A *partial deterministic automaton* is a non-deterministic automaton with no $\varepsilon$-transitions, exactly one initial state, no dead states, and at most one $x$-transition from any state.

We can recover a deterministic automation from a partial deterministic automaton by adding a failure state and making all arrows which are not already defined have the failure state as target. This does not affect the language accepted. Conversely, given a (normalized) deterministic automaton, we obtain a partial deterministic automaton accepting the same language by by dropping the failure state and all arrows leading to it (see Figure 1.5).

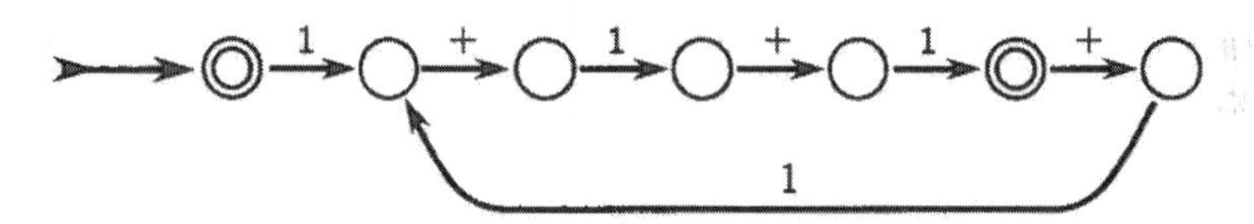


**Figure 1.5. Omitting the failure state.** This is the same automaton as in Figure 1.2, but with the failure state deleted.

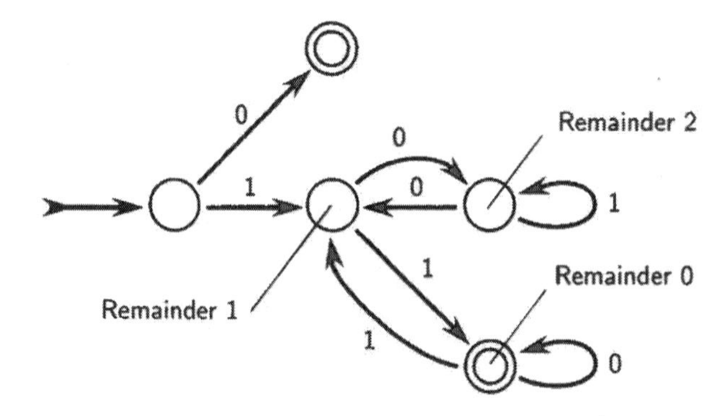


**Figure 1.6. Binary numbers mod 3.** Here we see an automaton that takes as input a string of zeroes and ones. It accepts the input if and only if the string is a legal representation of a binary number divisible by three. The failure state has been omitted. Note that the string "00" is not a legal binary number and is not accepted, but "0" is accepted. The labels on the states represent the remainder modulo 3 of the input so far.

It is sometimes useful to combine the definition of a finite state automaton with that of a regular expression.

**Definition 1.2.6 (generalized finite state automaton).** A *generalized finite state automaton* is the same as a non-deterministic finite state automaton, except that each arrow is labelled by a regular expression over $A$. Given a (possibly generalized) non-deterministic finite state automaton $M$, we will now define the language $L(M)$ *recognized*, or *accepted*, by $M$.

A *path of arrows* in $M$ is a sequence

$$(s_1, u_1, s_2, \ldots, u_n, s_{n+1}),$$

where $n \geq 0$ is the *length* of the path and each $u_i$ is an arrow with source $s_i$ and target $s_{i+1}$, for $1 \leq i \leq n$. We call $s_1$ the *source* and $s_{n+1}$ the *target* of the path of arrows. Let $w_i$ be the label of $u_i$; thus $w_i$ is $\varepsilon$ or a letter of $A$ in the case of a non-deterministic automaton, and a regular expression over $A$ in the case of a generalized automaton. We can regard $w_i$ as a regular expression in the first case as well, by giving $\varepsilon$ and the letters of $A$ the interpretation specified in Definition 1.1.3. Let $w$ be the concatenation $w_1 \ldots w_n$; if $n = 0$, we set $w = \varepsilon$. Then $w$ is called the *label* of the path, and it is a regular expression, with associated regular language $L(w)$. We define $L(M) \subset A^*$ as the union of the languages $L(w)$, where $w$ ranges over all labels of paths of arrows with source in $S_0$ and target in $Y$.

By considering a path of arrows of length zero, we see that if any element of $S_0$ lies in $Y$, the empty string is in $L(M)$.

We extend Convention 1.2.3 (normalized automaton) to non-deterministic and generalized automata in the obvious way.

It is perhaps surprising that deterministic and non-deterministic automata define the same class of languages:

**Theorem 1.2.7 (Kleene, Rabin, Scott).** *Let $A$ be a finite alphabet. The following four conditions on a language over $A$ are equivalent:*

(1) *The language is recognized by a deterministic finite state automaton.*

(2) *The language is recognized by a non-deterministic finite state automaton.*

(3) *The language is recognized by a generalized finite state automaton.*

(4) *The language is defined by a regular expression.*

*Proof of 1.2.7:* Clearly (1) implies (2) and (2) implies (3).

Next we show that (3) implies (4). We are given a generalized finite state automaton with a set $S$ of states, a set of arrows each labelled with a regular expression over $A$, a set $S_0$ of start states and a set $Y$ of accept states.

We say that a path of arrows $(s_1, u_1, \ldots, u_n, s_{n+1})$ *visits* the states $s_2, \ldots, s_n$. (Thus a path may or may not visit its own source and target.) For each pair of states $(s_1, s_2)$ and each subset $X$ of $S$, we define $L(s_1, s_2, X)$ to be the union of the languages defined by the set of paths with source $s_1$, target $s_2$ and visiting only states in $X$. We prove by induction on the size of $X$ that this language is defined by a regular expression.

First let $X = \varnothing$. There are a finite number of arrows going directly from $s_1$ to $s_2$. Let their labels be $r_1, \ldots, r_k$, where each $r_i$ is a regular expression. Then $L(s_1, s_2, X)$ is equal to $L(r_1 \vee \ldots \vee r_k)$ if $k > 0$, is empty if $k = 0$ and $s_1 \neq s_2$, and is equal to $\{\varepsilon\}$ if $k = 0$ and $s_1 = s_2$. Therefore $L(s_1, s_2, X)$ is defined by a regular expression.

Otherwise we choose some $s \in X$, and define $L_1 = L(s_1, s_2, X \backslash \{s\})$, $L_2 = L(s_1, s, X \backslash \{s\})$, $L_3 = L(s, s, X \backslash \{s\})$ and $L_4 = L(s, s_2, X \backslash \{s\})$. Then

$$L(s_1, s_2, X) = L_1 \cup (L_2 L_3^* L_4).$$

By induction, each of the terms on the right of the equation is defined by a regular expression. The same is therefore true for $L(s_1, s_2, X)$.

The proof that (3) implies (4) is completed by taking $X = S$, and then taking the union as $s_1$ varies over $S_0$ and $s_2$ varies over $Y$.

To see that (4) implies (2), we construct, by induction on the length of a regular expression, a non-deterministic finite state automaton accepting the language defined by the regular expression. The automaton will have exactly one start state and one accept state, which we can assume are distinct. If the regular expression is (), the corresponding language is the empty language, and the automaton has no arrows and no states beyond the start and accept states. If the regular expression consists of a single symbol $x \in A \cup \{\varepsilon\}$, the automaton has a single arrow labelled $x$, going from the start state to the accept state; there are no other states.

If $r$ is a regular expression corresponding to machine $M$, a machine corresponding to $r^*$ is obtained by creating a new $\varepsilon$-arrow from the accept state of $M$ to its start state, as shown in Figure 1.7. Concatenation corresponds to connecting the machines in series, with an $\varepsilon$-transition in between, as shown in Figure 1.8. Finally, to construct an automaton that accepts the union of two regular languages, we take the corresponding automata, create a new start state and a new accept state, and use $\varepsilon$-arrows to go from the new start state to the old ones, and from the old accept states to the new ones. This is illustrated in Figure 1.9.

Finally we show that (2) implies (1). Let $S$ be the set of states of our non-deterministic finite state automaton. If $X$ is a subset of $S$, we define $\overline{X}$, the $\varepsilon$-closure of $X$, to be the set of all possible targets of paths of arrows

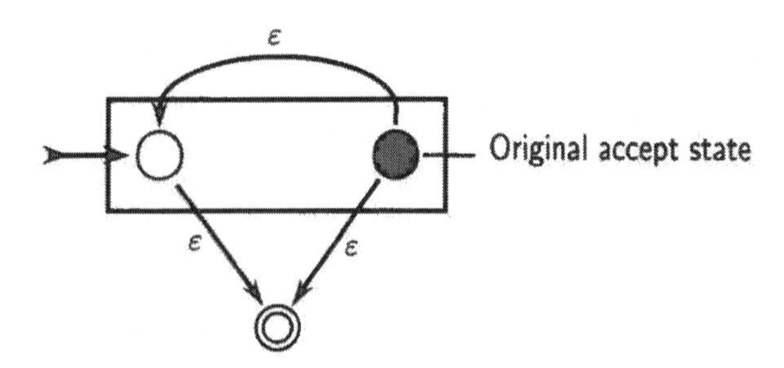


**Figure 1.7. Star closure of a language.** Given a non-deterministic automaton with one start state and one accept state, accepting a language $L$, we can form a new automaton, also with one start state and one accept state, which accepts $L^*$.

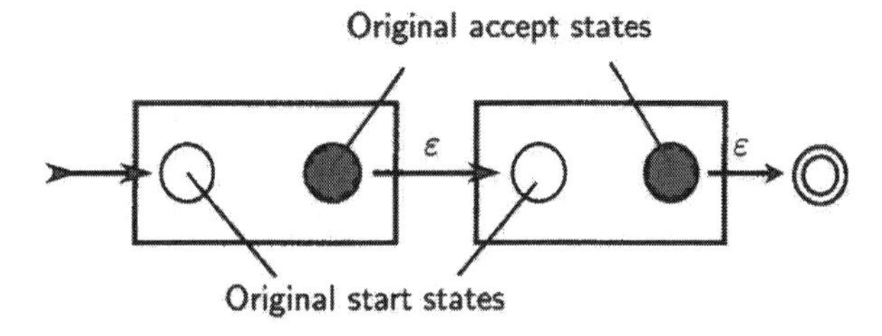


**Figure 1.8. Concatenation of two languages.** Given non-deterministic automata accepting languages $L_1$ and $L_2$, each with one start state and one accept state, we form a non-deterministic automaton also with one start state and one accept state, which accepts the concatenation $L_1L_2$.

all labelled $\varepsilon$ and such that the source of the path is in $X$. By considering paths of length zero, we see that $X \subset \overline{X}$; in addition, $\overline{\overline{X}} = \overline{X}$. We define a deterministic finite state automaton as follows: The new set of states is the set of all $\varepsilon$-closed subsets of $S$. If $X$ is a new state and $x \in A$, we define $Xx$ to be the $\varepsilon$-closure of the set of all targets of arrows labelled with $x$ and with source in $X$. The new start state is the $\varepsilon$-closure of the original start state, and the new accept states are those containing an old accept state. It is easy to check that the deterministic automaton defined in this way accepts the same language as the original automaton. 1.2.7

Notice that in the last part of the proof we replaced a non-deterministic automaton with $n$ states by a deterministic one with possibly as many as $2^n$ states. This exponential blowup is not due to lack of ingenuity, but is, in fact, unavoidable: see Figure 1.10.

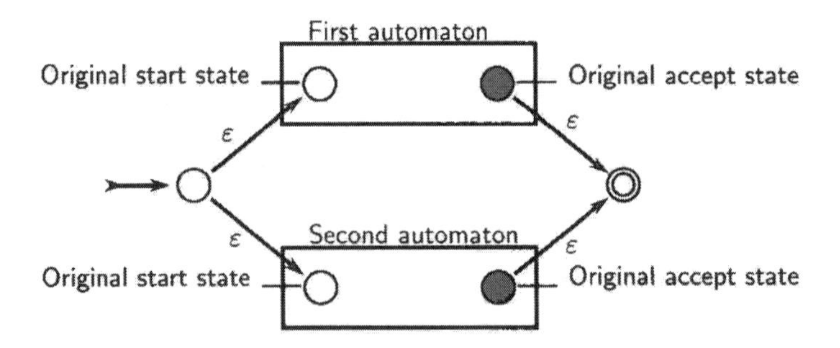


**Figure 1.9. Union of two languages.** Given two non-deterministic automata accepting languages $L_1$ and $L_2$, each with one start state and one accept state, we can put them together to form a new non-deterministic automaton, also with one start state and one accept state, which accepts $L_1 \cup L_2$.

We conclude this section with several classical results about regular languages and finite state automata, many (but not all) of which we will need later.

**Theorem 1.2.8 (reversal).** *If $L$ is a regular language over an alphabet $A$, the language consisting of the strings of $L$ written in the reverse order is also regular.*

*Proof of 1.2.8:* Take a non-deterministic automaton accepting $L$, reverse all the arrows, and interchange the set of start states with the set of accept states. [1.2.8]

**Theorem 1.2.9 (Myhill–Nerode).** *Given a language $L$ over an alphabet $A$, consider the equivalence relation on $A^*$ defined as follows: two strings $w_1$ and $w_2$ are equivalent if, for each string $u$ over $A$, $w_1u \in L$ if and only if $w_2u \in L$. Then $L$ is regular if and only if there are only finitely many equivalence classes.*

*Proof of 1.2.9:* If the set of equivalence classes is finite, we define an automaton with one state for each equivalence class. If $x \in A$ and if $[w]$ is the equivalence class of $w$ in $A^*$, we define $[w]x = [wx]$. We define the start state to be $[\varepsilon]$ and a state $[w]$ to be an accept state if and only if $w \in L$.

Conversely, suppose the language is regular. Then it is given by some deterministic finite state automaton, which we may assume has no inaccessible states. We assign to each state $s$ of the automaton an equivalence class by choosing a string $w$ which sends the start state to $s$, and then taking the equivalence class of $w$. Clearly, the class is independent of the choice of $w$. The map from states to equivalence classes is surjective, so the number of equivalence classes is finite. [1.2.9]

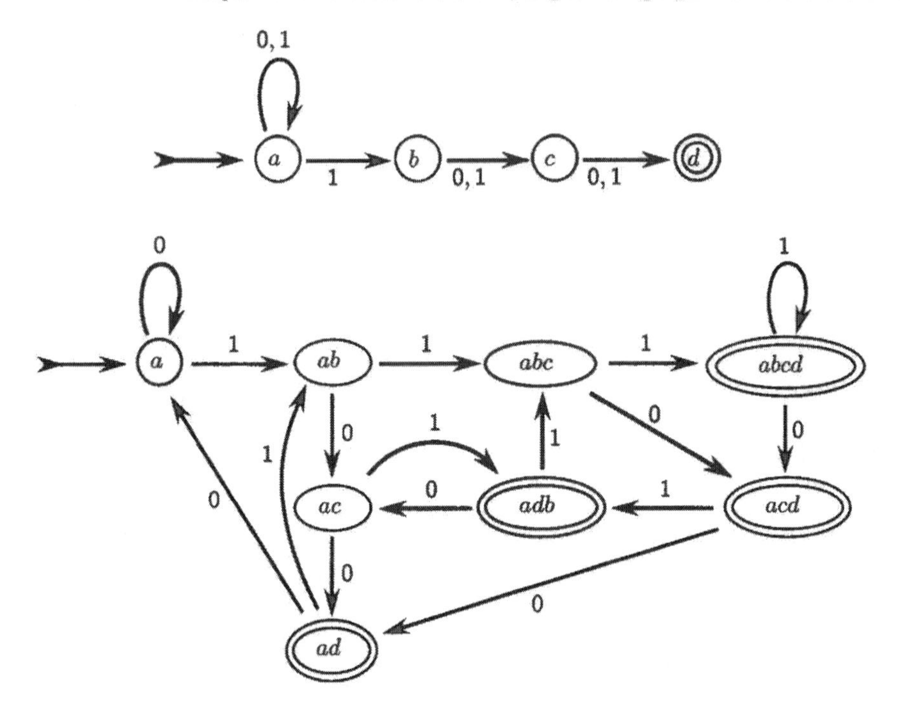


**Figure 1.10. Exponential blowup of states.** The upper part of the diagram shows a non-deterministic automaton which checks whether a string of zeroes and ones has a 1 in the position third from the end. The lower part of the diagram shows the deterministic version of the automaton. A non-deterministic automaton similar to this, checking whether the 1 occurs in position $n$ from the end, uses $n+1$ states; the smallest deterministic machine that does the same job has $2^n$ states.

This proof shows that there is a unique minimal deterministic finite state automaton accepting a given regular language.

**Definition 1.2.10 (prefix closure).** If $L$ is a language over an alphabet $A$, the *prefix closure* of $L$ is the set of all prefixes of strings in $L$. A *prefix-closed* language is one that equals its prefix closure.

**Theorem 1.2.11 (prefixes).** *Let $L$ be a regular language. Then the prefix closure of $L$ and the largest prefix-closed sublanguage of $L$ are regular languages.*

*Proof of 1.2.11:* Let $W$ be a deterministic automaton accepting $L$, which we can assume is normalized (see Convention 1.2.3). If we turn every non-failure state of $W$ into an accept state, the language accepted by the new automaton is the prefix closure of $L$. If, instead, we remove from $W$ all non-accept states and all arrows to and from such states, we get an automaton that accepts the largest prefix-closed subset of $L$. (This procedure may leave us without an initial state, if the initial state was not an accept state. In this case the nullstring is not in $L$, so $L$ has no nonempty prefix-closed subsets.) 1.2.11

Let $A$ and $B$ be alphabets, and let $f : A \to B$ be a map. If $L_A$ is a language over $A$, the *image* of $L_A$ under $f$ is the image of $L_A$ under the unique semigroup homomorphism $A^* \to B^*$ extending $f$. If $L_B$ is a language over $B$, the *inverse image* of $L_B$ under $f$ is the inverse image of $L_B$ under the same homomorphism. The definition remains the same if $f$ is a map $f : A \to B^*$.

More generally, if $f$ is a map from $A$ to the set of regular expressions over $B$, we say that $f$ is a *substitution*. In this case we define the image $f(L_A)$ as follows: we extend $f$ to a semigroup homomorphism from $A^*$ to the semigroup of regular languages over $B$ (see Proposition 1.1.4). We then set $f(L_A) = \bigcup_{u \in L_A} f(u)$. Notice that in this case there is no analogue for the inverse image.

**Lemma 1.2.12 (effect of map).** *Let $L_A$ be a regular language over $A$ and let $L_B$ be a regular language over $B$. For any $f : A \to B$ or $f : A \to B^*$, the image $f(L_A)$ and the inverse image $f^{-1}(L_B)$ are regular languages over $B$ and $A$, respectively. For any substitution $f$, the image $f(L_A)$ is a regular language over $B$.*

*Proof of 1.2.12:* Let $M_A$ be a deterministic automaton over $A$ accepting $L_A$. We replace every arrow labelled $a$ in $M_A$ by an arrow labelled $f(a)$. This gives a (possibly non-deterministic or generalized) new automaton over $B$, whose accepted language is clearly $f(L_A)$.

Let $M_B$ be a finite state automaton over $B$ accepting $L_B$. We define a finite state automaton over $A$ with the same states as $M_B$. Given a state $s$ and $a \in A$, we define $sa = sf(a)$. (This does not work if $f$ is a substitution.) Clearly the new automaton accepts $f^{-1}(L_B)$. 1.2.12

The next result is often used in the theory of automata to show that a language is not regular. Also, the technique used in its proof is fruitful in similar contexts (see Lemma 2.3.9, for example).

**Theorem 1.2.13 (pumping lemma).** *Let $L$ be a regular language. Then there is a number $n > 0$ such that any $x \in L$ of length at least $n$ is of the form $x = uvw$, where $|v| > 0$ and $uv^iw \in L$ for all $i \geq 0$.*

*Proof of 1.2.13:* Let $n$ be the number of states in a finite state automaton accepting $L$. If an accepted string $x$ has length at least $n$, it must visit some state twice, and therefore it has a substring $v$ whose corresponding path of arrows forms a loop. Clearly this loop can be traced any number of times. [1.2.13]

The following result is included for general reference, but is not used in this book except in the trivial case where $L'$ has just one element. Its proof is left to the reader.

**Theorem 1.2.14.** *Let $L$ and $L'$ be languages over $A$. If $L$ is regular, so are the languages $\{u \mid (\exists v \in L')(uv \in L)\}$ and $\{u \mid (\exists v \in L')(vu \in L)\}$.* [1.2.14]

This is only really useful when $L'$ is regular, because in this case automata accepting the derived languages can be constructed from the automata accepting $L$ and $L'$.

## 1.3. Simply Starred Languages

In this section, whose results are mainly due to Paterson, we investigate the sizes of regular languages. John Sullivan has greatly improved the exposition and the pattern of proof.

Only regular expressions involving the Kleene star represent infinite languages. In fact, there is a further distinction in size between applying star-closure to languages containing a single string and applying it to those with distinct strings of the same length. Our characterization of languages with polynomial growth as being simply starred will be important when we prove Theorem 8.2.8.

**Definition 1.3.1 (language size).** The *growth function* of a language $L$ over $A$ is a function $g_L : \mathbf{N} \to \mathbf{N}$ which counts the elements of $L$ of different lengths: $g_L(n) = \#(L \cap A^n)$. Sometimes is it most convenient to encode this information (which is really a sequence of natural numbers) as a formal power series $f_L$, which (despite being formal) is called the *generating function* for $g_L$:

$$f_L(t) = \sum_{n=0}^{\infty} g_L(n) t^n.$$

We speak of one growth function or generating function as being bounded by another if the bound holds for each $n$. If $L = A^*$, we have $g_L(n) = |A|^n$, so $|A|^n$ is an upper bound for every language. We say that $L$ has *polynomial growth* if its growth function is bounded (above) by some polynomial function of $n$, and *exponential growth* if its growth function is bounded below by an exponential function of the form $\lambda^n$, with $\lambda > 1$. We will see that every regular language has either polynomial or exponential growth. (Confusingly, we say that $L$ has *rational growth* if its generating function is the power series for a rational function of $t$; this is a comment not on the size of $L$ but on the simplicity of its growth function. Every regular language has rational growth, but we will not prove this standard result.)

**Lemma 1.3.2 (polynomial growth concatenation).** *Let $f_1$ and $f_2$ be the generating functions of two languages $L_1$ and $L_2$. The generating function for $L_1 \cup L_2$ is bounded above by $f_1 + f_2$, and below by $\max(f_1, f_2)$. The generating function for $L_1L_2$ is bounded above by $f_1f_2$, and, assuming $L_1$ contains a string of length $k$, is bounded below by $t^k f_2$. Thus the operations of union and concatenation preserve the class of languages with polynomial growth.*

*Proof of 1.3.2:* The bounds follow from the definitions: any string of length $n$ in $L_1 \cup L_2$ is in $L_1$ or in $L_2$; similarly any string of length $n$ in $L_1L_2$ can be written as $u_1u_2$ with $u_1 \in L_1$, $u_2 \in L_2$ and $|u_1| + |u_2| = n$, and this matches the definition of multiplication for power series. These bounds fail to be sharp when strings have several such expressions. If the growth functions for $L_1$ and $L_2$ are bounded by polynomials of degree $n_1$ and $n_2$, it is easy to check from these formulas that the growth functions for $L_1 \cup L_2$ and $L_1L_2$ are bounded by polynomials of degree $\sup(n_1, n_2)$ and $n_1 + n_2 + 1$, respectively. [1.3.2]

The lower bounds in this lemma show that a union or concatenation of two languages has exponential growth if at least one of the starting languages does (except in the case of concatenation with the empty language, which results in the empty language). Since $L \subset L^*$, star closures also do not reduce the size of a language.

**Definition 1.3.3 (simply starred).** A regular expression over $A$ in which star closure is applied only to languages with a single element (that is, to strings over $A$) is called *simply starred*. We also say that the regular language represented by such an expression is simply starred.

**Lemma 1.3.4 (one-letter automata).** *Any regular language over a one-letter alphabet, or, more generally, any regular language consisting entirely of powers of a single string, is simply starred.*

*Proof of 1.3.4:* Any finite language $L$ can be described by a regular expression without Kleene stars, for example in the form $w_1 \vee w_2 \vee \cdots \vee w_k$, where the $w_i$ are the elements of $L$. Any language consisting of powers of a single string $w$ is the image of a language $L$ over $A = \{x\}$ under the map $x \mapsto w$, and thus is simply starred if $L$ is, as we can see by applying the same map to a regular expression for $L$.

The behaviour of a deterministic finite state automaton over the alphabet $\{x\}$ is given by the single transition function from the finite state set to itself. This behaviour is therefore ultimately periodic—see Figure 1.11. If $n$ is its period, the language $L$ accepted by the automaton is the union $L_1 \cup (x^n)^* L_2$ for some finite languages $L_1$ and $L_2$, which can each be written in the standard form without Kleene stars as noted above. Thus $L$ is simply starred. 1.3.4

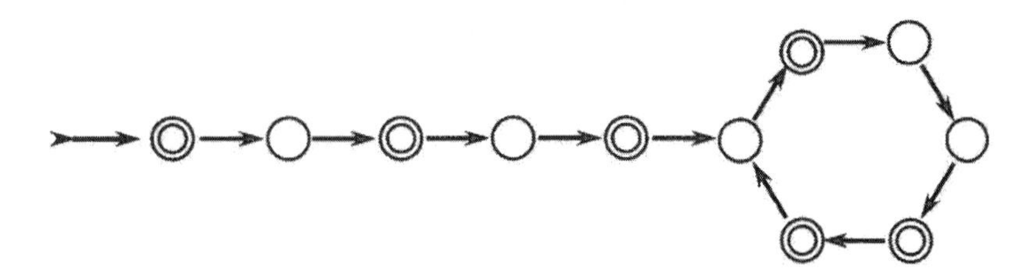

**Figure 1.11. One-letter automaton.** Every automaton over an alphabet with exactly one letter $x$ looks something like this (though there may not be a loop at the end). The language $L$ accepted by such an automaton is of the form $L_1 \cup (x^n)^* L_2$, for $L_1$ and $L_2$ finite; here $L_1 = \{\varepsilon, x^2, x^4\}$, $L_2 = \{x^6, x^9, x^{10}\}$ and $n = 6$.

Note that a language $L$ may be both simply starred and star-closed (meaning that $L = L^*$), although the two properties cannot be reflected in a single regular expression. For example, $(x^3 \vee x^4)^*$ can also be written as $(x^3 \vee x^4) \vee x^*(x^6)$, but not as $(w)^*$ for any string $w$.

Given two strings $u$ and $v$ over $A$, we can view them as defining a map from the language $\{u, v\}^*$ to $A^*$. This map is most often injective: distinct words in $\{u, v\}$ are still distinct when the strings $u$ and $v$ are substituted. The next lemma, due to John Sullivan, shows that the only exception is when $u$ and $v$ are powers of some common substring.

**Lemma 1.3.5 (string powers).** *Let $D$ be an alphabet of two letters $u$ and $v$, and let $f : D \to A^*$ be a map. The induced homomorphism $f^* : D^* \to A^*$ is not injective if and only if $f(u)$ and $f(v)$ are powers of a single string $z \in A^*$.*

*Proof of 1.3.5:* If $f(u) = z^r$ and $f(v) = z^s$, then $f^*(u^s) = f^*(v^r)$, so $f^*$ is not injective.

Suppose conversely $f^*$ is not injective. We will prove, by induction on $|f(u)| + |f(v)|$ that they are both powers of some $z \in A^*$. Let $w_1$ and $w_2$ be distinct strings over $D$ whose images in $A^*$ are equal. We may remove any common prefixes, and assume that $w_1$ starts with $u$ and $w_2$ starts with $v$. We may also assume that $|f(u)| \leq |f(v)|$. Then $f(v) = f(u)v'$ for some $v' \in A^*$. If $v' = \varepsilon$, then $f(u) = f(v) = z$ and the induction is proved. Otherwise, the substitution $v = uv'$ in $w_1$ and $w_2$ gives us words $w_1'$ and $w_2'$ in the formal variables $u$ and $v'$. Moreover $w_1'$ and $w_2'$ are distinct, since $w_2'$ starts with $uv'$ and $w_1'$ starts with $u$, but not with $uv'$. By induction, $f(u)$ and $v'$ are powers of some string $z$; therefore the same is true of $f(v)$. 1.3.5

This lemma has two special cases that we will find most useful. First, using the words $uv$ and $vu$, we see that $u$ and $v$ commute in the concatenation semigroup if and only if they are powers of a common substring. Second, using $u^p$ and $v^q$, we see that a word $w = u^p = v^q$ can be written as a power of a substring $z$ in a unique maximal way.

**Lemma 1.3.6 (polynomial star).** *If a language $L = L^*$ is star-closed, $L$ has polynomial growth if it has at most one string of any given length, and exponential growth otherwise.*

*Proof of 1.3.6:* If $L$ has two distinct strings $s$ and $t$ of the same length $n$, then $L = L^* \supset (s \vee t)^*$ contains at least $2^k$ strings of length $kn$, for each $k \in \mathbf{N}$, so $L$ has exponential growth. Conversely, a language with at most one string of any given length has polynomial growth by definition. 1.3.6

**Lemma 1.3.7 (distinct strings in star).** *For any language $L$, its star closure $L^*$ has at most one string of any given length if and only if $L$ (or equivalently $L^*$) consists entirely of powers of a single string.*

*Proof of 1.3.7:* Suppose first that there are two distinct strings of the same length in $L$ (or $L^*$). Then they cannot both be powers of a single string. Therefore $L$ cannot consist entirely of powers of a single string.

Conversely, suppose there is at most one string of any given length in $L^*$. Let $u \neq \varepsilon$ and $v$ be distinct strings in $L$. Then $uv = vu$ and Lemma 1.3.5 (string powers) shows that $u$ and $v$ are powers of a common string $z$; we may assume that $z$ is not itself a proper power. (We are not claiming that $z \in L^*$.) Fixing $u$ and varying $v$, Lemma 1.3.5 (string powers) shows that every string in $L$ is a power of $z$. 1.3.7

**Proposition 1.3.8 (polynomial growth condition).** *A regular language $L$ has polynomial growth if it is simply starred, and exponential growth otherwise.*

*Proof of 1.3.8:* If $L$ is star closed, this follows from Lemmas 1.3.4, 1.3.6 and 1.3.7.

If $L$ is any simply starred language, it can by definition be constructed by union and concatenation from finite languages and from languages which are star closed and simply starred, so Lemma 1.3.2 shows that it also has polynomial growth.

Conversely, if $L$ is not simply starred, any regular expression for $L$ by definition involves some $E^*$ such that $L(E^*)$ is a star closed language, which is also not simply starred. By Lemma 1.3.4, $L(E^*)$ does not consist of powers of a single string. By Lemma 1.3.7, there are distinct strings $u$ and $v$ of the same length, matching $E^*$. By Lemma 1.3.6, the growth of $L(E^*)$ is exponential. Since $L$ is not empty, the note after Lemma 1.3.2 shows that the further operations producing $L$ from $E^*$ do not reduce the size of the language below exponential growth. 1.3.8

## 1.4. Predicate Calculus and Regular Languages

In this section we will define the operators of first-order predicate calculus on the class of regular languages, and show that, if the languages are presented in terms of finite state automata, the operators can be realized constructively as operations on the automata.

Let $A$ be an alphabet. A *predicate* $P$ over $A$ is a *boolean valued function* on $A^*$, that is, a function that takes only the values T (for "true") and F (for "false"). There is a one-to-one correspondence between predicates and languages over $A$, in which a predicate $P$ corresponds to the language $P^{-1}(\mathtt{T})$. A *regular predicate* is a predicate whose corresponding language is regular. Examples of regular predicates on a variable $w$ are:

- "$w$ is the binary representation of an integer modulo 3" (see Figure 1.6).
- "$w$ consists of digits in non-decreasing order."
- "$w$ is a legal identifier in C."

The truth of a regular predicate, when values are specified for the variables, can be checked by a finite state automaton. The third example above is a typical example of the use of finite state automata in practical use in compiler programs which convert a program written in a high-level computer language into machine code.

The operators of first-order predicate calculus are $\neg$, $\wedge$, $\vee$, $\exists$ and $\forall$, pronounced "not", "and", "or", "there exists" and "for all". By thinking of a language as a predicate on a variable, the string, we can extend these operators to languages. Thus, given languages $L$ and $L'$ over $A$, we define $\neg L$ as $A^* \setminus L$, $L \wedge L'$ as $L \cap L'$, and $L \vee L'$ as $L \cup L'$. (We will generally not use the notations $L \wedge L'$ and $L \vee L'$, however.)

**Lemma 1.4.1 (NOT, AND and OR).** *Let $A$ be an alphabet, and let $L$ and $L'$ be regular languages over $A$. Then $\neg L$, $L \cup L'$ and $L \cup L'$ are regular languages over $A$.*

*Proof of 1.4.1:* Let $M$ be a finite state automaton accepting $L$ (see Theorem 1.2.7 (Kleene, Rabin, Scott)). An automaton accepting $\neg L$ is obtained by interchanging accept and non-accept states of $M$. That $L \cup L'$ is regular follows immediately from Definition 1.1.3 (regular expression). For $L \cup L'$ we use de Morgan's law, writing $L \cup L' = \neg(\neg L \cap \neg L')$. [1.4.1]

A common situation in practice is when we have two automata accepting languages $L$ and $L'$, and want an automaton accepting the intersection or union of $L$ and $L'$. Although in theory we could find the solution by using Lemma 1.4.1 and Theorem 1.2.7 (Kleene, Rabin, Scott) (for which we gave a constructive proof), in practice there are much better ways. If $S$ and $S'$ are the state sets of the two automata, we construct an automaton with state set $S \times S'$ and transitions defined by $(s, s')a = (sa, s'a)$ for $a \in A$, $s \in S$ and $s' \in S'$. The initial state is the pair of initial states of the original automata. An intersection automaton is obtained by defining as accept states all states $(s, s')$ such that $s \in S$ and $s' \in S'$ are accept states; a union automaton has as accept states all states $(s, s')$ such that $s$ or $s'$ is an accept state.

To find a counterpart for $\exists$ and $\forall$, we need to extend Definition 1.1.1 (language) to include *many-variable languages*, which correspond to predicates on $n$ variables, instead of one. This extension could conceivably cause confusion, but we will avoid that by using the term *one-variable language* when we wish to stress that we are using Definition 1.1.1.

**Definition 1.4.2 (many-variable language).** Let $A_1, \ldots, A_n$ be alphabets. By a *language over* $(A_1, \ldots, A_n)$ we mean a set of $n$-tuples of strings $(w_1, \ldots, w_n)$, where $w_i \in A_i^*$, together with the $n$-tuple of alphabets $(A_1, \ldots, A_n)$. A language over an $n$-tuple of alphabets is called an *$n$-variable language*. A *predicate over* $(A_1, \ldots, A_n)$ is a boolean valued function on $A_1^* \times \ldots \times A_n^*$; predicates over $(A_1, \ldots, A_n)$ are in one-to-one correspondence with languages over $(A_1, \ldots, A_n)$.

In the special case $n = 0$, a language is a set of zero-tuples of strings. Since there is exactly one zero-tuple of anything (the cartesian product of no sets has cardinality one), there are two languages over $()$: the empty language, and the language $\{()\}$. The corresponding predicates are denoted simply by F and T.

To define the remaining operators of predicate calculus, $\exists$ and $\forall$, we quantify over the last variable: $\exists(L)$ is the language over $(A_1, \ldots, A_{n-1})$

consisting of $(a_1, \ldots, a_{n-1})$ such that $(a_1, \ldots, a_{n-1}, a_n) \in L$ for some $a_n \in A_n$, and similarly for $\forall(L)$.

Sometimes we will want to think of $n$-tuples of strings as strings of $n$-tuples. To account for the case when the strings have different lengths, we make some technical definitions:

**Definition 1.4.3 (padding).** Let $A_1, \ldots, A_n$ be alphabets. We adjoin to each $A_i$ an end-of-string or *padding symbol*, denoted by $\$_i$, which is assumed not to lie in $A_i$, and we define $B_i = A_i \cup \{\$_i\}$. The *padded alphabet* associated with $(A_1, \ldots, A_n)$ is the set

$$B = B_1 \times \ldots \times B_n \setminus \{(\$_1, \ldots, \$_n)\}.$$

A *padded string* over $B$ is a string $w = (w_1, \ldots, w_n)$, such that, once an end-of-string symbol occurs in one of the component strings $w_i$, all subsequent letters of that string are end-of-string symbols. A *padded language* is a language over $B$ all of whose elements are padded strings. Given any language $G$ over $B$, the *padded restriction* of $G$ is the set of the padded strings in $G$; clearly, a language is padded if and only it equals its padded restriction.

**Definition 1.4.4 (padded extension).** Given a language $L$ over $(A_1, \ldots, A_n)$, we define a one-variable language $L^\$$ over the padded alphabet $B$ associated with $(A_1, \ldots, A_n)$, as follows: For each $n$-tuple $(w_1, \ldots, w_n) \in L$, let $m$ be the maximal length of the $w_i$, for $1 \leq i \leq n$. We pad each $w_i$ with $\$_i$'s at the end, so as to make its length $m$. The resulting $n$-tuple of strings is a string of length $m$ in the alphabet $B$; these are the strings in $L^\$$. We call $L^\$$ the *padded extension* of $L$.

For example, the padded string corresponding to the pair (hello, goodbye) is

$$(\mathrm{h}, \mathrm{g})(\mathrm{e}, \mathrm{o})(\mathrm{l}, \mathrm{o})(\mathrm{l}, \mathrm{d})(\mathrm{o}, \mathrm{b})(\$, \mathrm{y})(\$, \mathrm{e}).$$

For any language $L$ over $(A_1, \ldots, A_n)$, the padded extension $L^\$$ is a padded language over the corresponding padded alphabet $B$. $L$ and $L^\$$ determine each other uniquely.

**Definition 1.4.5 (regular many-variable language).** We say that $L$ is a *regular language* over $(A_1, \ldots, A_n)$ if $L^\$$ is a regular language over the padded alphabet $B$ associated with $(A_1, \ldots, A_n)$. A finite state automaton over $B$ accepting the language $L^\$$ is said to be an *$n$-variable automaton* over $(A_1, \ldots, A_n)$ accepting $L$.

If $B$ and $C$ are padded alphabets, the image of a padded language $L_B$ under a map $f : B \to C$ (see page 17) is not a padded language. In order to make it into one, we can take its padded restriction, that is, its intersection with the language of all padded strings over $C$; the result is called the *padded image* of $L_B$, and is denoted by $f_*(L_B)$. If $L_B$ is a regular language, so is $f_*(L_B)$; this follows immediately from Lemma 1.2.12 (effect of map), Lemma 1.4.1 (NOT, AND and OR) and the fact that the language of all padded strings over $C$ is regular. Similarly, the *padded inverse image* of a regular language $L_C$ over $C$ is a regular language over $B$.

**Theorem 1.4.6 (predicate calculus).** *Let $L$ and $L'$ be regular languages over $(A_1, \ldots, A_n)$.*

(1) *The languages $\neg L$, $L \cap L'$ and $L \cup L'$ are regular languages over $(A_1, \ldots, A_n)$.*

(2) *The languages $\exists(L)$ and $\forall(L)$ are regular languages over $(A_1, \ldots, A_{n-1})$.*

(3) *For any alphabet $A_{n+1}$, the language*

$$\{(w_1, \ldots, w_n, w_{n+1}) \mid (w_1, \ldots, w_n) \in L\}$$

*is a regular language over $(A_1, \ldots, A_{n+1})$.*

(4) *For any permutation $\sigma$ of $\{1, \ldots, n\}$, the language*

$$L_\sigma = \{(w_1, \ldots, w_n) \mid (w_{\sigma(1)}, \ldots, w_{\sigma(n)}) \in L\}$$

*is a regular language over $(A_1, \ldots, A_{n+1})$.*

The remark following the proof of Lemma 1.4.1 (NOT, AND and OR) applies equally well here: although we give a constructive proof for this theorem, one would, in practice, use more sophisticated constructions to do predicate calculus on automata.

*Proof of 1.4.6:* Part (1) follows from the one-variable case (Lemma 1.4.1), by taking intersections with the language of all padded strings.

The $\exists$ operator is just $f_*$, where $f : A_1 \times \ldots \times A_n \to A_1 \times \ldots \times A_{n-1}$ is the obvious projection. The construction for $\forall$ follows from this, since $\forall$ is equal to the composition $\neg \circ \exists \circ \neg$. This proves (2).

Parts (3) and (4) follow immediately from the definition of a finite state automaton. 1.4.6

**Corollary 1.4.7 (predicates closed).** *The class of regular predicates is closed under the operators $\neg$, $\wedge$, $\vee$, $\exists$, $\forall$, $\Rightarrow$, $\Leftrightarrow$.* 1.4.7

It is obvious, but worth pointing out, that part (4) of Theorem 1.4.6 allows us to shuffle the order of the variables (and in particular to quantify over any variable, not just the last), and that part (3) allows us to consider variables that do not figure explicitly in a predicate. For instance, suppose we want to show that the language

$$\{(w_1, w_2, w_3) \mid w_3 < w_2 \land (w_2, w_3, w_1) \in L\}$$

is regular, knowing that $L$ is regular and that the relation $<$ can be checked by a finite-state automaton ($<$ might be lexicographical order, say). We can write this language as

$$\{(w_1, w_2, w_3) \mid w_3 < w_2\} \cap \{(w_1, w_2, w_3) \mid (w_2, w_3, w_1) \in L\},$$

where $w_1$ is "missing" from the first predicate $w_3 < w_2$ and the variables are "out of order" in the second. We apply parts (4) and (3) of Theorem 1.4.6 to the first operand of $\cap$ and part (4) to the second, and conclude with an application of part (1).

One consequence of Theorem 1.4.6 (predicate calculus) is that the truth or falsity of a fully quantified predicate defined by means of regular languages can be checked by an algorithmic process. Members of the Geometry Group at Warwick have written computer programs to carry out this process. One of the difficulties which arises is the exponential increase in the size of the automata, as the various operators of predicate calculus are applied. However, when dealing with particular types of predicates, more efficient methods may be available, or else the given automata may be sufficiently small that the exponential increase in size does not create insuperable problems.

# Chapter 2

# Automatic Groups

In this chapter we show how the idea of a regular language can be relevant in group theory. We define the central notion of this book, that of an automatic group (Definition 2.3.1), and find a characterization for such groups (Theorem 2.3.5). Roughly speaking, an automatic group is a finitely generated group for which one can check, by means of a finite state automaton, whether two words in a given presentation represent the same element or not, and whether or not the elements they represent differ by right multiplication by a single generator. The totality of these data—the generators and the automata—constitute an automatic structure for the group.

An automatic structure determines the group uniquely, and we should regard it as a particularly good presentation of the group. Presentations of groups by generators and relators are notoriously difficult to handle; in contrast, presentations by means of finite state automata possess feasible algorithms for solving many problems. For example, there is a quadratic algorithm which has as its input a string in the generators and their inverses, and as its output a unique "normal form" for the corresponding group element—see Theorem 2.3.10. Such algorithms can be used to search systematically for homomorphisms or isomorphisms between groups; programs developed by Holt and Sarah Rees have had a certain amount of success in this regard [HR00]. The finite state automata can also be used to enumerate the group elements, listing a unique word for each group element. The nice thing here is that the first $n$ words can be listed in time $O(n \log n)$, with the constant depending on the group and its automatic structure. This enables one to do fast computations of objects invariant under an automatic group, notably limit sets of Kleinian groups.

Before introducing automatic groups, we define the weaker notion of a regularly generated group, and prove that the word problem can be solved for such groups (Theorem 2.1.9). In Section 2.2 we cover some standard material on combinatorial group theory, including the notion of the Cayley graph of a group, which is used throughout the book, and isoperimetric inequalities.

The definition of an automatic group is Thurston's formulation of the results of Cannon about cocompact groups of hyperbolic isometries [Can84]. The details were worked out jointly by the authors.

In Section 2.4 we show that the property of a group being automatic does not depend on the choice of generators. This is in contrast with other properties of groups found in the combinatorial group theory literature, where the choice of a presentation is crucial.

There are a number of interesting variants on the concept of an automatic structure. In Chapter 7 we will discuss asynchronous automatic groups, which have less structure than automatic groups. In Section 2.5, on the other hand, we consider automatic structures that satisfy various additional conditions. Some of these conditions, such as uniqueness or prefix closedness, can always be achieved after a suitable modification to the structure: see Theorems 2.5.1 and 2.5.9. Others, like representation by geodesics or invariance of the language of accepted words under inversion, cannot. In this case the extra structure often allows us to prove additional results.

## 2.1. Groups As Languages

We consider the relationship between groups and languages over an alphabet of generators. Although the material down to Theorem 2.1.9 is elementary, we present it carefully because is includes some tricky points (see in particular Definition 2.1.1 and the subsequent paragraph).

Let $G$ be a group and $A \subset G$ a finite set of elements of $G$. By interpreting concatenation as multiplication in $G$, we define a semigroup homomorphism $\pi : A^* \to G$. If $w$ is a string over $A$, we say that $\pi(w)$ is the element of $G$ *represented* by $w$; we also denote this element by $\overline{w}$. If the semigroup homomorphism $\pi : A^* \to G$ obtained in this way is surjective, that is, if every element of $G$ is represented by a string over $A$, we say that $A$ is a set of *semigroup generators* for $G$, or that *$A$ generates $G$ as a semigroup*. By contrast, *$A$ generates $G$ as a group* if $A \cup A^{-1}$ generates $G$ as a semigroup. For example, the set $\{1\}$ generates the additive group of integers $\mathbf{Z}$ as a group, but not as a semigroup, while $\{3, -4\}$ does generate $\mathbf{Z}$ as a semigroup.

For our purposes, it is important to allow a set of generators to have repeated elements, that is, to allow different symbols in the alphabet to represent the same element of $G$. Formally, we have the following definition:

**Definition 2.1.1 (generators).** Let $G$ be a group, $A$ an alphabet and $p : A \to G$ a map, which need not be injective. We extend $p$ to a semigroup homomorphism $\pi : A^* \to G$. If this homomorphism is surjective, that is, if $\pi(A)$ generates $G$ as a semigroup, we say that $A$ is a *set of semigroup generators* for $G$. If $\pi(A)$ generates $G$ as a group, we say that $A$ is a *set of group generators* for $G$. If $A$ has a total order, we talk about an *ordered set* of semigroup or group generators. If $x \in A$ and $g \in G$, we generally write $gx$ for $g\,p(x)$, $x^{-1}g$ for $(p(x))^{-1}g$, and so on. If we extend $A$ by adjoining an end-of-string symbol \$, we also extend $p$ so it takes \$ to the identity element in $G$.

Notice that we normally suppress mention of the map $p : A \to G$ in this definition; a "set" of semigroup or group generators is really a set together with a map. If $p$ is injective, of course, we can identify $A$ with $p(A) \subset G$, and we have a set in the traditional sense. When mathematicians talk about group or semigroup generators, they usually mean the setup described in Definition 2.1.1, even when they don't realize it.

Given a language $L$ over $A$, we also use $\pi$ to denote the restriction of $\pi : A^* \to G$ to $L$. The most interesting case is when $\pi : L \to G$ is surjective, that is, when every group element is represented by at least one string in $L$. We think of the elements of the language as providing a kind of normal, or standard, form for the elements of the group. In the best situation the normal form is unique, that is, $\pi : L \to G$ is one-to-one; we then say that $L$ has the *uniqueness property*. But we will also work with languages that do not have this property.

**Convention 2.1.2 (inverse generators).** We will often want our alphabet $A$ of semigroup generators to be *closed under inversion*, which means that there is an involution $\iota : A \to A$ such that $p(\iota(x)) = (p(x))^{-1}$ for all $x \in A$. We call $\iota(x)$ the *formal inverse* of the generator $x \in A$, and we will generally write $x^{-1}$ instead of $\iota(x)$, even though $x$ is an element of $A$, not of $G$. It is also convenient to denote specific pairs of formal inverses by the uppercase and lowercase variants of the same letter: thus the inverse of $x$ will be represented by $X$, and so on.

Here are some examples where there is a reasonable way of reflecting some aspect of the structure of a group $G$, using the idea of a language $L$ over an alphabet $A$ of semigroup generators.

**Example 2.1.3 (free group).** Let $G$ be the free group on two (group) generators $x$ and $y$, and $A = \{x, y, X, Y\}$, where $X = x^{-1}$ and $Y = y^{-1}$ under Convention 2.1.2. Let $L$ be the language of all reduced strings, that is, strings in which $x$ and $X$ are never adjacent and $y$ and $Y$ are never adjacent. This language is regular; Figure 2.1 shows an automaton accepting it. The map $\pi : L \to G$ is a bijection.

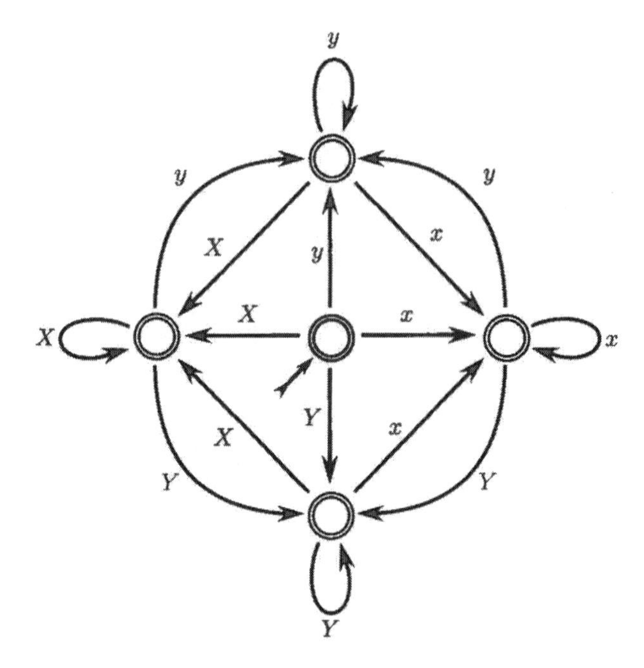


**Figure 2.1. Automaton for the free group.** This partial automaton accepts precisely the reduced strings in the free group on two generators $x$ and $y$.

**Example 2.1.4 (shortest).** The previous example can be generalized to any group $G$ and any set $A$ of semigroup generators. We set $L$ equal to the set of strings over $A$ which are shortest representatives for elements of $G$. The map $\pi : L \to G$ is usually not bijective, but it is surjective and finite-to-one, because, for any $g$, all elements of $L$ in $\pi^{-1}(g)$ have the same length, and therefore they are finite in number.

**Example 2.1.5 (free abelian).** Let $G$ be the free abelian group on three (group) generators $x$, $y$ and $z$, and let $A = \{x, y, z, X, Y, Z\}$. If $L$ is the language defined by the regular expression $(x^* \vee X^*)(y^* \vee Y^*)(z^* \vee Z^*)$, the map from $L$ to $G$ is bijective.

**Example 2.1.6 (nilpotent).** Let $G$ be the group of $3 \times 3$ matrices with integer coefficients, ones down the diagonal, and zeroes above the diagonal. The matrices

$$x = \begin{pmatrix} 1 & 0 & 0 \\ 1 & 1 & 0 \\ 0 & 0 & 1 \end{pmatrix}, \quad y = \begin{pmatrix} 1 & 0 & 0 \\ 0 & 1 & 0 \\ 0 & 1 & 1 \end{pmatrix}, \quad z = \begin{pmatrix} 1 & 0 & 0 \\ 0 & 1 & 0 \\ -1 & 0 & 1 \end{pmatrix}.$$

generate $G$ as a group; we have $[x, y] = z$, and $z$ commutes with $x$ and $y$. Taking the same alphabet and language as in Example 2.1.5, we again obtain a bijective map from $L$ to $G$.

The last two examples show that, even if the map from the language to the group is bijective, the language is far from determining the group.

We have mentioned that the existence of a regular language $L$ such that $\pi : L \to G$ is bijective gives a sort of *normal form* for elements of $G$. Normal forms usually lead to a solution of the word problem for $G$, which we now define.

**Definition 2.1.7 (word).** Let $A$ be an alphabet. We form an alphabet $A' = A \cup A^{-1}$ closed under inversion (Convention 2.1.2) as follows: $A^{-1}$ is simply a set of the same cardinality as $A$ and disjoint from $A$, and $\iota : A' \to A'$ is an involution interchanging $A$ and $A^{-1}$. A *word over* $A$ is a string over $A'$; the word is *reduced* if it has no substring of the form $xx^{-1}$ (recall that $x^{-1}$ is a shorthand for $\iota(x)$, the formal inverse of $x$). The set $F(A)$ of reduced words over $A$ has a natural group structure: the product of two words is obtained by concatenating them and cancelling out pairs of the form $xx^{-1}$, and the inverse of a word is obtained by replacing each letter with its inverse, then reversing the resulting string. We call $F(A)$ the *free group* on $A$.

**Definition 2.1.8 (generators and relators).** Let $A$ be an alphabet and $R \subset F(A)$ a set of words over $A$. The smallest normal subgroup of $F(A)$ that contains $R$ consists of all the words of the form

$$\prod_{i=1}^{n} w_i r_i^{\pm 1} w_i^{-1},$$

with $r_i \in R$ and $w_i \in F(A)$ for all $i$. Given a group $G$ such that $A$ is a set of group generators for $G$, we extend the map $p$ to $A \cup A^{-1}$ by setting $p(\iota(x)) = (p(x))^{-1}$ (see Definition 2.1.1 and Definition 2.1.7), and thus obtain a group homomorphism $\pi : F(A) \to G$. If the kernel of this homomorphism is the smallest normal subgroup of $F(A)$ containing $R$, we say that $R$ is a set of *relators* for $G$, and that $A$ and $R$ together form a *group presentation* for $G$; this is denoted by $G = \langle A/R \rangle$.

The *word problem* in $G$ consists in finding an algorithm that takes as its input a word $w$ over $A$ and answers Yes or No, depending on whether or not $w$ represents the identity in $G$.

**Theorem 2.1.9 (regularly generated).** *Let $G$ be a finitely presented group, with semigroup generators $A$. Assume that $G$ is regularly generated, which means that there is a regular language $L$ over $A$ such that $\pi : L \to G$*

*is surjective, and that the inverse image $L_0$ in $L$ of the identity under $\pi$ is also a regular language. Then the word problem in $G$ is solvable.*

The statement and proof of Theorem 2.1.9, as well as Corollary 2.1.10 below, remain valid under the weaker assumption that $G$ has a presentation with a *recursively enumerable* set of relations, that $L$ is a recursively enumerable language and that $L_0$ is a *recursive* language. Intuitively, a recursive set is one for which membership can be checked algorithmically. A recursively enumerable set is one whose elements can be enumerated algorithmically; this is a weaker condition, because in general there is no way to know if an object that has not appeared yet in the enumeration belongs to the set or not. For more details on these important concepts, see [HU69]. (The reason we state the results for regular languages is to give an intermediate point between general groups and automatic groups.)

*Proof of 2.1.9:* Given a word $w$ over $A$ (Definition 2.1.7), we want to check whether or not $\overline{w} \in G$ is equal to the identity. Let $F(A)$ be the free group on $A$ and let $K$ be the kernel of $\pi : F(A) \to G$, that is, the set of reduced words which represent the identity element of $G$. Since the group is finitely presented, $K$ is recursively enumerable (see Definition 2.1.8). Reducing the elements of $Kw$ by cancelling inverses, we get the full inverse image $X$ of $\overline{w}$ in $F(A)$. This means we can enumerate the set $Y$ of all words over $A$ which represent $\overline{w}$ by inserting into words which are elements of $X$ cancelling pairs of generator and inverse. Since the intersection of the two recursively enumerable sets is recursively enumerable, we can recursively enumerate the set $Y \cap L$, which consists of all elements of $L$ that map to $\overline{w}$. By hypothesis this set is non-empty, so our computation will eventually produce such an element in $L$. We then check whether it is in $L_0$ or not. 2.1.9

**Corollary 2.1.10 (finite-to-one).** *Let $G$ be a finitely presented group, with semigroup generators $A$. Let $L$ be a regular language over $A$ and let $\pi : L \to G$ be surjective and finite-to-one. Then the word problem in $G$ is solvable.* 2.1.10

These two results also display the usual ambiguity in the statements of non-intuitionistic mathematics. To say that some object *exists* does not say that we know how to construct that object. For example, there is an algorithm which tells us whether the Riemann hypothesis is true: it is either the algorithm that says Yes, or the algorithm that says No. (If the Riemann hypothesis were independent of the axioms of set theory, this would not be the case, but it is reasonable to believe that the Riemann hypothesis or its negation can be deduced from the axioms of set theory.) In Theorem 2.1.9 we know how to construct the algorithm in question if $L$ and $L_0$ are given constructively (for example, by specifying finite state automata). This fact is a bit more hidden in Corollary 2.1.10; we only know how to use the knowledge that $L_0$ is finite if we are given $L_0$ constructively.

Also, the algorithm for the word problem derived here is horrendously long, and therefore impracticable. Theorem 2.3.10 (quadratic algorithm) will give a fast algorithm to solve the word problem, which works when the group is automatic. A group can be automatic with respect to one language, but only regularly generated with respect to another language (see Figure 2.2). The word problem can be solved quickly using the first structure, but only slowly using the second. Also, a group can be regularly generated with respect to some language, but not automatic with respect to any language: see Example 2.1.6 and Theorem 8.2.8 (nilpotent implies not automatic).

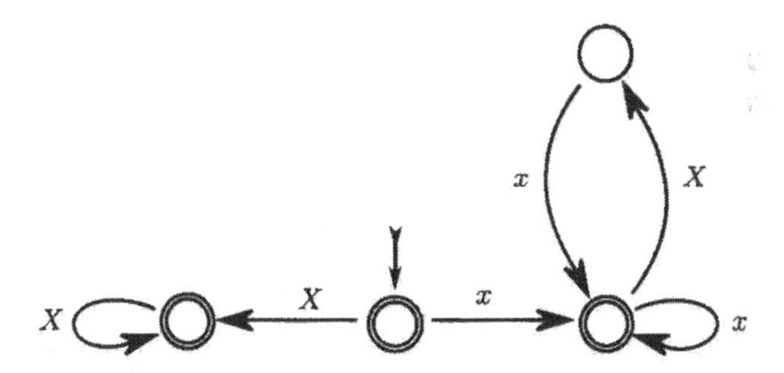


**Figure 2.2. A finite state automaton for the integers.** This automaton accepts the language $XX^* \vee x((Xx) \vee x)^*$, where $x$ is a generator of the group $\mathbf{Z}$ of the integers, and $X$ is its inverse. With this language, $\mathbf{Z}$ is regularly generated, but not automatic (since Lemma 2.3.2 is not satisfied).

## 2.2. Cayley Graphs and Isoperimetric Inequalities

The choice of a set of (group or semigroup) generators $A$ for a group $G$ gives a notion of distance in $G$, where two distinct elements are at distance one if they can be obtained from one another by right multiplication by a generator. By connecting such neighbouring elements with an edge labelled with the corresponding generator, we obtain the so-called *Cayley graph* of $G$. Figure 2.3 illustrates this for the symmetric group on three elements.

More generally, let $G$ be a group and let $H$ be a subgroup of $G$. We denote by $H\backslash G$ the set of *right cosets* $Hg$ of $H$, where $g$ ranges over $G$. The group $G$ acts on $H\backslash G$ on the right.

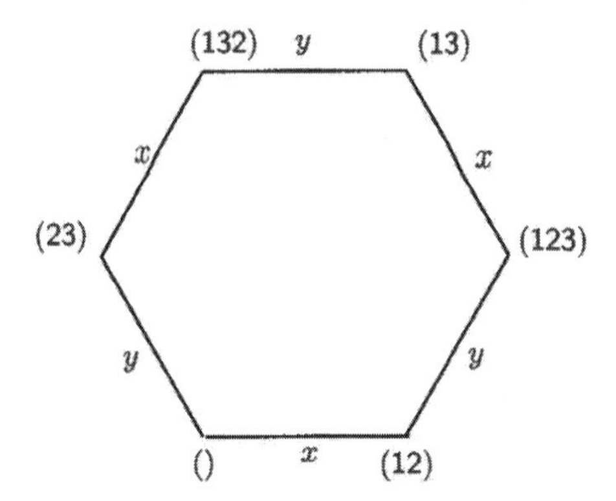


**Figure 2.3. The Cayley graph of the symmetric group** $S(3)$. The generators are $x = (12)$ and $y = (23)$.

**Definition 2.2.1 (Cayley graph).** Let $G$ be a group, $A$ a set of semigroup generators, and $H$ a subgroup of $G$. The *Cayley graph* $\Gamma(H\backslash G, A)$ of $G$ with respect to $A$ and $H$ is a directed, labelled graph defined as follows: The set of vertices is $H\backslash G$, and the set of labels is $A$. There is a directed edge or *arrow* with label $x$, source $Hg_1$ and target $Hg_2$ if and only if $Hg_1x = Hg_2$. We sometimes write such an arrow as $(Hg_1, x, Hg_2)$. The Cayley graph $\Gamma(G, A)$ of $G$ with respect to $A$ is the same as the Cayley graph of $G$ with respect to $A$ and the trivial subgroup. The *basepoint* of a Cayley graph is the coset of the identity element.

If $A$ is closed under inversion (Convention 2.1.2), the two arrows $(Hg, x, Hgx)$ and $(Hgx, x^{-1}, Hg)$ are usually thought of as making up a single edge pointing both ways, and often only one or the other of the arrowheads is drawn, with the corresponding label. If $x$ and $x^{-1}$ have the same image in $G$, the generator has order two and the edge can be thought of as undirected (see Figures 2.3, 2.4 and 2.5.) Likewise, if two generators $x$ and $y$ give the same element in $G$, an $x$-arrow and a $y$-arrow having the same source can be thought of as sharing an edge. These "optimizations," however, make no significant difference to the construction.

If $H$ is a normal subgroup of $G$, left cosets and right cosets coincide, and there is an action of $G$ on $\Gamma(H\backslash G, A)$, coming from left multiplication by elements of $G$, which is a homeomorphism. Thus $\Gamma(H\backslash G, A) = \Gamma(G/H, AH/H)$ is *homogeneous*: any two vertices look alike. When $H$ is not normal, the Cayley graph is no longer homogeneous (see Figure 2.6). In fact, it is easy to see that $H$ is normal if and only if the group of symmetries of the Cayley graph preserving the labels on the arrows acts transitively on the vertices.

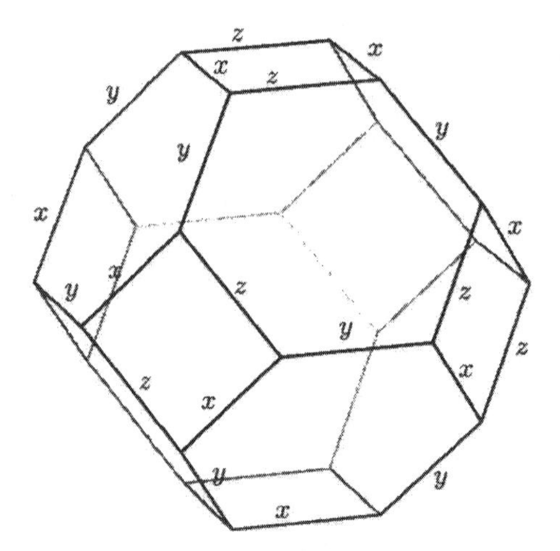


**Figure 2.4. Cayley graph of $S(4)$.** This truncated octahedron is the Cayley graph of the symmetric group $S(4)$ with respect to the generators $x = (12)$, $y = (23)$ and $z = (34)$. No arrowheads need to be drawn, because each generator has order two.

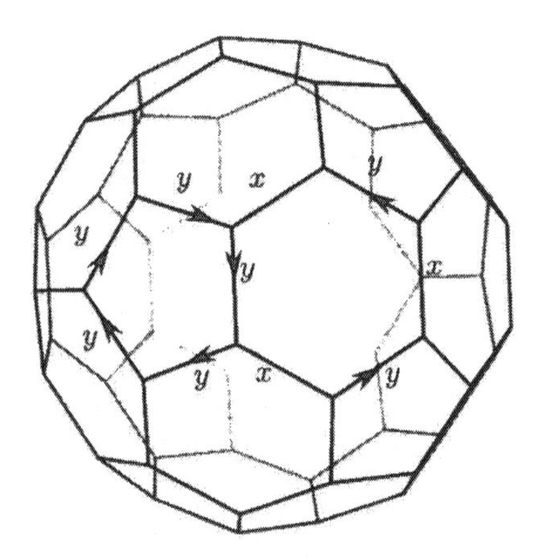


**Figure 2.5. The Cayley graph of $A(5)$.** This truncated dodecahedron is the Cayley graph of the alternating group $A(5)$, with respect to the generators $x = (12)(34)$ and $y = (12345)$. One can deduce from the fact that the surface of the truncated dodecahedron is simply connected that $x^2$, $y^5$ and $(xy)^3$ form a set of relators for a presentation of $A(5)$.

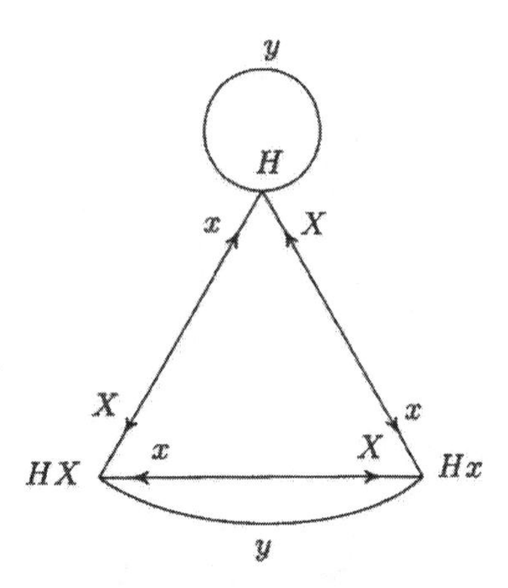


**Figure 2.6. Cayley graph of a set of cosets.** The Cayley graph $(H\backslash G, A)$, where $G$ is the symmetric group on three elements, $A$ consists of the generators $x = (123)$, $X = x^{-1}$ and $y = (12)$, and $H$ is the subgroup generated by $y$. The graph is not homogeneous because $H$ is not normal.

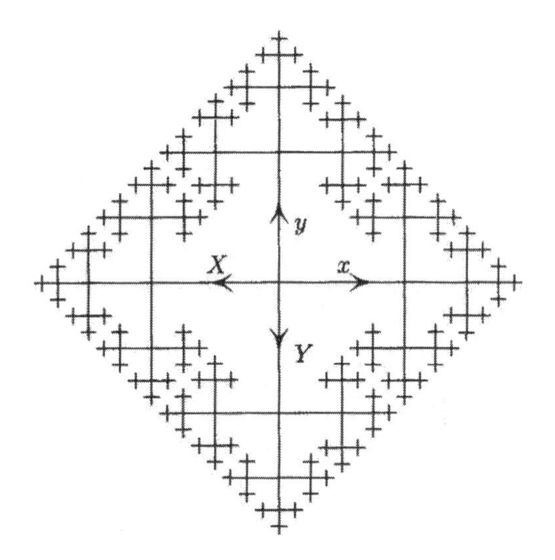


**Figure 2.7. Cayley graph of a free group.** Part of the Cayley graph of the free group on two group generators $x$ and $y$, with inverses $X$ and $Y$.

The Cayley graph is connected because $A$ generates $G$. We make it into a metric space by applying the following procedure, which works for any connected graph: The distance between two vertices is the minimum number of edges in any chain of edges connecting the two vertices (ignoring the directions of the edges). To extend the distance function from the set of vertices to the edge, we make each edge isometric to the unit interval, or, if the edge's endpoints are equal, to the circle of length one. The space thus obtained is a path metric space (see Section 3.1), and the distance between any two distinct points is realized by a geodesic, that is, a path whose length is equal to the distance.

When $H$ is the trivial subgroup, the metric thus obtained on $G$ is called the *word metric*. Clearly, the distance between two elements $g_1$ and $g_2$ in the word metric is $|g_1^{-1}g_2|$, where $|g|$ denotes the *word length* of $g \in G$, that is, the minimum length of a word over $A$ representing $g$ (remember from Definition 2.1.7 that a word can include inverses of generators).

Given a word $w$ over $A$, we define a path $\widehat{w} : [0, \infty) \to \Gamma(G, A)$ in the Cayley graph as follows: If $t$ is an integer, $\widehat{w}(t) = w(t)$ is the image in $G$ of the prefix of $w$ of length $t$ (recall from page 4 that $w(t) = w$ if $t$ is greater than the length of $w$). We extend $\widehat{w}$ to non-integral values of $t$ in the obvious way, by moving along the respective edges with unit speed. This is illustated in Figure 2.8. Thus $\widehat{w}$ travels at unit speed for parameter values in $[0, |w|]$, and then stops.

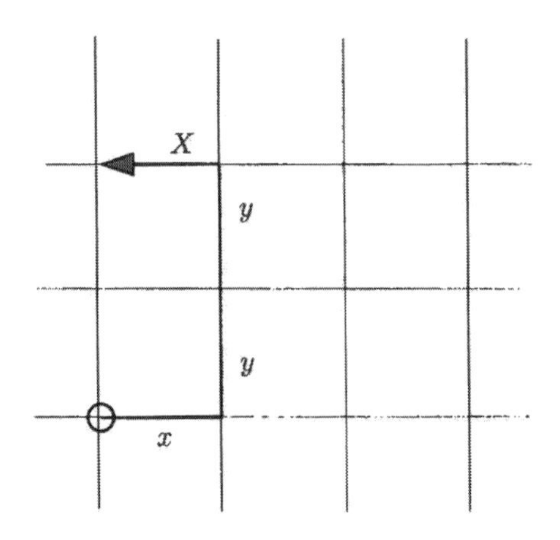


**Figure 2.8. A path in a Cayley graph.** The string $w = xyyX$ is represented by a path of length four in the Cayley graph of the free abelian group on two generators $x$ and $y$. We have $w(0) = \varepsilon$, $w(1) = x$, ..., $w(4) = w(5) = xyyX$.

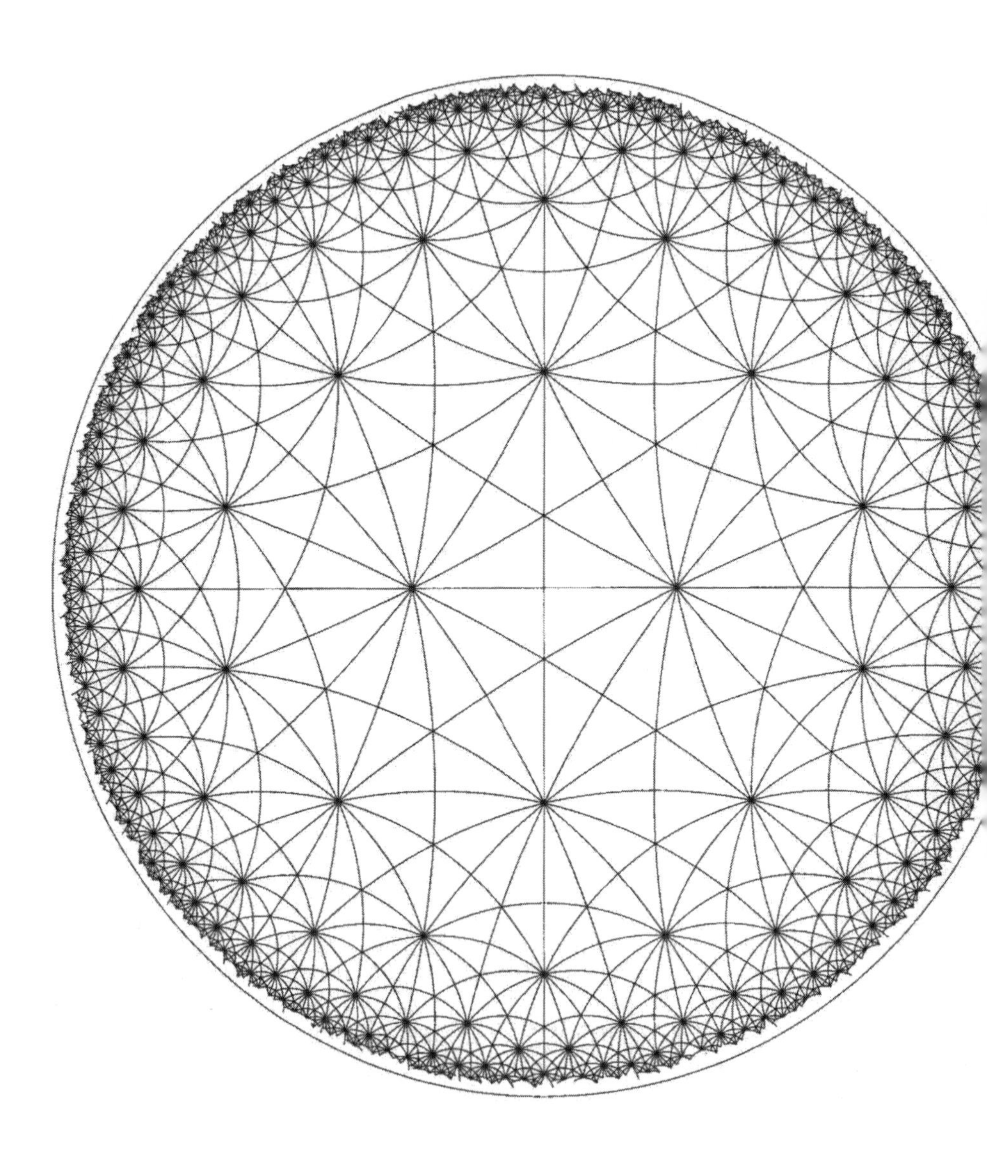

**Figure 2.9. Tesselation of the hyperbolic plane.** The tiles are triangles, with angles $\pi/2$, $\pi/3$ and $\pi/7$. The group of symmetries is generated by reflections in the sides of one of the triangles. This picture was drawn by Steve Rumsby, using *gm*, a program written by Epstein and Rumsby.

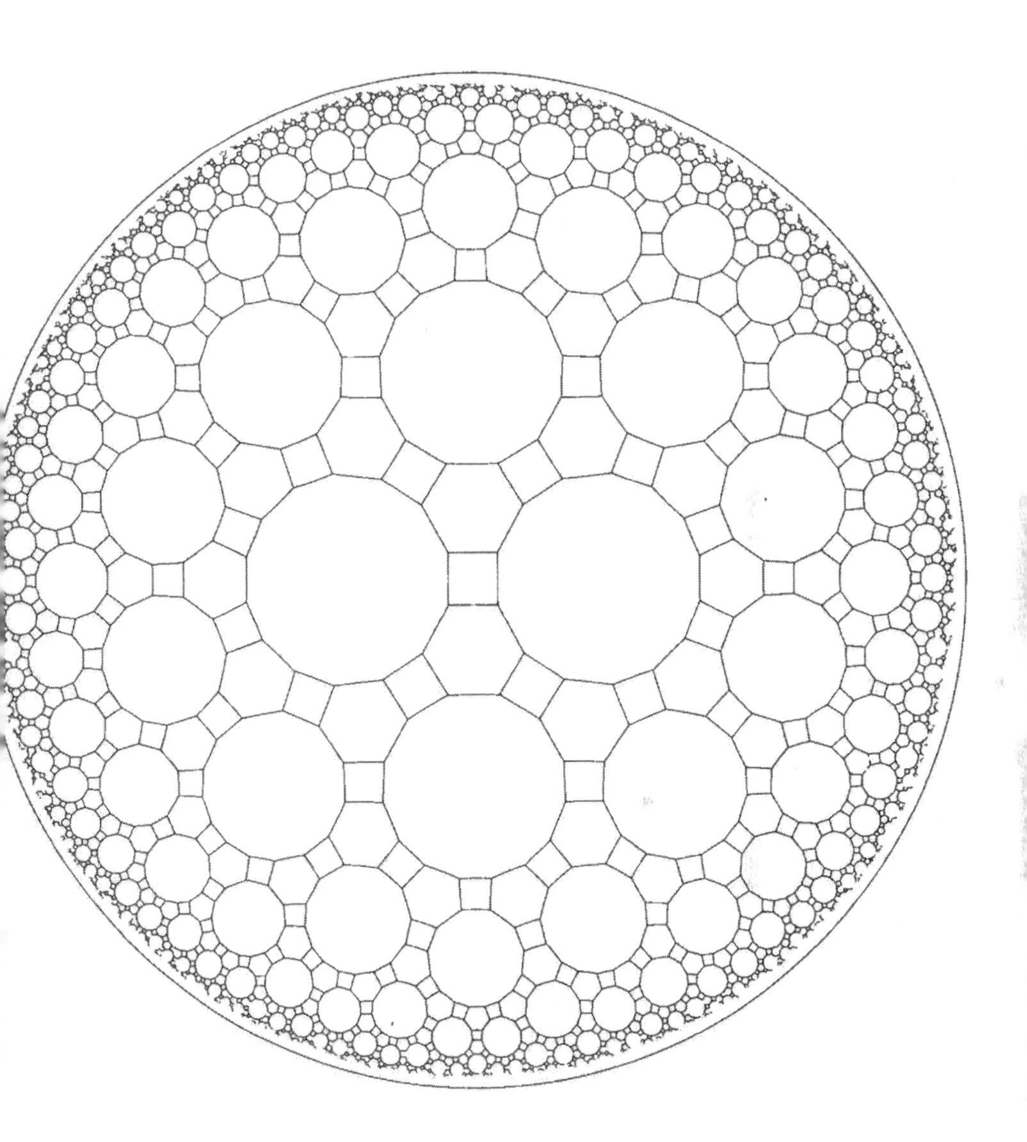

**Figure 2.10. The Cayley graph of the (2,3,7)-triangle group.** This is the Cayley graph of the group of symmetries of the pattern in Figure 2.9, generated by reflections in the sides of the triangles. This picture was drawn by Steve Rumsby, using *gm*, a program written by Epstein and Rumsby.

If we read off the labels as we go around a loop in the Cayley graph $\Gamma(G,A)$, the word we obtain represents the identity in $G$. As discussed in Definition 2.1.8 (generators and relators), if $\langle A/R\rangle$ is a presentation for $G$, any representative $w$ of the identity can be written in the form

**2.2.2.** $$w = \prod_{i=1}^{n} v_i r_i^{\pm 1} v_i^{-1}, \qquad \text{with } r_i \in R \text{ and } v_i \in F(A) \text{ for all } i.$$

In the Cayley graph, this corresponds to subdividing the loop labelled by $w$ into a number of loops labelled by the relators $r_i$; the $v_i$ are needed to translate the model relator loops, which are based at the basepoint (the identity element), to the place where they fit into the loop labelled by $w$. A *Dehn diagram* is a portion of the Cayley graph displaying this relationship.

**Example 2.2.3.** Let $G$ be the group $\langle\{x,y\}/\{r_1,r_2\}\rangle$, where $r_1 = x^3$ and $r_2 = xyx^{-1}y^{-1}$. (It is easy to see that $G = \mathbf{Z}\times\mathbf{Z}_3$.) The word $y^2x^{-1}y^{-1}x^{-1}y^{-1}x^{-1}$ is trivial in $G$, and can be decomposed into the product $(x^{-1}r_2x)(x^{-1}yr_2y^{-1}x)(x^{-2}r_2x^2)(r_1^{-1})$. Figure 2.11 shows a Dehn diagram for this word.

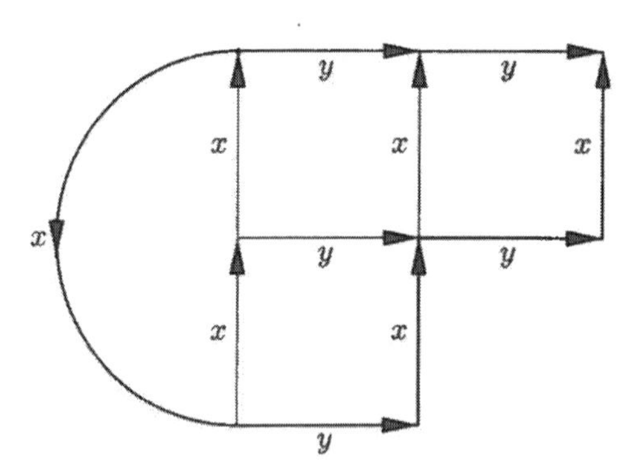


**Figure 2.11. The Dehn diagram.** In the group $\langle\{x,y\}/\{x^3, xyx^{-1}y^{-1}\}\rangle$, the word $y^2x^{-1}y^{-1}x^{-1}y^{-1}x^{-1}$ can be decomposed into the product of four conjugates of relators. The Dehn diagram illustrates this decomposition. The basepoint is marked on the upper left.

We often call a decomposition of the form 2.2.2 a *disk* with boundary $w$ and *combinatorial area* $n$. This terminology originates from the following construction, called a *dual Dehn diagram* or *picture* (see Figure 2.12). We start by drawing the conjugates of relators in clockwise order around a point, each $v_i r_i^{\pm1} v_i^{-1}$ appearing as a path (representing $v_i$) terminated by a loop (representing $r_i^{\pm1}$). We call this a *bouquet*. Reducing the product of conjugates

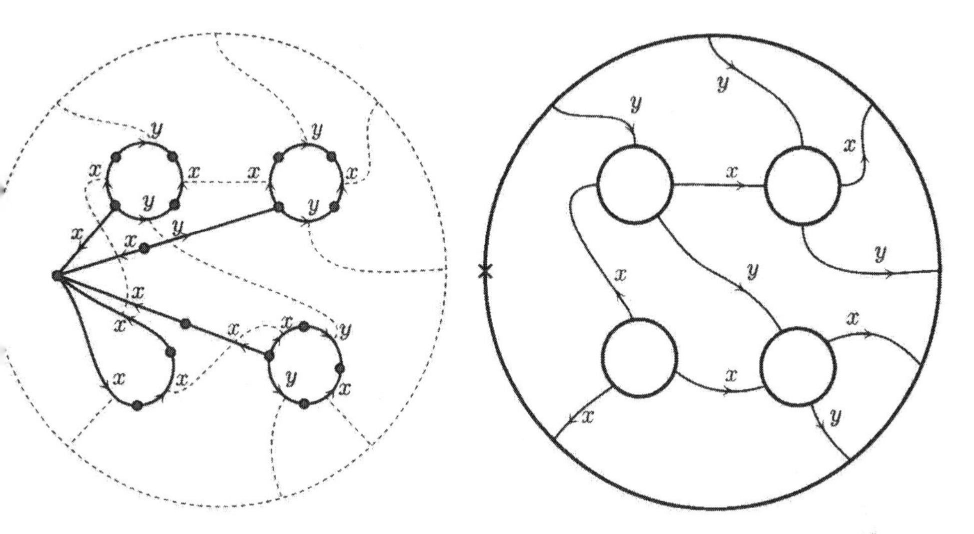

**Figure 2.12. Building the dual Dehn diagram.** (This construction does not take place in the Cayley graph.) To build the picture dual to Figure 2.11, start with a bouquet of conjugates of relators (left); each arrow between two nodes represents a symbol in a relator $r_i^{\pm 1}$, or two matching inverse symbols in a pair $v_i, v_i^{-1}$. Next draw curves (dashed on the left) connecting symbols that cancel each other. The left-over symbols form the reduced word, which we represent by an outer circle. The outer circle, the inner circles (relators) and the connecting curves make up the dual Dehn diagram (right).

in the free group means pairing off adjacent edges with inverse labels. We mark the pairing off with arcs connecting the midpoints of the edges. After edges can no longer be paired off, the remaining edges spell out $w$. We draw $w$ as a circle surrounding the diagram, and pair off the remaining edges of the bouquet with the edges of $w$.

We interpret the region between the outer circle and the inner relator loops as a disk with holes, and the identifications as arcs running between the various boundary components. Conceivably arcs might also join to make a closed loop, but this corresponds to a total cancellation of the relators contained in the interior of the loop, so we could have omitted these terms from the product in the first place.

Conversely, given a dual Dehn diagram for $w$ we can reconstruct an expression of the form 2.2.2 by drawing non-intersecting paths from the basepoint to each of the holes, and reading off the path labels from the arcs intersected, keeping track of orientation.

(Readers with some background in differential topology may prefer to visualize the dual Dehn diagram in a different way. Start with the standard way of describing a two-dimensional complex $K$ whose fundamental group is $\langle A/R \rangle$, namely: Take a basepoint, an interval for each element of $A$ and a two-dimensional disk for each element of $R$. Identify the two endpoints of each interval with the basepoint, forming a bouquet $K^1$. To this bouquet, attach each disk $D^2$ corresponding to a relator $r \in R$ by means of the obvious map $\gamma_r : S^1 \to K^1$ defined on the boundary $S^1 \subset D^2$. A general word $w$ corresponds, in the same standard way, to a map $\gamma_w : S^1 \to K^1$. Now $w$ represents the identity if and only if $\gamma_w$ can be extended to $D^2$. The dual Dehn diagram is a picture of one particular way to carry out this extension. In this picture, the small disks of Figure 2.12 are copies of one of the disks forming $K$, mapped to the disk homeomorphically; the inverse image of the interiors of the disks forming $K$ is otherwise disjoint from the picture. Each arc labelled by $x \in A$ in the picture is mapped to the midpoint of the corresponding interval in $K^1$, and these labelled arcs in the picture are the complete inverse image of these midpoints. The two sides of an arc labelled $x$ in the picture are mapped to different sides of the midpoint of the edge of $K^1$ labelled $x$. Since the complement in $K^1$ of the midpoints of the intervals is contractible, it does not matter, up to homotopy, how the rest of the picture is mapped into $K^1$.)

This construction makes it easy to prove the following result:

**Lemma 2.2.4 (bounding lengths).** *If $w$ is a reduced word expressed as a product of $n$ conjugates of relators as in Equation 2.2.2, we can rewrite the product so that each $v_i$ has length at most $(|w| + 2k)2^n$, where $k$ is the maximum length of the relators $r_i$.*

*Proof of 2.2.4:* The basepoint $*$ of $w$ can be connected to the basepoint of one of the loops by a path $u$ that crosses at most $\frac{1}{2}(|w|+k)$ arcs. To see this, cut the disk with holes along all arcs that begin or end at a hole, and look at the connected component $P$ containing $*$. The only remaining arcs are those running from $w$ to itself, and there are at most $\frac{1}{2}|w|$ of them. To connect $*$ with some hole, therefore, we need to cross at most this many arcs. We might need to cross another $\frac{1}{2}k$ arcs to get to the basepoint of the relator.

We now cut the diagram open along the path $u$, getting a disk with one fewer holes (Figure 2.13). The boundary of the new disk is $w' = wuru^{-1}$, where $r$ is equal to some relator or its inverse and $u$ has length at most $\frac{1}{2}(|w| + k)$. Thus $w'$ has length at most $2(|w| + k)$. Since $w'$ can be written as the product of $n-1$ conjugates of relators and $w = w'ur^{-1}u^{-1}$, the result follows by induction on $n$. $\boxed{2.2.4}$

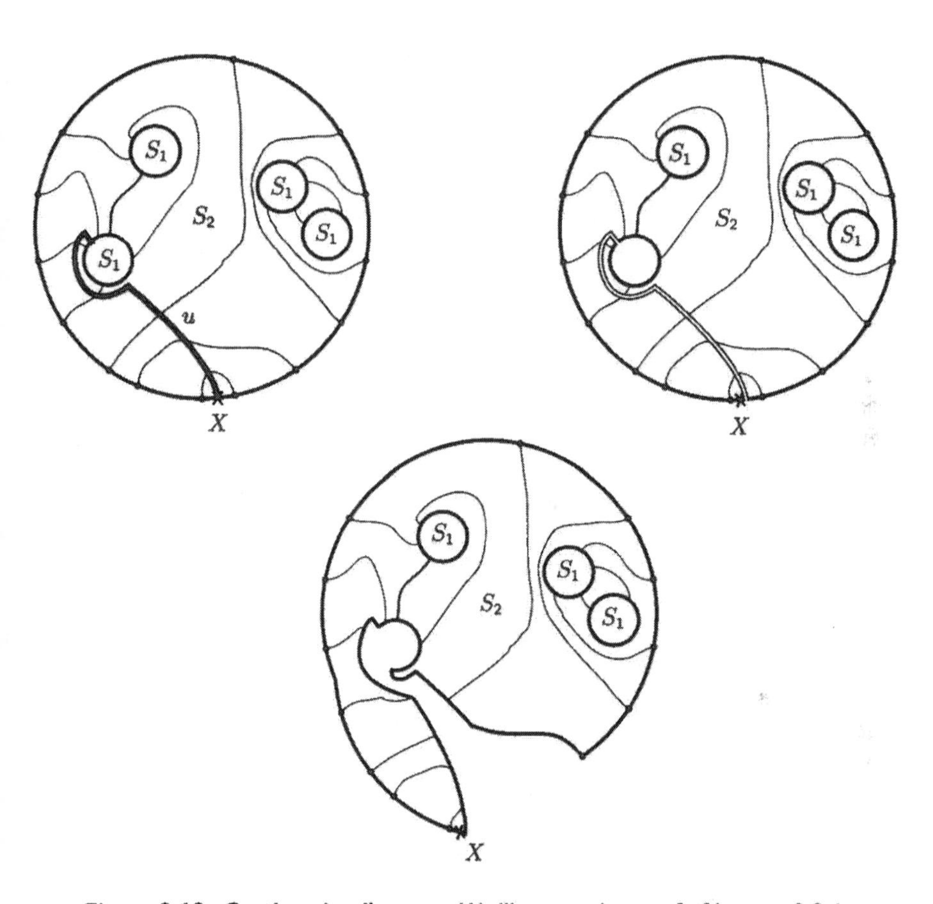


**Figure 2.13. Cutting the diagram.** We illustrate the proof of Lemma 2.2.4 (bounding lengths). We start with a word $w$ of length $|w| = 14$, bounding a disk of combinatorial area $p = 4$. The shaded area in the upper left diagram denotes the piece $P$. The length $k$ of the longest relator is 5. The point $X$ denotes the basepoint. After cutting along $u$ and reducing, we get a word of length 21, bounding a disk of combinatorial area 3.

There are many ways to decompose a given trivial word $w$ in the form 2.2.2; that is, there are many disks with boundary $w$. The *combinatorial area* of $w$, denoted by $\text{area}(w)$, is the minimum combinatorial area of any disk spanning $w$. The *isoperimetric function* of a presentation is defined by

$$\phi(i) = \max\{\text{area}(w) \ : \ |w| \leq i, \overline{w} = e\},$$

where $e$ is the identity.

In general, $\phi$ is not a recursive function, that is, there is no finite procedure to compute $\phi(i)$ given $i$ (see also the discussion following Theorem 2.1.9). We will see shortly that $\phi$ is recursive if and only if the word problem is solvable.

Of course, the isoperimetric function depends on the presentation; but if $\phi$ and $\phi'$ are isoperimetric functions for different presentations of the same group, it is easy to see that $\phi'(i) \leq N_1\phi(N_2 i)$, where $N_1$ is the maximum area of an old relator in terms of new ones, and $N_2$ is the maximum length of a new generator in terms of old ones. Thus if $\phi$ is bounded by a polynomial, or exponential, or recursive function, so is $\phi'$. In this case we say that the group satisfies a polynomial, or exponential, or recursive *isoperimetric inequality*. We will see in the next section that automatic groups satisfy a quadratic isoperimetric inequality, a fact that can sometimes be used to prove that a group is not automatic.

**Theorem 2.2.5 (solvable word problem).** *A group has a decidable word problem if and only if it has an isoperimetric inequality in the class of recursive functions. This happens if and only if the isoperimetric function is itself recursive.*

*Proof of 2.2.5:* Let $f$ be a recursive function bounding the isoperimetric function from above. Given a reduced word $w$, we want to check if $w$ can be written in the form 2.2.2. If it can, the value of $n$ can be chosen not to exceed $f(|w|)$, and Lemma 2.2.4 gives a bound for the lengths of the $v_i$. We can therefore try all possibilities.

Conversely, if the word problem is solvable, we can compute the isoperimetric function $\phi(i)$ for a given value of $i$. For we can list all words of length at most $i$, select those that are equivalent to the identity, and for each one find a disk of minimal combinatorial area by exhaustively trying all possibilities in order of increasing area. Lemma 2.2.4 guarantees that only a finite numbers of possibilities with a given combinatorial area need to be tried.

2.2.5

## 2.3. Automatic Groups: Definition

The reader may want to recall Definitions 1.4.5 (regular many-variable language) and 2.1.1 (generators) at this point.

**Definition 2.3.1 (automatic group).** Let $G$ be a group. An *automatic structure* on $G$ consists of a set $A$ of semigroup generators of $G$, a finite state automaton $W$ over $A$, and finite state automata $M_x$ over $(A, A)$, for $x \in A \cup \{\varepsilon\}$, satisfying the following conditions:

(1) The map $\pi : L(W) \to G$ is surjective.

(2) For $x \in A \cup \{\varepsilon\}$, we have $(w_1, w_2) \in L(M_x)$ if and only if $\overline{w_1 x} = \overline{w_2}$ and both $w_1$ and $w_2$ are elements of $L(W)$.

We call $W$ the *word acceptor*, $M_\varepsilon$ the *equality recognizer*, and each $M_x$, for $x \in A$, a *multiplier automaton* for the automatic structure. An automatic structure is sometimes called an *automation*. An *automatic group* is one that admits an automatic structure.

Note that, in a certain sense, $W$ is redundant, because $L(W)$ is the same as the diagonal subset of $L(M_\varepsilon)$.

We will prove in Theorem 2.4.1 that if $G$ is automatic for one set of generators it is automatic for any other set.

In Definition 7.2.1 we will introduce the more general notion of an *asynchronous automatic group* and we will sometimes use the word "synchronous" to refer to automatic groups as just defined; however, if the word is omitted, Definition 2.3.1 will be the one intended.

Recall from page 4 that $w(t)$ denotes the prefix of $w$ of length $t$, or $w$ itself if $t$ exceeds the length of $w$. Recall also from page 37 that the word metric on a group $G$ (with respect to a given set $A$ of generators) is the metric induced on $G$ from the Cayley graph $\Gamma(G, A)$.

**Lemma 2.3.2 (lipschitz property).** *If $G$ has an automatic structure, there is a constant $k$ (depending on the structure) with the following property: If $(w_1, w_2)$ is accepted by one of the automata $M_x$, for $x \in A \cup \{\varepsilon\}$, the distance between $\overline{w_1(t)}$ and $\overline{w_2(t)}$ in the word metric is bounded by $k$, for any integer $t \geq 0$. Such a $k$ is called a lipschitz constant for the structure.*

The name of this lemma comes from analysis. We recall that if $X$ is any set, and $Y$ is a metric space with metric $d_Y$, the *uniform metric* on the set of maps from $X$ to $Y$ is the metric given by $d(f, g) = \sup_{x \in X} d_Y(f(x), g(x))$, where $f, g : X \to Y$. If $Y$ is not bounded, $d(f, g)$ may be infinite. Thus Lemma 2.3.2 says that the paths $\widehat{w_1}$ and $\widehat{w_2}$ are at a distance at most $k$ from each other, in the uniform metric on paths in the Cayley graph, if $\overline{w_1}$ and $\overline{w_2}$ are at a distance at most one from each other.

Now the word acceptor assigns to each $g \in G$ an acceptable string representing $g$ (or possibly several strings all uniformly close to each other). We can regard this assignment (ignoring for the moment the multiple representatives) as a map $G \to A^*$, which is a section (right inverse) to the map $\pi : A^* \to G$. If we move $g$ a distance 1, the string is moved by at most $k$, in the uniform metric on paths in the Cayley graph. It follows that if $g$ is moved a distance $d$, the path is moved a distance at most $kd$; that is to say, the assignment of strings to group elements is $k$-lipschitz. (Again, this is not quite right because group elements which are equal can be assigned to distinct strings. What we really have is a $(k, k)$-lipschitz pseudomap: see page 71.)

*Proof of 2.3.2:* We assume all automata are normalized (Convention 1.2.3).

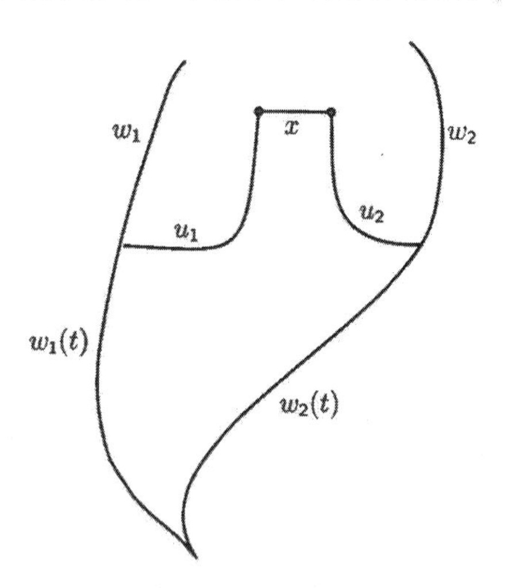


**Figure 2.14. The lipschitz property for automatic groups.** The notation in the picture matches that in the proof of Lemma 2.3.2.

Let $c$ be the maximum number of states in any of the automata $M_x$, for $x \in A \cup \{\varepsilon\}$. For any integer $t \geq 0$, suppose that $M_x$ has read the prefixes $w_1(t)$ of $w_1$ and $w_2(t)$ of $w_2$, and has arrived at state $s$. Let $(u_1, u_2) \in A^* \times A^*$ represent a shortest path of arrows to an accept state of $M_x$; then $u_1$ and $u_2$ have length less than $c$, and $(w_1(t)u_1, w_2(t)u_2)$ is accepted by $M_x$. It follows that $\overline{w_1(t)u_1x} = \overline{w_2(t)u_2}$ (see Figure 2.14), and that the distance between $\overline{w_1(t)}$ and $\overline{w_2(t)}$ in the Cayley graph is at most $2c - 1$. $\boxed{2.3.2}$

**Definition 2.3.3 (standard automata).** Let $G$ be a group and let $A$ be a set of semigroup generators of $G$. Let $W$ be a deterministic finite state automaton over $A$ and let $N$ be a fixed finite neighbourhood of the identity in $G$ (for example, the set of elements of $G$ with word length at most $k$). Assume that $N$ contains $A$. For each $x \in A \cup \{\varepsilon\}$, we define a finite state automaton $M_x$ over $(A, A)$, called the *standard automaton* $M_x$ *based on* $(W, N)$. The idea is that $M_x$ keeps track of the action of $W$ on both input strings, and also of the *word difference* between the prefixes read so far. The precise definition is the following:

Let $W'$ be a finite state automaton over $A \cup \{\$\}$ accepting the language $L(W)\$^*$, and let $S$ be the set of states of $W'$. The automaton $M_x$ has $S \times S \times N$ as its set of states, but we lump together into a single failure all states of the form $(s_1, s_2, g)$, where $s_1$ or $s_2$ is a failure state of $W'$. The initial state is $(s_0, s_0, e)$, where $s_0$ is the initial state of $W'$ and $e$ is the identity element of $G$. If $M_x$ is in state $(s_1, s_2, g)$ and reads a letter $(y_1, y_2) \in B$, where $B$ is the padded alphabet associated with $(A, A)$, it goes into state $(s_1y_1, s_2y_2, h)$, where $h = y_1^{-1}gy_2$ (Figure 2.15); but if $h \notin N$, $M_x$ goes into the failure state. The accept states of $M_x$ are the states of the form $(s_1, s_2, x)$, where $s_1$ and $s_2$ are accept states of $W'$. (Thus the only difference among the various $M_x$'s is the set of accept states.)

Notice that this definition is constructive only when we know $N$ constructively. This may seem an obvious remark, but it is worth keeping in mind. Even if $N$ consists only of the generators plus the identity, we may not really "know" it, because different generators in $A$ may correspond to the same element of $G$, or we may not know how to connect the generators in the Cayley graph.

If $G$ has an automatic structure, we can solve the word problem and therefore know $N$: see Theorem 2.3.10. However, one often wants to build the standard automata without prior knowledge of an automatic structure, but *in order* to find one. One may know from other sources what a neighbourhood of the identity looks like, and then it is possible to bootstrap the procedure from this much more restricted knowledge.

**Theorem 2.3.4 (standard automata theorem).** *If $G$ has an automatic structure with alphabet $A$ and word acceptor $W$, and $N$ is a sufficiently large, but finite, neighbourhood of the identity, the standard automata based on $(W, N)$, together with $A$ and $W$, also give an automatic structure on $G$.*

*Proof of 2.3.4:* By Lemma 2.3.2 (lipschitz property), there is a number $k > 0$ such that, if $(w_1, w_2)$ is accepted by any of the original automata, the uniform distance between the paths $\widehat{w_1}, \widehat{w_2} : [0, \infty) \to \Gamma$ is less than $k$. Let $N$ contain

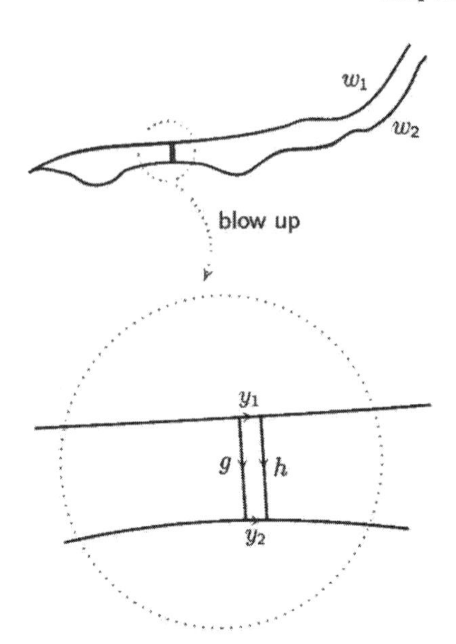


**Figure 2.15. Standard automata.** This is a diagram showing the main idea of a standard automaton. The diagram is drawn in the Cayley graph of the group and the relationship of $g$ with $h$ is shown (using the notation of Definition 2.3.3). At time $t$, the component $g$ of the automaton is the word difference $\overline{w_1(t)}^{-1}\overline{w_2(t)}$; if this gets too far from the identity, the automaton fails.

the ball of radius $k$ about the identity element in the Cayley graph. Let $M_x$, for $x \in A \cup \{\varepsilon\}$, be the corresponding standard automata.

The first condition for an automatic structure (Definition 2.3.1) is trivially satisfied. The second is also easy to check: Let $(w_1, w_2)$ be a string over $(A, A)$. If $w_1$ and $w_2$ are accepted by $W$ and $\overline{w_1 x} = \overline{w_2}$, for $x \in A$, the pair $(w_1, w_2)$ is accepted by the original $x$-multiplier, and therefore the distance between $w_1(t)$ and $w_2(t)$ in the Cayley graph never exceeds our chosen value of $k$. This implies that $\overline{w_1(t)}^{-1}\overline{w_2(t)} \in N$ for all $t \geq 0$, by the definition of $N$, so $(w_1, w_2)$ is accepted by $M_x$. The converse is trivial. $\boxed{2.3.4}$

The following characterization of an automatic structure follows from the existence of standard automata.

**Theorem 2.3.5 (characterizing synchronous).** *Let $G$ be a group and let $A$ be a finite set of semigroup generators for $G$. Let $W$ be a finite state automaton over $A$ and suppose that $\pi : L(W) \to G$ is surjective. Then $A$ and $W$ are part of a (synchronous) automatic structure on $G$ if and only if* *there is a number $k$ with the property that whenever two strings $w_1$ and $w_2$ accepted by $W$ are such that $\overline{w_1 x} = \overline{w_2}$ for some $x \in A$, the corresponding paths $\widehat{w_1}$ and $\widehat{w_2}$ are a uniform distance less than $k$ apart.* $\boxed{2.3.5}$

Using this theorem, we can readily construct examples of automatic structures. Examples 2.1.3 (free group) and 2.1.5 (free abelian) are straightforward. Here is another example.

**Example 2.3.6 (finite).** Let $G$ be a finite group. Then we may take the alphabet $A$ to equal $G$. For the word acceptor $W$ we take an automaton which accepts all strings. This shows that in an automatic structure the map $\pi : L(W) \to G$ may be far from being bijective.

**Convention 2.3.7 (multipliers dropped).** In some situations the structure of the multipliers is of no significance: we only need to know that they exist. It is then convenient to talk more loosely of the "automatic structure $(A, L)$" or the "automatic structure $(A, W)$" on a group, where $L$ is a regular language and $W$ is a finite state automaton over $A$.

Theorem 2.3.5 has the following immediate consequence:

**Corollary 2.3.8 (restriction is automatic).** *If $(A, L)$ is an automatic structure for a group $G$, and $L' \subset L$ is a regular language over $A$ that maps onto $G$, the pair $(A, L')$ is also an automatic structure for $G$.* $\boxed{2.3.8}$

We conclude this section with two related and important results: automatic groups satisfy a quadratic isoperimetric inequality (see end of Section 2.2), and the word problem in such a group is solvable in quadratic time (see end of Section 2.1). We will need the following simple fact, which we state as a lemma for future reference:

**Lemma 2.3.9 (bounded length difference).** *Let $G$ be an automatic group and let $(A, L)$ be an automatic structure for $G$. There is a constant $N$ such that, if $w \in L$ is an accepted word and $g \in G$ is a vertex of the Cayley graph at a distance at most one from $\overline{w}$, we have the following situation:*

(a) *$g$ has some representative of length at most $|w| + N$ in $L$; and*

(b) *if some representative of $g$ in $L$ has length greater than $|w| + N$, there are infinitely many representatives of $g$ in $L$.*

*It follows from* (a) *that any word $w$ over $A$ is equivalent to an accepted word of length at most $N|w| + n_0$, where $n_0$ is the length of an accepted representative of the identity.*

*Proof of 2.3.9:* Let $N$ be greater than the number of states in any of the automatic structure's automata, let $w$ and $g$ be as in the statemenent of the theorem, and let $w'$ be any representative of $g$ in $L$. One of the pairs $(w, w')$ or $(w', w)$ is accepted by the appropriate automaton $M_x$, for some $x \in A \cup \varepsilon$. If $|w'| > |w| + N$, the automaton $M_x$ undergoes more than $N$ transitions after reading all of $w$, and therefore visits the same state twice. One can then shorten $w'$ by eliminating the loop between two visits to this repeated state, and still get an accepted word representing the same group element; this proves (a). Alternatively, one can lengthen $w'$ arbitrarily by going over the loop repeatedly; this proves (b). [2.3.9]

**Theorem 2.3.10 (quadratic algorithm).** *Let $G$ be an automatic group and $(A, L)$ an automatic structure for $G$. For any word $w$ over $A$, we can find a string in $L$ representing the same element of $G$ as $w$, in time proportional to the square of the length of $w$.*

In particular, knowing some word $e$ representing the identity element, we can solve the word problem in quadratic time, by finding a representative in $L$ for the desired word, and feeding it and $e$ to the equality recognizer.

*Proof of 2.3.10:* We assume that the multiplier automata $M_x$, for $x \in A$, are normalized. Suppose we are given an accepted string $u$ and a generator $x$, and we want to find an accepted representative $v$ for $\overline{ux}$. The idea is that $M_x$ accepts some pair $(u', v')$, where $u'$ equals $u$, possibly followed by some number of $'s, and $v'$ is an accepted word representing $\overline{ux}$, also possibly followed by $'s. If we ignore the second element of its labels, $M_x$ becomes a non-deterministic automaton in one variable accepting $u'$. We find a path of arrows corresponding to $u'$, and read off $v'$ by looking at the second element of the labels in this path of arrows.

More precisely, let $S_0$ be the set whose only element is the initial state $s_0$ of $M_x$. For $i > 0$, we inductively define $T_i$ as the set of arrows with source in $S_{i-1}$ and label $(x_i, y_i)$, where $y_i \in A \cup \{\$\}$ is arbitrary, and $x_i$ equals either the $i$-th character in $u$ (if $i \leq |u|$) or $\$$ (if $i > |u|$). We also define $S_i$ as the set of targets of arrows in $T_i$. We let $n$ be the smallest number such that $n \geq |u|$ and $S_n$ includes an accept state of $M_x$. Working backwards, we read off the labels $y_n, \ldots, y_1$ on a path of arrows joining the initial state to an accept state, and we form the string $v' = y_1 \ldots y_n$, which represents $\overline{ux}$. We finally discard trailing $'s to obtain $v$.

Since there are only finitely many arrows and states, the time taken by each step in this induction is bounded by a constant. Therefore the overall time is proportional to the number $n$ defined above. By Lemma 2.3.9, $n$ can only exceed $|u|$ by a bounded amount $N$, otherwise it would not be minimal.

This shows that, given an accepted string $u$, we can find a representative for $\overline{ux}$ in time $O(|u|)$, and the representative's length is at most $|u| + N$. Replacing $(x_i, y_i)$ by $(y_i, x_i)$ in the previous paragraph, we see that similar estimates apply when we multiply $u$ by $x^{-1}$. Repeating this process, we see that we can find a representative of $\overline{w}$ of length at most $N|w|+n_0$, where $n_0$ is the length of a representative of the identity. The time taken is $O\bigl(\sum_{i=1}^{|w|}(iN+n_0)\bigr) = O(|w|^2)$. Finally, to make sure the entire process is constructive, notice that we can find explicitly an accepted representative of the identity from the automatic structure: we start with any accepted string $x_1 \dots x_n$ and use the procedure above to multiply successively by $x_n^{-1}, \dots, x_1^{-1}$. 2.3.10

**Open Question 2.3.11 (isomorphism problem).** Is the isomorphism problem solvable for synchronous automatic groups? That is, is there an algorithm that takes as its input two automatic groups and finds out whether they are isomorphic or not? (We conjecture that there is no such algorithm.)

**Theorem 2.3.12 (quadratic isoperimetric inequality).** *Every automatic group $G$ is finitely presented and satisfies a quadratic isoperimetric inequality, that is, if $\langle A/R \rangle$ is a presentation of $G$, any word $w$ representing the identity can be written in the form 2.2.2 with $n = O(|w|^2)$.*

*Proof of 2.3.12:* Let $L$ be the language of accepted words of an automatic structure over $A$. Lemma 2.3.9 (bounded length difference) gives constants $N$ and $n_0$ such that, for any word $w$, the prefixes $w(t)$ have representatives $u_t \in L$ of length at most $N|w| + n_0$. For each $t$, the elements $\overline{u_t}$ and $\overline{u_{t+1}}$ are one generator apart, say $x_t$. By Lemma 2.3.2 the uniform distance between the paths $\widehat{u_t}$ and $\widehat{u_{t+1}}$ is bounded by the lipschitz constant $k$ of $L$, so the loop $u_t x_t u_{t+1}^{-1}$ can be decomposed into at most $N|w| + n_0$ loops of length at most $2k+2$ (see Figure 2.16). Then $w$ itself (if it is a loop) can be decomposed into at most $N|w|^2 + n_0|w|$ loops of at most this length. We get a finite set of relators for $G$ by taking all loops of up to this length; in terms of this presentation, $G$ clearly satisfies a quadratic isoperimetric inequality. As we remarked at the end of Section 2.2, this implies a quadratic isoperimetric inequality under any presentation. 2.3.12

Theorem 2.3.12 is due to Thurston. Epstein had previously used the idea of cutting up the disk (without any counting argument) to show that automatic groups are finitely presented; this itself was a development of an argument by Cannon in his discussion of almost convex groups.

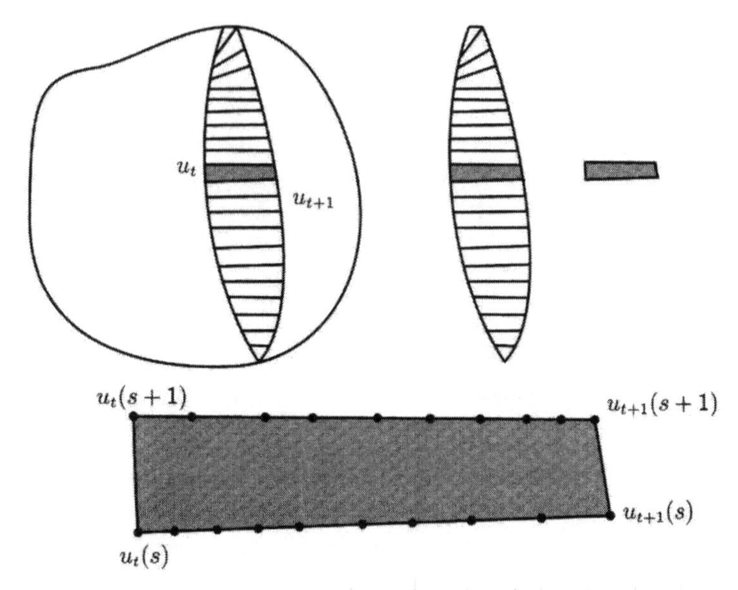


**Figure 2.16. Every loop bounds a disk.** To find a disk for a word $w$ representing the identity, we first take representatives $u_t$ for the prefixes $w(t)$; the $u_t$ can be chosen with length bounded by a multiple of $|w|$ (upper left). For two consecutive values of $t$, this gives us a wedge (upper centre), which can be divided into four-sided and three-sided regions (three-sided regions occur if $u_t$ and $u_{t+1}$ have different lengths). These regions have perimeter at most $2k+2$, where $k$ is a lipschitz constant for the structure (bottom).

## 2.4. Invariance under Change of Generators

In this section we show that the property of a group being automatic is invariant under change of generators. This is one of the main features that distinguishes this concept from many earlier approaches to combinatorial properties of groups, where the particular presentation is a vital ingredient. For example, the property of being a small cancellation group is not invariant under change of generators. Gromov's word hyperbolic groups (Definition 3.4.4) also have a definition independent of the choice of generators.

**Theorem 2.4.1 (changing generators).** *Let $(A_1, W_1)$ be an automatic structure on a group $G$ (see Convention 2.3.7). If $A_2$ be any other finite set of semigroup generators of $G$, there is a finite state automaton $W_2$ over $A_2$ such that $(A_2, W_2)$ is an automatic structure on $G$.*

Taken at its face value, this result gives somewhat too rosy a picture. Some groups may have a nice, very easily described, automatic structure with respect to one set of generators, but a very cumbersome one with respect to another. The situation is not well understood.

*Proof of 2.4.1:* We start by showing that we can add or delete a generator equal to the identity. Let $A_2 = A_1 \cup \{e\}$, where $e$ maps to the identity in $G$ (see Definition 2.1.1). We make $W_2$ equal to $W_1$, except that we take the underlying alphabet to be $A_2$; this means that every string containing $e$ fails. Then $(A_2, W_2)$ is an automatic structure on $G$, by Theorem 2.3.5 (characterizing synchronous), because the word metric does not change.

Next assume that $A_1 = A_2 \cup \{e\}$, where $e$ maps to the identity. If $A_2$ is empty, we have $A_1 = \{e\}$ and $G$ is the trivial group. The language $\{\varepsilon_\varnothing\}$ over the empty alphabet gives an automatic structure on $G$ by Theorem 2.3.5. Therefore we many suppose that $A_2$ is non-empty.

To obtain $L(W_2)$, we cannot simply delete $e$ whenever it appears in a string in the language $L(W_1)$, because Theorem 2.3.5 requires that a pair $(\widehat{w_1}, \widehat{w_2})$ of accepted paths whose endpoints are within distance one of each other in the Cayley graph should be uniformly near each other. Deletion of $e$'s may cause these paths to cease to keep pace with each other. Instead, we find a string $z$ over $A_2$ that also maps to the identity, and has non-zero length $m$. (For instance, we can take $x \in A_2$ and $u \in A_2^*$ that $\overline{u} = \overline{x}^{-1}$, and set $z = xu$.) We then define the new language $L_2$ of accepted strings as follows: for each string of $L(W_1)$, we replace every $m$-th occurrence of $e$ by $z$, and delete all other occurrences of $e$ (Figure 2.17).

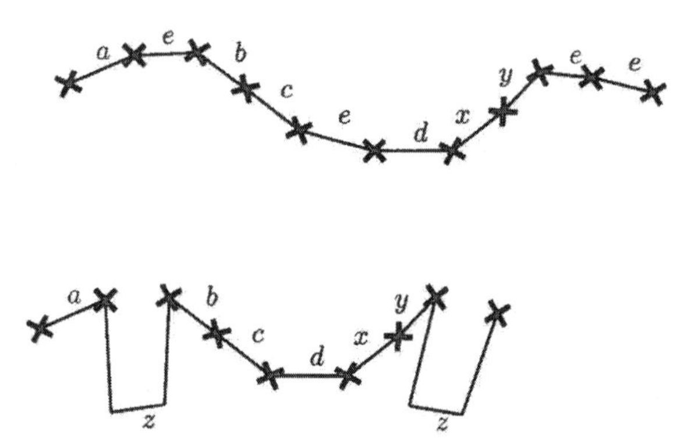


**Figure 2.17. Removing an identity.** To eliminate occurrences of the generator $e$, which is equal to the identity in $G$, we replace every third occurrence of $e$ by the string $z$ of length three, which also represents the identity.

To show that the $L_2$ is regular, we construct a generalized nondeterministic finite state automaton over $A_2$ that accepts $L_2$. A state of this automaton is of the form $(s, i)$, where $s$ is a state of $W_1$ and $i$ is an integer satisfying $0 \leq i < m$, which indicates by how much the modified string is out of phase with respect to the original one. The start state is $(s_0, 0)$, where $s_0$ is the start state of $W_1$. The accept states in $W_2$ are any states of the form $(s, i)$, where $s$ is an accept state of $W_1$. Each arrow $(s, x, t)$ of $W_1$ gives rise to $m$ arrows in the new automaton, as follows: If $x \neq e$, the new arrows are $((s, i), x, (t, i))$ for $0 \leq i < m$. If $x = e$, the new arrows are $((s, 0), z, (t, m-1))$ and $((s, i), \varepsilon, (t, i-1))$ for $1 \leq i < m$.

A string in $L_2$ that leaves the new automaton in the accept state $(s, i)$ is $i$ characters longer than some string accepted by $W_1$ and represents the same element in $G$. Since $i$ is bounded by $m$, the conditions of Theorem 2.3.5 are satisfied.

Having proved the case where $A_2$ is obtained from $A_1$ by adjoining or removing a representative of the identity element, we tackle the case of $A_2$ arbitrary. We assume, without loss of generality, that $A_1$ and $A_2$ contain an

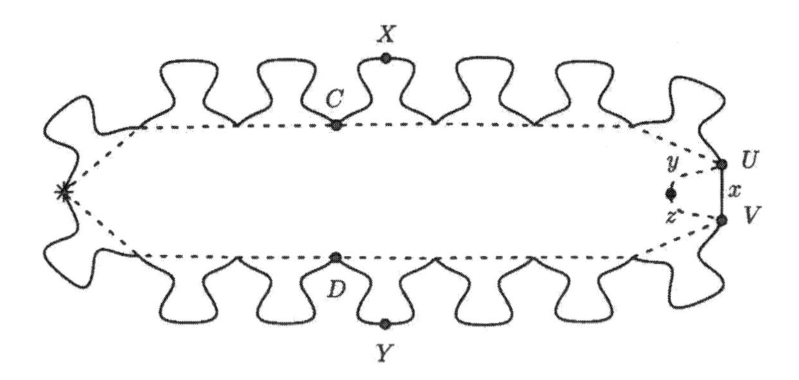


**Figure 2.18. Change of generators I.** We have two sets of generators $A_1$ and $A_2$ for $G$, giving rise to Cayley graphs $\Gamma_1$ and $\Gamma_2$. We indicate edges in $\Gamma_1$ with dotted lines and edges in $\Gamma_2$ with solid lines. We show a pair of accepted strings in $\Gamma_2$ (solid lines), representing elements $U$ and $V$ of $G$ that differ by right multiplication by the element $x \in A_2$. These strings are obtained, by substitution, from strings over $A_1$ (dotted lines). The points $X$ and $Y$ on the solid paths correspond to prefixes of equal length: we must show that their distance in $\Gamma_2$ is bounded. We start by finding points $C$ and $D$ on the dotted paths, also corresponding to prefixes of equal length, and such that the distances $d(X, C)$ and $d(Y, D)$ are at most than $c/2$. We also express $x \in A_2$ as a product of at most $c'$ generators in $A_1$; here $x = yz$.

identity element. We choose $c > 0$ such that any element of $A_1$ can be written as a string of length at most $c$ over $A_2$, and $c'$ such that any element of $A_2$ can be written as a string of length at most $c'$ over $A_1$. For each $x \in A_1$, we take a string $u_x \in A_2$ of length $c$ representing $x$ (we use the trivial generator to pad the length as needed).

We define the new language $L_2$ by replacing $x$ by $u_x$ in the strings of $L(W_1)$. By Lemma 1.2.12 (effect of map), $L_2$ is regular. Clearly the map from $L(W_2)$ to $G$ is surjective. Let $k$ be the bound given by Theorem 2.3.5 for $(A_1, W_1)$. As explained in the captions of Figures 2.18, 2.19 and 2.20, the condition of Theorem 2.3.5 is then satisfied for $(A_2, L_2)$ with bound $c + kcc'$. 2.4.1

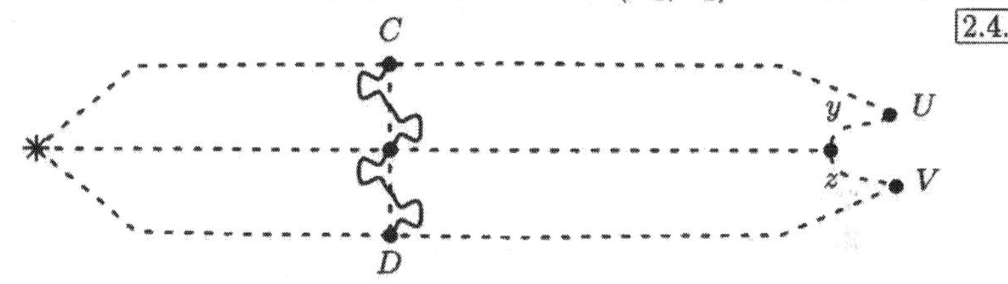


**Figure 2.19. Change of generators II.** We bridge the gap between $U$ and $V$ in $\Gamma_1$ by a sequence of at most $c' - 1$ points adjacent to one another, using the expression of $x$ in terms of generators in $A_1$. We draw auxiliary paths from the basepoint to these points, and mark off points corresponding to the prefixes of same length as $C$ and $D$. By assumption, the distance in $\Gamma_1$ between each of these auxiliary points and the next is at most $k$, so the distance in $\Gamma_1$ between $C$ and $D$ is at most $kc'$. Therefore the distance in $\Gamma_2$ between $C$ and $D$ is at most $kcc'$.

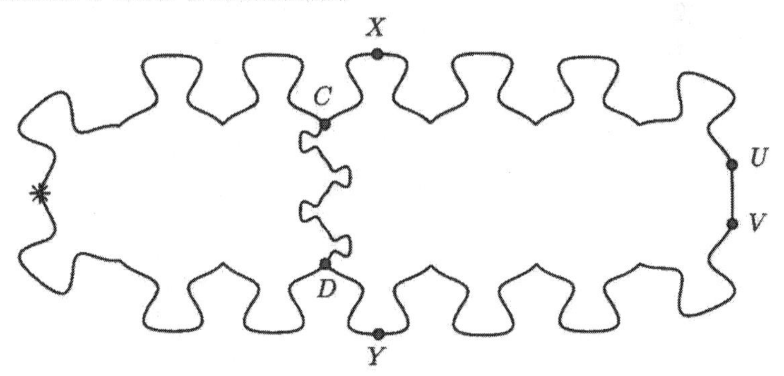


**Figure 2.20. Change of generators III.** Adding the bounds from the previous two figures, we conclude that the distance between $X$ and $Y$ in $\Gamma_2$ is at most $c + kcc'$.

## 2.5. Improving the Automatic Structure

The exposition in this section owes a lot to John Sullivan.

By itself, the language of accepted strings of an automatic group may tell one nothing about the group (see Example 2.3.6). In this section we introduce several interesting properties that the language may satisfy. These properties tend to make theoretical arguments easier to carry out and computer programs easier to implement. For example, it is generally desirable for the language of accepted words to have unique representatives for group elements (we recall that such a language is said to have the uniqueness property).

Let $A$ be an *ordered* alphabet. Recall that *lexicographic order* ranks strings of the same length in $A^*$, by comparing the letters in the first position where the strings differ. We define three order relations on $A^*$:

(1) In the *length order*, $v < w$ if and only if $v$ is shorter than $w$. This is a partial order, and does not depend on the order in $A$.
(2) In the *ShortLex order*, $v < w$ if and only if $v$ is shorter than $w$, or they have the same length and $v$ comes before $w$ in lexicographical order. ShortLex order is a well-ordering.
(3) In the *dictionary order*, $v < w$ if and only if $v$ is a proper prefix of $w$, or $v$ and $w$ have prefixes $v'$ and $w'$ of the same length such that $v'$ comes before $w'$ in lexicographical order. Dictionary order is a total order, but not a well-ordering.

Now let $G$ be a group and $A$ a set of semigroup generators for $G$. A string $w \in A^*$ is called a *geodesic* if it has minimal length among all strings representing the same element as $w$. We saw in Example 2.1.4 that the language of all geodesic strings maps finite-to-one onto $G$, but in general this language does not have to be regular, or even recursively enumerable. If the language of all geodesics is part of an automatic structure for $G$, we say that $G$ is *strongly geodesically automatic*; an example is the free group on $n$ generators $x_1, \ldots, x_n$, with alphabet $\{x_1, \ldots, x_n, X_1, \ldots, X_n\}$. If some language consisting only of geodesics is part of an automatic structure for $G$, we say that $G$ is *weakly geodesically automatic*. Theorem 2.5.1 (uniqueness) below implies that a group with an automatic structure containing a geodesic representative for each element is weakly geodesically automatic.

A string $w \in A^*$ is a *ShortLex-geodesic* if it is the minimum in the ShortLex order among all strings representing the same element as $w$. Again, the language of all ShortLex-geodesics may be very badly behaved. If this language is part of an automatic structure, we say the group is *ShortLex-automatic*.

**Theorem 2.5.1 (uniqueness).** *Let $G$ be an automatic group with automatic structure $(A, L)$. Let $L' \subset L$ be the set of strings $w \in L$ such that $w$ is ShortLex-minimal in $L \cap \pi^{-1}(\overline{w})$. Then $(A, L')$ is an automatic structure for $G$. In particular, $G$ has an automatic structure over $A$ with the uniqueness property.*

*Proof of 2.5.1:* Let $M_\varepsilon$ be the equality recognizer for the given structure. We have

$$L' = \{w \in L : (\forall v)((w, v) \in L(M_\varepsilon) \Rightarrow w \leq v \text{ in the ShortLex order})\}.$$

Since the predicate "$w \leq v$ in the ShortLex order" can be checked by a finite state automaton, Corollary 1.4.7 (predicates closed) implies that $L'$ is a regular language, and therefore part of an automatic structure for $G$ by Corollary 2.3.8 (restriction is automatic). [2.5.1]

Notice that any total order which can be recognized by some finite state automaton would work equally well in this theorem.

**Corollary 2.5.2 (geodesic hierarchy).** *A strongly geodesically automatic group is ShortLex-automatic, for any ordering of the generators. If a group is ShortLex-automatic, for some ordering of the generators, it is weakly geodesically automatic.*

*Proof of 2.5.2:* The first assertion follows from the proof of Theorem 2.5.1. The second is trivial. [2.5.2]

The hierarchy in Corollary 2.5.2 is mostly strict. The free abelian group on two generators $x$ and $y$ cannot be strongly geodesically automatic with respect to the generators $\{x, y, X, Y\}$, because the strings $x^m y^n$ and $y^n x^m$ contradict the criterion in Theorem 2.3.5. But it is easy to check that the group is ShortLex-automatic with respect to any ordering of these generators; indeed, Theorem 4.3.1 will show that any finitely generated abelian group is ShortLex-automatic with respect to any generators. Next, Example 6.2.2 (confluent but not ShortLex-automatic) shows that a group can be ShortLex-automatic with respect to one ordering of the generators, but not with respect to another ordering. Finally, we will see in Section 3.5 that a group may be automatic, but not weakly geodesically automatic.

**Open Question 2.5.3.** Find a group that is weakly geodesically automatic, but not ShortLex-automatic for any ordering of the generators.

The requirement that an automatic structure $(A, L)$ be strongly geodesic is very strong. If the set $A$ is closed under inversion (see Convention 2.1.2), it implies that the automatic structure is *symmetric*, that is, it satisfies $L = L^{-1}$, where $L^{-1}$ is the set of formal inverses of strings in $L$. This means that each pair of elements of $G$ is assigned a set of canonical paths in the Cayley graph from one element to the other; the paths in this set are uniformly close to each other. If the pair is interchanged, the paths are reversed. In addition, the set of labels of these paths must give a regular language, and the paths must vary in a uniform way as the two endpoints are changed. The set of all such paths, as the endpoints vary, is required to be invariant under left translation by elements of $G$.

Word hyperbolic groups (Theorem 3.4.5), finitely generated abelian groups (Theorem 4.1.3) and braid groups (Theorem 9.3.6) can all be given symmetric automatic structures, but only in the case of word hyperbolic groups is the structure also strongly geodesic.

A weaker concept of symmetry is provided by the following definition:

**Definition 2.5.4 (biautomatic).** Let $G$ be an automatic group with automatic structure $(A, L)$, where $A$ is closed under inversion (Convention 2.1.2). We say that the structure is *biautomatic* if $(A, L^{-1})$ is also an automatic structure.

**Lemma 2.5.5 (characterizing biautomatic).** *Let $G$ be a group and let $A$ be a finite set of generators closed under inversion. Let $L$ be a regular language over $A$ that maps onto $G$. Then $L$ is biautomatic if and only if, for each $w \in L$ and each pair of generators $x, y \in A$, the uniform distance between the path $xwy$ and the path corresponding to any element of $L$ representing $\overline{xwy}$ is bounded by a fixed constant $k > 1$.*

*Proof of 2.5.5:* By Theorem 1.2.8 (reversal), the language obtained by reversing each string of $L$ is regular. Thus $L^{-1}$ is also regular, since we can obtain it from the reversal of $L$ by replacing each generator $x$ by $x^{-1}$. The result now follows from Theorem 2.3.5 (characterizing synchronous). 2.5.5

This result is similar to Theorem 2.3.5 (characterizing synchronous), except that we allow multiplication by generators on the left as well as on the right. It is clear from this lemma and from the proof of Theorem 2.5.1 that any biautomatic structure can be replaced by one with uniqueness. The lemma also implies that any automatic structure on an abelian group is biautomatic.

**Open Question 2.5.6.** Example 4.4.1 (order matters) gives an example of an automatic structure that is not biautomatic. But we don't know of any automatic group that can be proved to have no biautomatic structure (compare Open Question 4.1.5 and the last paragraph of Chapter 8).

The property of a group being biautomatic does not depend on the choice of generators. The proof is very similar to that of Theorem 2.4.1 (changing generators); details are left to the reader.

The concept of a biautomatic group is due to Thurston. S. M. Gersten and H. Short have used it extensively; some of their results are discussed in see Section 8.3. Here we give only the following result from [GS00b]:

**Theorem 2.5.7 (biautomatic implies solvable conjugacy problem).** *If $G$ has a biautomatic structure, the conjugacy problem is solvable in $G$, that is, one can algorithmically determine whether or not two words represent conjugate elements in $G$.*

*Proof of 2.5.7:* Let $A$ be the set of generators and let $L$ be the language of accepted words. Given a word $p$ over $A$, we can construct a finite state automaton $M_p$ in two variables that accepts a pair of strings $(u, v)$ over $A$ if and only if $\overline{up} = \overline{v}$ and $u$ and $v$ are both in $L$. We do this by induction: if $p = p'x$, where $x \in A$, the language

$$\{(u,v) : u \in L \wedge v \in L \wedge \overline{up} = \overline{v}\}$$

is equal to the language

$$\{(u,v) : (\exists w)((u,w) \in L(M_{p'}) \wedge (w,v) \in L(M_x))\},$$

where $M_x$ is a multiplier automaton for $x$ and we assume that $M_{p'}$ has already been constructed. By Theorem 1.4.6 (predicate calculus), the latter language is regular, which completes the induction.

Since $G$ is biautomatic, we can also find in a similar way a finite state automaton $M^q$ that accepts a pair $(u, v)$ if and only if $u$ and $v$ are both elements of $L$ and $\overline{u} = \overline{qv}$. It follows that there is a finite state automaton $M_p^q$ that accepts a pair $(u, v)$ if and only if $u$ and $v$ are both in $L$ and $\overline{up} = \overline{qv}$. The conjugacy problem is solved by asking whether $M_p^q$ accepts some string of the form $(u, u)$. 2.5.7

The algorithm above is expensive in both time and space, since the process to form $M_{px}$ from $M_p$ and $M_x$ is potentially exponential in the number of states of the two given automata, so the whole process is multiply exponential. Gersten and Short have a singly exponential algorithm (not yet published).

**Open Question 2.5.8.** Is the conjugacy problem solvable for automatic groups?

We now turn to another useful property of languages:

**Theorem 2.5.9 (prefix closure of automatic groups).** *Let $(A, L)$ be an automatic structure on a group $G$, and let $L'$ be the prefix closure of $L$. Then $(A, L')$ is an automatic structure on $G$. Furthermore, suppose there is $n > 0$ such that the number of representatives of any element of $G$ in $L$ is bounded by $n$. Then the number of representatives of any element of $G$ in $L'$ is bounded by $nc$, where $c$ is the number of states of a finite state automaton accepting $L$.*

*Proof of 2.5.9:* We know that $L'$ is regular by Theorem 1.2.11 (prefixes): if $W$ is a finite state automaton with language $L$, the automaton obtained by making every live state of $W$ into an accept state accepts $L'$. Let $c$ be the number of states in $W$.

To check that the hypotheses of Theorem 2.3.5 are verified, let $w_1$ and $w_2$ be prefixes of strings in $L$, representing elements of $G$ a distance at most one apart in the Cayley graph. Let $u_1$ and $u_2$ be strings of length less than $c$ such that $w_1u_1$ and $w_2u_2$ are in $L$. Since $\overline{w_1u_1}$ and $\overline{w_2u_2}$ are at most $2c+1$ apart, the paths $\widehat{w_1u_1}$ and $\widehat{w_2u_2}$ are a uniform distance at most $k(2c+1)$ apart, where $k$ is the constant in Theorem 2.3.5 applied to $L$. It follows that $\widehat{w_1}$ and $\widehat{w_2}$ are a uniform distance at most $k(2c+1)+2c$ apart.

To prove the last statement of the theorem, let $C_g = L' \cap \pi^{-1}(g)$, for $g \in G$. We partition $C_g$ into sets $C_{g,s}$, where $w \in C_{g,s}$ if $W$ is in state $s$ after reading $w$. For each $s$, we take a string $u_s$ leading from $s$ to an accept state; then the strings $wu_s$, for $w \in C_{g,s}$, lie in $L$ and represent the same element of $G$. By assumption, there are at most $n$ such strings. Therefore $C_g$ has at most $cn$ elements. 2.5.9

**Open Question 2.5.10.** We have shown that every automatic structure can be made to have the uniqueness property or to be prefix-closed. Can the two properties be achieved simultaneously? More precisely, given an automatic group $G$ with generators $A$, can one find an automatic structure over $A$ whose language is both prefix-closed and has the uniqueness property? How about if one is allowed to change the set of generators? The answer to each of these questions is probably negative.

Be warned that this seems to be a tricky question; several researchers have claimed proofs that later turned out to be flawed. Nonetheless, we have the following partial result:

**Proposition 2.5.11 (geodesic automaton ShortLex).** *Let $(A, L)$ be an automatic structure on a group $G$, and assume that $L$ includes* ShortLex$(G, A)$ *(this is certainly the case if $G$ is strongly geodesically automatic). Then $G$ has an automatic structure that is prefix-closed and has the uniqueness property.*

These hypotheses are fulfilled for word hyperbolic groups (see Theorem 3.4.5) and abelian groups (see Theorem 4.3.1), for example.

*Proof of 2.5.11:* We know from Theorem 2.5.1 that $(A, \mathsf{ShortLex}(G, A))$ is an automatic structure for $G$ having the uniqueness property. We need to show that $\mathsf{ShortLex}(G, A)$ is prefix-closed; but this is clear because the ShortLex order is preserved under left and right multiplication by elements of $A^*$.

2.5.11

We conclude this section with another corollary of Theorem 2.5.1 (uniqueness), which has been noted independently by a number of different researchers.

**Example 2.5.12 (infinite torsion).** An infinite group all of whose elements have finite order cannot be automatic. For assume, to the contrary, that it is; then it has an automatic structure with uniqueness. Let $L$ be the language of accepted words for this structure. By Theorem 1.2.13 (pumping lemma), there is some integer $n$ such that any string in $L$ having length greater than $n$ is of the form $uvw$, with $|v| > 0$ and $uv^iw \in L$ for all $i$. But this is impossible, because $v$ has finite order and $L$ has the uniqueness property. Therefore no strings in $L$ have length greater than $n$, and $L$ is finite.

The existence of a finitely generated infinite torsion group is highly nontrivial; a comprehensible construction appears in [Gri80]. The study of torsion groups, and the conditions under which they are infinite, is called the *Burnside problem*. It is still an open question whether there exists a finitely presented infinite torsion group.

# Chapter 3

# Quasigeodesics, Pseudoisometries and Combings

The origins of this book lie in the work of Cannon [Can84], who showed the existence of a recursive structure on the set of geodesics of a discrete group of isometries of hyperbolic space. In this chapter we investigate how an automatic structure can be derived from such properties of the geodesics, and we show that large classes of geometrically motivated groups have automatic structures.

We start by recapitulating some standard terminology in Section 3.1. In Section 3.2 we give two criteria for a group to be (strongly or weakly) geodesically automatic, one due to Cannon (Theorem 3.2.1) and another to Epstein (Theorem 3.2.2). (A further generalization is provided in Theorem 11.2.2.)

In applying these theorems, the central idea, which goes back to Dehn [Deh12, Deh87], is that many qualitative features of a group can be seen in any space on which the group acts properly discontinuously and with compact quotient. The Cayley graph is one such space. For a discrete subgroup of a Lie group, a more natural choice might be some homogeneous space. Making this point precise involves understanding in what sense two metric spaces can be "almost the same," and a path can be "almost geodesic;" hence the definitions of *pseudoisometries* and *quasigeodesics* (Section 3.3).

Section 3.4 uses these ideas and the criteria of Section 3.2 to show that fundamental groups of negatively curved manifolds (3.4.1) and word-hyperbolic groups (3.4.5) are automatic.

In Section 3.5 we present an example, due to Paterson, of an automatic group that is not even weakly geodesically automatic. (Section 4.4 will give an example of a group that is ShortLex-automatic but not strongly geodesically automatic).

Finally, in Section 3.6, we introduce the concept of *combable groups*, which are like automatic groups without the automata. A combing assigns to each vertex of the Cayley graph a set of paths from the basepoint to the vertex, so that nearby vertices are assigned nearby paths. The proximity condition is purely geometric in nature, in contrast with the situation for automatic structures. Although combable groups have less structure than automatic ones, many interesting results can be proved for them as well.

## 3.1. Metric Spaces, Path Metrics and Geodesics

In this section we define several well-known geometric concepts, to fix the terminology. The most important of them are geodesics in a metric space, path metric spaces and geodesic spaces.

By an *interval* we mean a non-empty connected subset of the real line. Thus, an interval may be closed or open at each end, finite or infinite at each end, and may also reduce to a single point, in which case the two endpoints are equal.

Let $(X, d)$ be a metric space. (We will use a slightly extended definition of a metric space where the distance between two points is allowed to be infinite.) A *path* in $X$ is a continuous map $\alpha : J \to X$, where $J$ is an interval. Let $J$ have endpoints $a$ and $b$, where $a$ and $b$ may be infinite. The *length* of $\alpha$ is defined to be zero if $a = b$; otherwise, it is the supremum over all partitions $a < t_0 \leq t_1 \leq \cdots \leq t_k < b$ of $J$, as $k \geq 0$ varies, of the number $\sum_{i=1}^{k} d(\alpha(t_{i-1}), \alpha(t_i))$. The triangle inequality shows that this sum cannot decrease as the partition gets finer. For many paths the supremum will be infinite, even if the domain is compact. If, on each compact subinterval of the domain of $\alpha$, the supremum is finite, we say that $\alpha$ is *rectifiable*. For example, the path $t \mapsto (t, t\sin(1/t))$ in $\mathbf{R}^2$, defined on $(0, 1)$, is rectifiable, but its continuous extension to $[0, 1)$ is not. The path $t \mapsto (t, 0)$, defined on $(-\infty, \infty)$, is rectifiable.

Given a metric space $(X, d)$, we define the *path metric* $d'$ on $X$ by setting $d'(x_1, x_2)$ equal to the infimum of lengths of rectifiable paths from $x_1$ to $x_2$ (Figure 3.1). (The infimum is infinity if there is no such path.) The $d$-topology and the $d'$-topology on $X$ may differ, even if $d'$ is finite.

We have $d \leq d'$ and the $d$-length of every path is equal to its $d'$-length. It follows that a path is $d$-rectifiable if and only if it is $d'$-rectifiable and that $d' = d''$. If $d = d'$, we say that $(X, d)$ is a *path metric space*. Therefore, for any metric space $(X, d)$, the new metric space $(X, d')$ is automatically a path metric space. A common and fundamental example of a path metric space is a riemannian manifold.

The word "geodesic" is used in more than one sense in riemannian geometry. Sometimes it means a path with certain local properties, and sometimes

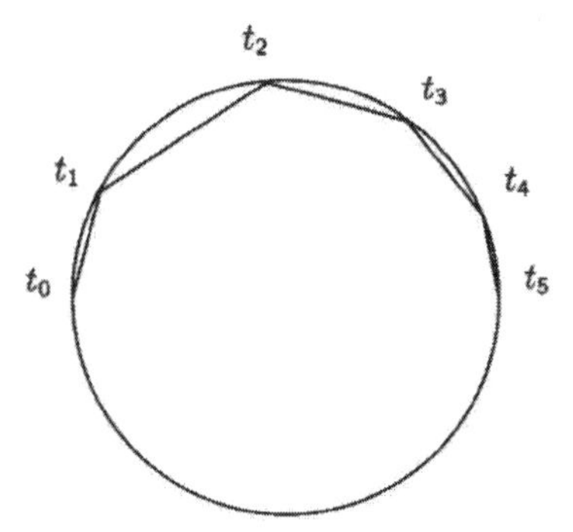


**Figure 3.1. The path metric on a circle.** We start with the metric $d$ induced on the circle (the curve, not the disk) as a subspace of the euclidean plane. The limit of the lengths of paths, as the subdivision gets finer and finer, is the path metric $d'$. We have $d'((1,0),(-1,0)) = \pi$, while $d((1,0),(-1,0)) = 2$.

a subset of the riemannian manifold which is the image of a path with certain local properties. In this book, we will restrict the meaning so that it can only refer to a path with certain global properties.

A *shortest path* in a metric space $(X, d)$, with domain an interval $J$, is a map $\alpha : J \to X$ such that, for any subinterval $[a, b] \subset J$, the length of $\alpha|_{[a,b]}$ is $d(\alpha(a), \alpha(b))$. If this equality holds for some $[a, b]$, it clearly holds also for each closed subinterval. We say that $\alpha$ is a *geodesic* if there is a constant $s$, called the *speed* of the geodesic, such that $d(\alpha(a), \alpha(b)) = s(b - a)$ for each subinterval $[a, b]$ of $J$. It is easy to see that a geodesic is a shortest path.

By a change of parametrization, a shortest path may be parametrized by path length, giving rise to a unit speed geodesic. A unit speed geodesic is determined by its image, up to direction and the starting point of the parameter.

A metric space $X$ is said to be a *geodesic space* if it is path metric and if any pair of points in $X$ can be joined by a geodesic. It is easy to prove [CEG87] that a complete, locally compact, connected, path metric space is a geodesic space and that a closed metric ball in such a space is compact.

We have already seen on page 37 how to metrize the Cayley graph of a group. In this metric the Cayley graph is a geodesic space. In fact, any graph can be metrized in the same way, yielding a geodesic space.

## 3.2. Shortest Strings

If a group is automatic, it will have many different automatic structures, even if the generators are fixed. The most important of these have the property that each accepted string is geodesic; we recall from Section 2.5 that a group having such an automatic structure is called weakly geodesically automatic. A stronger concept is that the language of *all* geodesics (see Example 2.1.4) should give an automatic structure; then the group is called strongly geodesically automatic. The reader should be clear that, in general, the language of all geodesics is not even regular—if it is, the group has a solvable word problem by Theorem 2.1.9 (regularly generated). However, it is a beautiful feature of certain important groups that this language is regular for some choice of generators.

In fact, the very concept of an automatic group originated from Cannon's study of the language of geodesics of discrete groups of hyperbolic transformations [Can84]. For the groups investigated by Cannon, as for many others, the language of geodesic strings is regular, no matter what the choice of generators. In this section we give a criterion, Theorem 3.2.1 (geodesic automaton 1), for a group to satisfy this condition. This result is implicit in the work of Cannon, and his original proof is given. Theorem 3.2.1 can also be derived from a more general criterion, Theorem 3.2.2 (geodesic automaton 2), due to Epstein. The proof of Theorem 3.2.2 is adapted from a proof by Holt of Theorem 3.2.1.

Let $G$ be a group and let $A$ be a non-empty finite set of generators for $G$. We will assume throughout this section that $A$ is closed under inversion (see Convention 2.1.2), so that geodesic strings are effectively shortest paths in the Cayley graph. For any string $\alpha$, the *cone type* $C(\alpha)$ of $\alpha$ is defined to be the set of strings $\gamma$ such that $\alpha\gamma$ is a geodesic. Clearly $C(\alpha)$ is non-empty if and only if $\alpha$ is a geodesic, for it must then include the nullstring. Also, if $\alpha$ is a geodesic, $C(\alpha)$ depends only on $\overline{\alpha}$, so we can define the *cone type* $C(g)$ of an element $g \in G$ as $C(g) = C(\alpha)$, where $\alpha$ is a geodesic string representing $g$. All non-geodesic strings $\alpha$ define the same cone type, namely the nullset, which we call the *failure type*.

We will now show how to form an automaton whose set of states is the set of cone types. If the number of cone types is finite, this will be a finite state automaton. Suppose $X$ is a cone type and $x \in A$. We write $X = C(\alpha)$ for some string $\alpha$, and define $\mu(X, x) = C(\alpha x)$. To see that this is well-defined, note that $\gamma \in C(\alpha x)$ if and only if $\alpha x\gamma$ is a geodesic string; this happens if and only if $x\gamma \in X$. So $\mu(X, x) = \{\gamma \mid x\gamma \in X\}$, which is independent of the choice of an $\alpha$ defining $X$.

The start state of the automaton is $C(e)$, the cone type of the identity element. Every cone type, except the failure type, is defined to be an accept state. The language accepted by this automaton is clearly the set of all geodesic representatives in $A^*$ of elements of $G$. For many groups, the automaton we have constructed is infinite. If there are only a finite number of cone types, we may apply Theorem 2.1.9 (regularly generated) to see that the word problem is solvable.

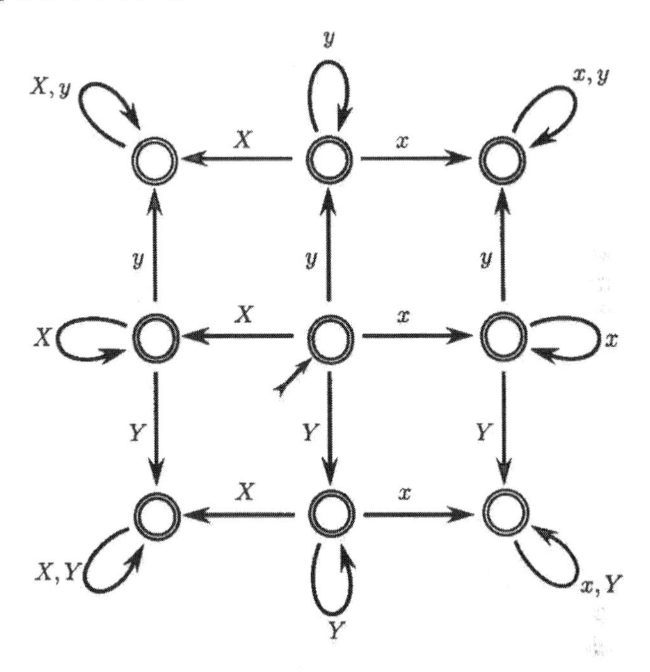


**Figure 3.2. Automaton for free abelian group.** This is a partial automaton for the free abelian group on generators $x$ and $y$ with inverses $X$ and $Y$ respectively. The automaton accepts all strings which are shortest representatives of some group element. Each group element $g$ is represented by some finite number of distinct strings, but there is no upper bound to this number as $g$ varies. The failure state has been omitted.

In the case of the free abelian group on two generators, we get a finite state automaton with ten states (if we include the failure state), which recognizes the language of geodesics—see Figure 3.2. However, this automaton is not part of an automatic structure, because two geodesics from the basepoint to another vertex of the Cayley graph can be arbitrarily far from each other and this contradicts Lemma 2.3.2 (lipschitz property).

In the case of the free group on two generators, we get a finite state automaton with six states, and this time it is part of an automatic structure (Figure 2.1).

Our next result is fundamental: it says that if geodesics going to nearby points are close in the hausdorff metric, the set of geodesics gives an automation. We recall that if $X$ is a metric space and $A, A' \subset X$ are closed subsets, the *hausdorff distance* between $A$ and $A'$ is the infimum of the values of $r$ such that $A \subset \bigcup_{x \in A'} B_r(x)$ and $A' \subset \bigcup_{x \in A} B_r(x)$, where $B_r(x)$ is the ball of radius $r$ around $x$. In words, every point of $A$ is at a distance of at most $r$ from some point in $A'$, and vice versa.

**Theorem 3.2.1 (geodesic automaton 1).** *Let $G$ be a group and $A$ a finite set of semigroup generators, closed under inversion. Suppose that there is a number $k > 1$ such that, for any two geodesic strings $u$ and $v$ over $A$ whose images under $\pi : A^* \to G$ are at distance at most one from each other, the hausdorff distance between the paths $\widehat{u}$ and $\widehat{v}$ in the Cayley graph $\Gamma(G, A)$ is at most $k$. Let $L_0$ be the language given by the set of geodesic strings over $A$. Then $(A, L_0)$ is an automatic structure for $G$ (and is even symmetric: see page 58).*

*Proof of 3.2.1:* This will follow from the next result, where we take $L = A^*$. [3.2.1]

This criterion can be used to prove that many different groups are automatic. In particular, Cannon's original results can be recovered if in addition we apply some standard geometric arguments: see Theorem 3.4.1 (negative curvature).

**Theorem 3.2.2 (geodesic automaton 2).** *Let $G$ be a group and $A$ a finite set of semigroup generators, closed under inversion. Let $L$ be a prefix-closed regular language over $A$. Suppose that there is a number $k > 1$ such that, for any two geodesic strings $u$ and $v$ in $L$ whose images under $\pi : A^* \to G$ are at distance at most one from each other, the hausdorff distance between the paths $\widehat{u}$ and $\widehat{v}$ in the Cayley graph $\Gamma(G, A)$ is at most $k$. Let $L_0$ be the language given by the set of geodesic strings in $L$, and suppose that $L_0$ maps onto $G$. Then $(A, L_0)$ is an automatic structure for $G$.*

This generalization has independent applications which cannot be deduced from Theorem 3.2.1. We will give a proof of Theorem 3.2.2 that depends on the fact that the set of regular languages is closed under the operations of predicate calculus: see Corollary 1.4.7 (predicates closed). We will also give an independent proof of Theorem 3.2.1 which follows Cannon's original proof [Can84]. The second proof can be generalized to prove Theorem 3.2.2, but at the cost of complicating the definition of a cone type, so as to bring in the role of $L$.

In each of the proofs we will need the following lemma:

**Lemma 3.2.3 (hausdorff implies uniform).** *Let $\alpha$ and $\beta$ be two elements of $L$ giving rise to geodesics $\widehat{\alpha}$ and $\widehat{\beta}$ in $\Gamma$ starting at the basepoint and ending a distance at most one apart. Then $\widehat{\alpha}$ and $\widehat{\beta}$ are uniformly at most $2k$ apart.*

*Proof of 3.2.3:* Let $|\alpha| \geq t \geq 0$. By hypothesis, there is an $s$ with $|\beta| \geq s \geq 0$, such that $d(\widehat{\alpha}(t), \widehat{\beta}(s)) \leq k$. Now $|s - t| \leq k$, for otherwise either $\widehat{\alpha}$ or $\widehat{\beta}$ would not be a geodesic. Hence $d(\widehat{\alpha}(t), \widehat{\beta}(t)) \leq 2k$ for all $t$. [3.2.3]

*Proof of 3.2.2:* By Theorem 2.3.5 (characterizing synchronous), we need only show that $L_0$ is a regular language. The problem is to express the condition of being a geodesic, using only finite state automata. The idea of the proof is to ensure that each prefix is a geodesic.

Let $V$ be an automaton accepting $L$. Let $N$ be the ball of radius $2k$ in $G$, with centre the identity element. We construct the standard automata $M_x$, for $x \in A \cup \{\varepsilon\}$, based on $(V, N)$. The language

$$L_x = \{w \in L : (w, w') \in L(M_x) \Rightarrow |w| < |w'|\}$$

is regular by Corollary 1.4.7 (predicates closed). Likewise,

$$L' = \left(\bigcup_{x \in A} L_x x \cup \{\varepsilon\}\right) \cap L$$

is regular. We let $L''$ be the largest prefix-closed subset of $L'$; by Theorem 1.2.11 (prefixes), $L''$ is also regular.

We claim that $L''$ coincides with the language $L_0$ of geodesic strings in $L$. The nullstring is geodesic and is contained in $L$ since $L$ is prefix-closed. Therefore $\varepsilon \in L_0$ and $\varepsilon \in L''$. Now take $w \in L''$ with $|w| > 0$, and suppose that $w = ux$, for some $x \in A$. Then $u \in L''$ since $L''$ is prefix-closed, and, by induction, $u$ is a geodesic string in $L$. Let $v \in L$ be a geodesic string with $\overline{v} = \overline{ux}$. By the hypotheses and Lemma 3.2.3, $v$ and $u$ are uniformly at most $2k$ distant from each other. Since $u$ and $v$ are both in $L$, the definition of $M_x$ shows that $(u, v)$ is accepted by $M_x$. By the definition of $L''$, we have $ux \in L'$ and so $u \in L_x$, which implies that $v$ is strictly longer than $u$. But this implies that $w = ux$ is also a geodesic.

Conversely, suppose that $w$ is a geodesic in $L$; we must show that $w \in L''$. We will show by induction on the length that any prefix $u$ of $w$ is in $L'$. The case of the null prefix being trivial, we assume that $u = vx$ for some $x \in A$. Since $L$ is prefix-closed, $v \in L$. We must show that $v \in L_x$, which is to say that, if $(v, v')$ is accepted by $M_x$, then $v'$ is longer than $v$. But $u$ is a geodesic representing the same element of $G$ as $v'$. Hence $|v'| \geq |u| > |v|$. 3.2.2

We now give the second proof, which is more geometric, and which follows Cannon's original discussion [Can84]. As remarked above, we will restrict ourselves to proving Theorem 3.2.1 (geodesic automaton 1).

*Proof of 3.2.1:* Let $n > 0$ be an integer. Let $B$ be the closed ball of radius $n$ in the Cayley graph of $G$, centred at the identity element. The *n-level* of an element $g \in G$ is defined to be the set of elements $h \in B$ such that $d(gh, e) < d(g, e)$, where $e$ is the identity element of $G$. Note that the $n$-level of $g$ is empty if and only if $g = e$.

**Lemma 3.2.4 (Cannon).** *If the hypotheses of Theorem 3.2.1 (geodesic automaton 1) are satisfied, the $(2k+1)$-level of an element determines its cone type.*

*Proof of 3.2.4:* Let $u$ and $u_1$ be geodesic strings, and suppose that $\overline{u}$ and $\overline{u_1}$

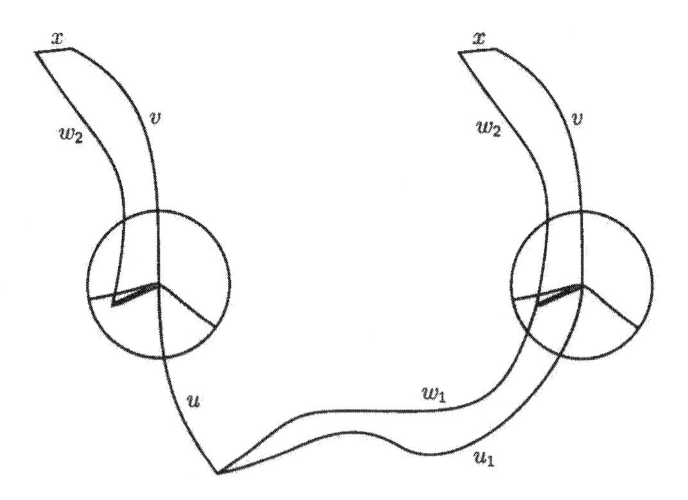


**Figure 3.3. The cone type determines the level.** This illustrates the proof of Lemma 3.2.4 (Cannon). The circles represent $(2k+1)$-neighbourhoods of $\overline{u}$ and $\overline{u_1}$. The thicker short interval in the circle represents the element $\overline{u_1}^{-1}\overline{w_1}$.

have the same $(2k+1)$-level. We prove by induction on the length of $v$, that if $uv$ is geodesic, then so is $u_1v$. This is true if $v$ has length zero. To prove the induction step, suppose that $uv$, $u_1v$ and $uvx$ are geodesic, where $x \in A$. We must show that $u_1vx$ is geodesic.

Assume, to the contrary, that there is a geodesic string shorter than $u_1vx$ and giving the same group element. Write it as $w_1w_2$, where $|w_1| = |u_1| - 1$. (We may assume that $u_1$ has length at least one, because if its length is zero, then the $(2k+1)$-level of $\overline{u_1}$ is empty. It would follow that both $u$ and $u_1$ are the nullstring, which would be fine.) Then $|w_1|+|w_2| = |w_1w_2| < |u_1|+|v|+1$ by assumption, so $|w_2| \leq |v| + 1$.

By Lemma 3.2.3 (hausdorff implies uniform), $d(\overline{w_1}, \overline{u_1}) \leq 2k+1$. Therefore $\overline{u_1}^{-1}\overline{w_1}$ is in the $(2k+1)$-level of $\overline{u_1}$, which equals the $(2k+1)$-level of $\overline{u}$—see Figure 3.3. It follows that $d(\overline{u}\,\overline{u_1}^{-1}\overline{w_1}, e) < d(\overline{u}, e) = |u|$, and therefore

$$d(uvx, e) \leq d(\overline{u}\,\overline{u_1}^{-1}\overline{w_1}, e) + |w_2| < d(\overline{u}, e) + |v| + 1 = |u| + |v| + 1 = |uvx|,$$

contradicting the assumption that $uvx$ is geodesic. $\boxed{3.2.4}$

Having proved this lemma, we see that the number of cone types in $G$ is finite. This shows that the language $L$ of geodesics is regular. The proof of Theorem 3.2.1 (geodesic automaton 1) is completed by referring to Lemma 3.2.3 (hausdorff implies uniform) and to Theorem 2.3.5 (characterizing synchronous). $\boxed{3.2.1}$

## 3.3. Pseudoisometries

Pseudoisometries and quasigeodesics are two of the basic ideas to have emerged from the last twenty years of geometry, and are largely to credit for the rapid progress witnessed in this period. They have been successfully exploited by Margulis and Mostow [Mar70, Mos68], and have played a fundamental role in the work of Gromov and of Thurston. We introduce these notions here for use in the next section and later on.

Let $X$ and $Y$ be metric spaces. We will be dealing with correspondences between $X$ and $Y$, in the sense of subsets of the product $X \times Y$. In set theory the word "relation" is used for such correspondences, but it would be awkward to follow this usage here because "relation" has a very different meaning in group theory, and has overtones of symmetry in everyday usage. To avoid this difficulty, we define a *pseudomap* as a relation between $X$ and $Y$ that projects onto $X$; that is, for every $x \in X$, there exists $y \in Y$ that is related to $x$.

**Convention 3.3.1 (pseudomap).** If $f$ is a pseudomap, we will generally use the shorthand $f(x)$ to refer to any point of $Y$ that is $f$-related to $x$. We call such a point an *image* of $x$. We also write $f : X \to Y$ as if $f$ were a map.

Let $k \geq 1$ and $\varepsilon \geq 0$ be real numbers. A *$(k,\varepsilon)$-lipschitz pseudomap*, or *$(k,\varepsilon)$-pseudomap*, from $X$ to $Y$ is a pseudomap such that

$$d(f(x_1), f(x_2)) \leq k\, d(x_1, x_2) + \varepsilon \quad \text{for } x_1, x_2 \in X.$$

Clearly, any $k$-lipschitz map is a $(k,0)$-pseudomap. A $(k,\varepsilon)$-pseudomap need not be continuous if $\varepsilon \neq 0$, though it looks continuous to a short-sighted person.

We say that $f$ is a *$(k,\varepsilon)$-pseudoisometric embedding* from $X$ to $Y$ if it is a $(k,\varepsilon)$-pseudomap and satisfies the further condition

$$d(x_1, x_2) \leq k\, d(f(x_1), f(x_2)) + \varepsilon \quad \text{for } x_1, x_2 \in X.$$

Any isometric embedding is a $(1,0)$-pseudoisometric embedding. A pseudoisometric embedding is not necessarily injective, but points far apart in the domain are sent to points far apart in the range.

The next notion is the analogue of an isometry. A $(k,\varepsilon)$-pseudoisometric embedding $f$ from $X$ to $Y$ is a *$(k,\varepsilon)$-pseudoisometry* if it projects onto $Y$; this is the same as saying that $f^{-1}$ is also a pseudomap (and a $(k,\varepsilon)$-pseudoisometric embedding). We sometimes also say that $f$ is an pseudoisometry if some "thickening" $f_\delta$ of $f$ is a pseudoisometry; the definition of a thickening is that $f_\delta$ relates $x$ and $y$ if and only if $f$ relates $x$ and $y'$, for some $y'$ in a $\delta$-neighbourhood of $y$. (The point is that $f_\delta$ has to be surjective. This can often be achieved by increasing $\delta$, since the image of $f_\delta$ in $Y$ increases as $\delta$ increases.) Under this definition the following result is immediate:

**Lemma 3.3.2 (subspace).** *Let $\varepsilon > 0$ be given. Let $X$ be a subspace of $Y$ such that, for each point $y \in Y$, there is a point $x \in X$ with $d(x,y) < \varepsilon$. Then the inclusion is a $(1, 2\varepsilon)$-pseudoisometry.* [3.3.2]

**Lemma 3.3.3 (change of generators is pseudoisometry).** *Let $A$ and $B$ be finite sets of generators for a group $G$, and let $k$ be such that any element in one set can be expressed as a word of length at most $k$ over the other set. Then the identity map on vertices is a $k$-pseudoisometry between the Cayley graphs $\Gamma(G,A)$ and $\Gamma(G,B)$.* [3.3.3]

A *$(k,\varepsilon)$-quasigeodesic* in $Y$ is a $(k,\varepsilon)$-pseudoisometric embedding of an interval in $Y$. In the literature, a quasigeodesic is generally assumed to be a continuous map, but this restriction seems never to be essential, and is sometimes inconvenient.

We will often omit the prefix $(k,\varepsilon)$, and talk about a lipschitz pseudomap, a pseudoisometric embedding, a pseudoisometry or a quasigeodesic without specifying the constants. Each of the first three classes of pseudomaps is closed under composition. Any compact metric space is pseudoisometric to a point. Any geodesic is an isometric embedding, and therefore a $(1,0)$-quasigeodesic.

Here is one connection between automatic groups and quasigodesics:

**Theorem 3.3.4 (accepted implies quasigeodesic).** *Let $G$ be a group with automatic structure $(A, L)$. If there is a bound for the number of accepted representatives in $L$ of an element $g \in G$ as $g$ varies, then there exists $N$ such that any path in the Cayley graph of $G$ corresponding to an accepted string is an $(N, N)$-quasigeodesic.*

Example 2.3.6 (finite) shows that the hypothesis of a bound is essential for the truth of this theorem.

*Proof of 3.3.4:* By Theorem 2.5.9 (prefix closure of automatic groups), we can replace $L$ by its prefix closure, which still satisfies the finiteness condition.

Let $w = ruv$ be an accepted string, where $r$, $u$ and $v$ are any strings

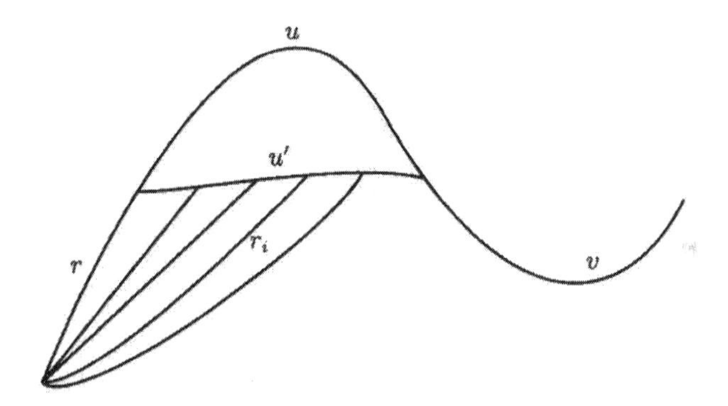


**Figure 3.4. An accepted string is quasigeodesic.** This picture illustrates the proof of Theorem 3.3.4. If $ruv$ is an accepted string, the geodesic between $\overline{r}$ and $\overline{ru}$ cannot be much shorter than $u$.

(Figure 3.4). Let $u'$ be a geodesic path in the Cayley graph from $\overline{r}$ to $\overline{ru}$; we must show that $|u'|$ is not much bigger or much smaller than $|u|$. Let $r_i$, for $0 \leq i \leq |u'|$, be an accepted word representing $\overline{ru'(i)}$; we can take $r_0 = r$ and $r_{|u'|} = ru$ because $L$ is prefix closed. By Lemma 2.3.9 (bounded length difference), the difference in length between $r = r_0$ and $ru = r_{|u'|}$ is at most $N \max(|u'|, 1)$, for some fixed $N$. Therefore $|r| + |u| = |ru| \leq |r| + N(|u'| + 1)$, which implies $|u'| \leq |u| \leq N(|u'| + 1)$. $\boxed{3.3.4}$

The spaces we will deal with will be almost always complete, locally compact path metric spaces. The next lemma shows that in such a space a quasigeodesic can be moved a little to give a piecewise geodesic continuous path.

**Lemma 3.3.5 (piecewise geodesic).** *Given $k \geq 1$ and $\varepsilon \geq 0$, there are constants $k' \geq 1$ and $\varepsilon' \geq 0$ with the following property: Let $X$ be a complete, locally compact path metric space and $\alpha$ a $(k, \varepsilon)$-pseudomap from an interval $[a, b]$ into $X$. Then there is a piecewise geodesic path $\beta : [a, b] \to X$ that is a $(k', \varepsilon')$-pseudomap within a uniform distance $\varepsilon'$ from $\alpha$. If, in addition, $\alpha$ is a $(k, \varepsilon)$-quasigeodesic, then $\beta$ is a $(k', \varepsilon')$-quasigeodesic, and remains one if we reparamatrize it by pathlength.*

*Proof of 3.3.5:* Set $\eta = 1+\varepsilon$. We can assume that the domain of $\alpha$ has length at least $\eta$; otherwise, the distance between any two points in the image of $\alpha$ is bounded by $k\eta + \varepsilon$, and the result follows by choosing $\beta$ to be constant.

Divide the domain $[a, b]$ of $\alpha$ into intervals with endpoints

$$a = t_0 < t_1 < \cdots < t_n = b,$$

where $t_{i+1} - t_i$ is between $\eta$ and $2\eta$. For each $i$, set $\beta(t_i)$ to some point $\alpha(t_i)$, and extend $\beta$ to $[a, b]$ by making it a constant-speed geodesic on $(t_i, t_{i+1})$. Because $\alpha$ is a $(k, \varepsilon)$-lipschitz pseudomap, we have

$$d(\beta(t_i), \beta(t_{i+1})) \leq k(t_{i+1} - t_i) + \varepsilon \leq 2k\eta + \varepsilon,$$

so the speeds of these geodesic pieces are bounded above. If $\alpha$ is a $(k, \varepsilon)$-quasigeodesic, the speeds are also bounded below, because

$$1 + \varepsilon = \eta \leq t_{i+1} - t_i \leq k\, d(\beta(t_i), \beta(t_{i+1})) + \varepsilon.$$

Now let $s_1 < s_2$ be any two points in $[a, b]$, and choose $i_1$ and $i_2$ so that $|s_1 - t_{i_1}| \leq \eta$ and $|s_2 - t_{i_2}| \leq \eta$. We have

$$\begin{aligned} d(\alpha(s_1), \beta(s_1)) &\leq d(\alpha(s_1), \alpha(t_{i_1})) + d(\beta(s_1), \beta(t_{i_1})) \\ &\leq (k\eta + \varepsilon) + (2k\eta + \varepsilon) = 3k\eta + 2\varepsilon, \end{aligned}$$

and similarly for $s_2$. In particular, the uniform distance between $\alpha$ and $\beta$ is bounded. We also have

$$\begin{aligned} d(\beta(s_1), \beta(s_2)) &\leq d(\beta(s_1), \alpha(s_1)) + d(\alpha(s_1), \alpha(s_2)) + d(\alpha(s_2), \beta(s_2)) \\ &\leq 2(3k\eta + 2\varepsilon) + k|s_1 - s_2| + \varepsilon, \end{aligned}$$

showing that $\beta$ is a lipschitz pseudomap. If $\alpha$ is a quasigeodesic, an upper bound for $|s_1 - s_2|$ in terms of $d(\beta(s_1), \beta(s_2))$ follows in a similar way, so that

$\beta$ is quasigeodesic. It remains one after reparametrization by pathlength because this is equivalent to composition with a pseudoisometry between intervals (recall that the speeds of the geodesic pieces are bounded above and below).

By construction, all the constants involved depend only on $k$ and $\varepsilon$.

3.3.5

We now discuss the fundamental idea that, for many purposes, studying a group and a space on which the group acts are equivalent activities. Let $X$ be a locally compact, connected, path metric space and let $G$ be a group of isometries of $X$, acting properly discontinuously. (This means that, for any compact subset $K$ of $X$, the set of $g \in G$ such that $K \cap gK \neq \varnothing$ is finite.) We suppose that $G \backslash X$ is compact, and fix a basepoint $*$ for $X$.

From these assumptions, we may deduce that $G$ is finitely generated and that $X$ is complete. To prove that $G$ is finitely generated, we notice that there is some open ball $U$ of radius $D$ and centered at $*$ such that $GU = X$, because the images of such balls in $G \backslash X$, for varying $D$, form an open cover of $G \backslash X$, which we have assumed compact. It can be shown [CEG87, Theorem 1.3.5] that a metric ball in a locally compact path metric space is relatively compact; since $G$ acts properly discontinuously, there are only finitely many elements $g \in G$ such that $gU \cap U \neq \varnothing$. A chaining argument shows that these elements generate $G$. The completeness of $X$ is proved by projecting a Cauchy sequence in $X$ to $G \backslash X$, where compactness is applied.

We now fix a finite set $A$ of generators for $G$ (not necessarily the ones found in the preceding paragraph). For each generator $\gamma \in A$, we fix a shortest geodesic $s_\gamma$ in $X$, from $*$ to $\gamma(*)$. Let $\Gamma$ be the Cayley graph of $G$. We define a map $f : \Gamma \to X$ by mapping the vertex corresponding to $g \in G$ to the point $g(*)$ and the edge $(g_1, \gamma, g_2)$ of $\Gamma$ to $g_1(s_\gamma) \subset X$.

**Theorem 3.3.6 (pseudoisometry of Cayley graph).** *Under the conditions of the previous three paragraphs, the map $f$ is a pseudoisometry of $\Gamma$ with $X$. It also induces a pseudoisometry of $G$ with $X$, where $G$ is given the word metric.*

*Proof of 3.3.6:* The statement about $\Gamma$ follows from the one about $G$. Let $D$ be the radius of an open ball centered at $*$ whose images under $G$ cover $X$. Then any point in $X$ is within a distance $D$ of $f(G)$, so by Lemma 3.3.2 we are reduced to proving that $f$ gives a pseudoisometric embedding of $G$ in $X$, when $G$ is endowed with the word metric.

To bound distances in $X$ in terms of distances in $G$, let $k_1$ be the maximum distance in $X$ between $*$ and $\gamma(*)$, as $\gamma$ runs over the set $A$ of generators. For $g_1, g_2 \in G$, it is clear that $d_X(g_1(*), g_2(*)) \leq k_1\, d_G(g_1, g_2)$, as indicated in Figure 3.5.

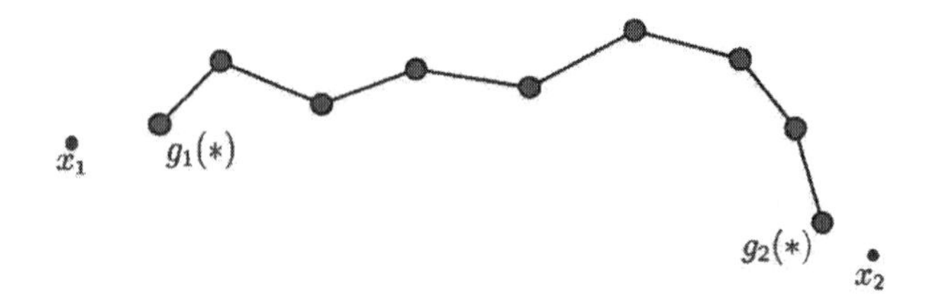


**Figure 3.5. Bounding the metric of $X$ from above.** Each straight segment represents a geodesic of the form $g(s_\gamma)$, which has length at most $k_1$.

To bound distances in $G$ in terms of distances in $X$, consider the subset $F \subset G$ of group elements that move the basepoint by less than $2D + 1$. This set is finite, again by the relative compactness of metric balls in $X$ and because $G$ acts properly discontinuously; let $k_2$ be the supremum of the word lengths of the elements of $F$.

Let $\sigma$ be a shortest geodesic in $X$ from $g_1(*)$ to $g_2(*)$. If the length of $\sigma$, that is, the distance between $g_1(*)$ to $g_2(*)$, is between $n-1$ and $n$, we choose points $z_0 = g_1(*), z_1, \ldots, z_n = g_2(*)$ along $\sigma$, with $d(z_{i-1}, z_i) \leq 1$ for $0 < i \leq n$. For $0 < i < n$, we also choose nearby points $h_i(*)$, where $h_i \in G$, that are within a distance $D$ of $z_i$; and we set $h_0 = g_1$ and $h_n = g_2$. Clearly, the distance between $h_i(*)$ and $h_{i+1}(*)$ is no more than $2D + 1$ (see Figure 3.6). Thus $d_G(h_i, h_{i+1}) \leq k_2$ for each $i$, and

$$d_G(g_1, g_2) \leq k_2 n \leq k_2(d_X(g_1(*), g_2(*)) + 1). \qquad \boxed{3.3.6}$$

**Corollary 3.3.7.** *The composition of a geodesic in the Cayley graph $\Gamma$ with the map $f$ defined above is a continuous quasigeodesic in $X$.* $\boxed{3.3.7}$

## 3.4. Applications to Fundamental Groups

In this section we show that some large classes of groups of great interest in geometry and topology satisfy the conditions of Theorems 3.2.1 and 3.2.2, and are therefore automatic. We have tried to define concepts that might be unfamiliar to readers from other areas, except for the most fundamental ones: riemannian manifold, completeness, and the like.

We say that a riemannian manifold $M$, possibly with boundary, is *convex* if any two points in the interior are joined by a geodesic in the interior. Although our geodesics, as defined on page 65, are not riemannian geodesics, our definition of convexity is equivalent to the definitions commonly used in riemannian geometry. Every compact manifold without boundary is convex. If $M$ is complete and has geodesic boundary it is convex. In hyperbolic space,

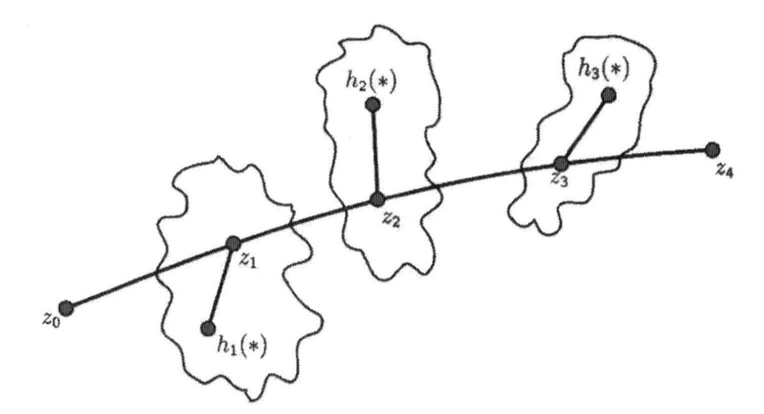


**Figure 3.6. Bounding the metric of $G$ from above.** We can follow the path of a geodesic in $X$, using a sequence of elements of the form $h_i(*)$, where the distances $d_G(h_i, h_{i+1})$ are bounded independently of the particular situation. This gives a corresponding path in the Cayley graph.

a horoball is an example of a convex manifold, and so is the interior of any metric ball.

A group action on a topological space is *properly discontinuous* if every point in the space has a neighbourhood that intersects only finitely many images of itself under the group. A group acting properly discontinuously on a simply connected complete convex riemannian manifold, possibly with boundary, is said to be *convex cocompact* if the quotient of the action is compact.

**Theorem 3.4.1 (negative curvature).** *Let $G$ be a group acting convex cocompactly on a simply connected manifold all of whose sectional curvatures are strictly negative. Then, for any finite ordered set of semigroup generators of $G$, the set of all shortest strings representing elements of $G$ provides an automatic structure on $G$. In particular, the fundamental group of a negatively curved compact manifold or orbifold or convex manifold with boundary is automatic.*

To prove the theorem, we need a well-known lemma, due to Mostow. We include a proof here for completeness.

**Lemma 3.4.2 (quasigeodesics near geodesics).** *Given $k \geq 1$, $\varepsilon \geq 0$ and $c > 0$, there is a constant $L$ with the following property: Let $f : [a, b] \to X$ be a $(k, \varepsilon)$-quasigeodesic, where $X$ is a complete convex simply connected riemannian manifold, with all sectional curvatures bounded above by $-c^2$.*

*Then an L-neighbourhood of a geodesic arc from (any point of) $f(a)$ to (any point of) $f(b)$ contains the image of $f$.*

*Proof of 3.4.2:* By Lemma 3.3.5 (piecewise geodesic), there is no loss of generality in supposing that $f$ is a piecewise geodesic parametrized by path-length. Let $\alpha$ be a geodesic arc joining $f(a)$ and $f(b)$, and let $N$ be the $s_0$-neighbourhood of $\alpha$, for some value of $s_0$ to be determined later. Let $(u, v)$ be a maximal interval mapped by $f$ into the complement of $N$, and let $R$ be its length in $X$, as shown in Figure 3.7.

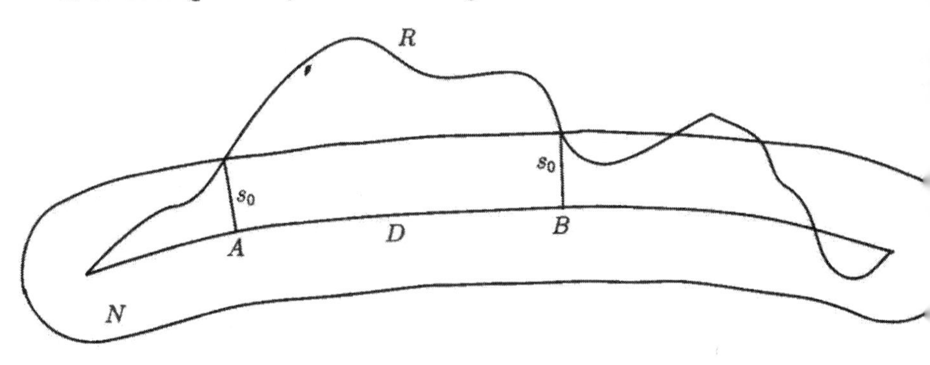


**Figure 3.7. Quasigeodesics are near geodesics.** This figure illustrates the proof of the fundamental fact that, in a negatively curved riemannian manifold, a quasigeodesic is near a geodesic.

Now it can be shown that nearest point projection $p$ onto $\alpha$ reduces length by a factor of at least $\cosh(cs)$, where $s$ is the distance from $\alpha$. We set $A = p(f(u))$, $B = p(f(v))$ and $D = d(A, B)$; thus $R \geq D\cosh(cs_0)$. On the other hand,

$$R = (v-u) \leq k\,d(fu, fv) + \varepsilon \leq k(2s_0 + D) + \varepsilon \leq k\left(2s_0 + \frac{R}{\cosh(cs_0)}\right) + \varepsilon.$$

If we now choose $s_0$ so that $\cosh(cs_0) > 2k$, we get $R \leq 4ks_0 + 2\varepsilon$; by setting $L = s_0 + 4ks_0 + 2\varepsilon$ we have proved the lemma. 3.4.2

**Corollary 3.4.3.** *Let $X$ and $c$ be as in the preceding lemma. Let $\alpha$ be a $(k, \varepsilon)$-quasigeodesic joining $x_1$ to $x_2$, and let $\beta$ be a $(k, \varepsilon)$-quasigeodesic joining $y_1$ to $y_2$. Then the hausdorff distance from $\alpha$ to $\beta$ can be estimated in terms of $c$, $k$, $\varepsilon$, $d(x_1, y_1)$ and $d(x_2, y_2)$.*

*Proof of 3.4.3:* We construct unit speed geodesics from $x_1$ to $y_1$ and from $x_2$ to $y_2$. Adjoining these geodesics to $\alpha$, we obtain a quasigeodesic $\gamma$ from $y_1$ to

$y_2$, whose constants can be expressed in terms of $k$, $\varepsilon$, $d(x_1, y_1)$ and $d(x_2, y_2)$. Both $\beta$ and $\gamma$ can be compared with the geodesic from $y_1$ to $y_2$, using the preceding lemma. 3.4.3

The proof of the Theorem 3.4.1 (negative curvature) will now follow easily.

*Proof of 3.4.1:* Let $\alpha$ and $\beta$ be two geodesics in the Cayley graph of $G$ that start at the basepoint and end at two vertices a distance at most one from each other. According to Theorem 3.2.1 (geodesic automaton 1), we need only find a constant $k$, independent of $\alpha$ and $\beta$, such that $\alpha$ and $\beta$ are in $k$-neighbourhoods of each other. According to Theorem 3.3.6 (pseudoisometry of Cayley graph), there is a pseudoisometry $f$ mapping the Cayley graph into $X$, so there are constants $K$ and $\varepsilon$ such that $f(\alpha)$ and $f(\beta)$ are $(K, \varepsilon)$-quasigeodesics in $X$ starting at the same point and ending a distance at most $K + \varepsilon$ apart. By the corollary just proved, there is a constant $L$, independent of $\alpha$ and $\beta$, such that the two quasigeodesics in $X$ are each inside an $L$-neighbourhood of the other. Once again applying the fact that $f$ is a pseudoisometry, we obtain our desired constant $k$, for which the hypotheses of Theorem 3.2.1 (geodesic automaton 1) can be verified. 3.4.1

Theorem 3.4.1 (negative curvature) is in fact a particular case of a more general result involving the concept of *word hyperbolic groups*. The introduction by Gromov [Gro87] of this class of groups, which are very well-suited to geometric analysis, proved to be a major development in combinatorial group theory. Gromov's paper is hard to read; the proofs are missing and some of the definitions are not correct in every detail. But much of the theory has now been worked out with detailed proofs, and there are several very good expositions [GdlH89, CDP90, Bow00], each of which has valuable aspects absent from the others. Here we content ourselves with defining word hyperbolicity and proving that word hyperbolic groups are automatic, quoting the necessary background results without proof.

**Definition 3.4.4 (word hyperbolic).** A geodesic space (see the end of Section 3.1) is said to be *hyperbolic* (in the sense of Gromov) if there is a number $\delta > 0$ such that, given any triangle with geodesic sides, the distance from a point on one side to the union of the other two sides is bounded by $\delta$. A group $G$ with a set of generators $A$ is called *word hyperbolic* if the Cayley graph $\Gamma(G, A)$ is hyperbolic. It can be shown [GdlH89, Chapter 5, Section 2, Theorem 12] that if $X$ and $Y$ are geodesic spaces and $f : X \to Y$ is a pseudoisometry, $X$ is hyperbolic if and only if $Y$ is. Therefore the definition of a word hyperbolic group does not depend on the choice of generators.

**Theorem 3.4.5 (word hyperbolic implies automatic).** *Let $G$ be a word hyperbolic group and let $A$ be a set of semigroup generators closed under* *inversion. The geodesics over A form a regular language, which is part of a (symmetric) automatic structure.*

*Proof of 3.4.5:* We show that the hypotheses of Theorem 3.2.1 (geodesic automaton 1) are satisfied. Given two geodesic strings $u$ and $v$ such that $d(\overline{u}, \overline{v}) \leq 1$, we obtain a geodesic triangle of sides $|u|$, $|v|$ and 1 (or 0). The fact that the Cayley graph is a hyperbolic space immediately implies that the paths $u$ and $v$ are within a hausdorff $(\delta+1)$-neighbourhood of each other. 3.4.5

Theorem 3.4.5 has been independently noted by a number of people, and was certainly known to Gromov before he published his work on word hyperbolic groups. To geometers, word hyperbolic groups are probably the most important general class of automatic groups. Although cocompact discrete groups of hyperbolic isometries are just a particular case of word hyperbolic groups, they are often the immediate object of interest, and this justifies the independent proof of Theorem 3.4.1 (negative curvature) given earlier in this section.

One of the striking aspects of Gromov's theory is the large number of equivalent definitions that can be given for a hyperbolic space. Definition 3.4.4 is merely the most convenient one for our current applications. [GdlH89, Chapter 2, Section 1, Proposition 21] provides several equivalent conditions. Also, Brian Bowditch [Bow00] has given a complete proof of Gromov's assertion that hyperbolicity is equivalent to the existence of a linear isoperimetric inequality, and in particular that a group is word hyperbolic if and only if it satisfies a linear isoperimetric inequality. Giving a good concept of area in appropriate generality is part of Bowditch's achievement in making Gromov's claim comprehensible in the context of general hyperbolic spaces. The best proof we know when one restricts oneself to Cayley graphs is due to Short [ABC+00].

## 3.5. Counterexample to the Use of Shortest Strings

In Theorem 3.4.5 we saw that every word hyperbolic group is strongly geodesically hyperbolic for any choice of generators. In Theorem 4.3.1 we will see that every abelian group is ShortLex-automatic, again for any choice of generators. These examples might lead one to conclude that any automatic group is, for any choice of generators, at least weakly geodesically automatic. But this is far from being the case: the following example, due to Paterson, gives a set of generators for a certain automatic group, under which it is impossible to obtain not only a weakly geodesic automatic structure, but even a regular language consisting only of geodesics.

**Example 3.5.1 (surfeit).** Let $P = F(a,b) \times F(c,d)$, where $F(a,b)$ is the free group on the group generators $\{a,b\}$ (see page 29) and similarly for $F(c,d)$. Now free groups are automatic by Theorem 3.2.1 (geodesic automaton 1), because each element is expressed by a unique geodesic. Direct products of automatic groups are automatic by Theorem 4.1.1, so $P$ is automatic (with respect to any set of generators: see Theorem 2.4.1).

Consider the set of generators

$$\Sigma = \{a,b,c,d,f,g,A,B,C,D,F,G\},$$

where $f = add$ and $g = bbc$, and uppercase letters, as usual, denote the inverses of the corresponding lowercase letters. Let $L$ be any regular language over $\Sigma$ mapping onto $P$. We will show that, for any $\varepsilon > 0$ and any $N > 0$, there is a string $u \in L$ with $|u| \geq N$ and $|u| > (\frac{3}{2} - \varepsilon)|\overline{u}|$ (see page 37 for the notation $|\overline{u}|$). In words, there are arbitrarily long elements of $L$ that are about 50% or more longer than their geodesic representatives.

The basic idea is that strings are abbreviated well by grouping $b$ and $c$ and by grouping $a$ and $d$, but typically groupings of one type get in the way of groupings of the other. We take a group element where both types of abbreviation can be applied several times, but not both at the same time. If we choose a representative with lots of $f$'s but no $g$'s, then because we are dealing with a finite state machine, we can find a substring that is also in $L$ and where not many $f$'s occur; we then show that this string could be written more economically with $g$'s.

Let $\Sigma_{ab} = \{a,b,A,B\}$, $\Sigma_{cd} = \{c,d,C,D\}$ and $\Sigma_{fg} = \{f,g,F,G\}$. For any string $w$ over $\Sigma$, we define a string $w_{ab}$ over $\Sigma_{ab}$ as follows: We first apply to $w$ the substitution

$$\{f \to add, F \to DDA, g \to bbc, G \to CBB\}$$

to get a string over $\Sigma_{ab} \cup \Sigma_{cd}$. We then delete all occurrences of symbols in $\Sigma_{cd}$, and finally cancel pairs of adjacent inverses. The result of the cancellation does not depend on the order in which adjacent pairs are cancelled, but we agree to remove the rightmost pair of inverses each time, so each symbol of $w_{ab}$ comes from a well defined symbol of $w$. We define $w_{cd} \in \Sigma_{cd}^*$ analogously. Note that $\overline{w} = \overline{w_{ab}w_{cd}}$, and that $|\overline{w}| \geq \frac{1}{3}|w_{ab}w_{cd}|$ because any symbol in $\Sigma$ can be replaced by at most three symbols in $\Sigma_{ab} \cup \Sigma_{cd}$.

We say that a symbol $x$ of $w$ *survives* if either one (for $x \in \Sigma_{ab} \cup \Sigma_{cd}$) or three (for $x \in \Sigma_{fg}$) symbols in $w_{ab}w_{cd}$ come from $x$; in other words, if it takes part in no cancellation. If $x$ survives and $w = w'xw''$, we have $w_{ab} = w'_{ab}x_{ab}w''_{ab}$ and $w_{cd} = w'_{cd}x_{cd}w''_{cd}$, since there is no cancellation between symbols on opposite sides of $x$. An elementary analysis of possible cancellations shows that, if $|w| \le |w_{ab}w_{cd}| - 2k$, then at least $k$ symbols $x \in \Sigma_{fg}$ survive. This is because the inequality implies that some abbreviation must take place, and this can only happen with elements of $\Sigma_{fg}$.

Let $W$ be a deterministic automaton accepting $L$, and let $W$ have $q-1$ states. Take $n \ge \max(N, q/\varepsilon)$ and let $w \in L$ represent

$$\overline{a^n b^{2n} c^n d^{2n}} = \overline{c^n a^n d^{2n} b^{2n}} = \overline{a^n g^n d^{2n}} = \overline{c^n f^n b^{2n}}.$$

Then $w_{ab} = a^n b^{2n}$ and $w_{cd} = c^n d^{2n}$, so $|w| \ge |\overline{w}| \ge \frac{1}{3}6n > N$. From the representative $a^n g^n d^{2n}$ we get $|\overline{w}| \le 4n$. If $w$ has length greater than $6n - 2q$, it satisfies the desired conditions $|w| > (\frac{3}{2} - \varepsilon)|\overline{w}|$ and $|w| > N$, and we are done.

We therefore assume that $|w| \le 6n - 2q$. Then $|w| \le |w_{ab}w_{cd}| - 2q$, so at least $q$ symbols in $\Sigma_{fg}$ must occur in $w$ and survive. None of them can be $F$ or $G$, because that would mean one of $A$, $B$, $C$ or $D$ appearing in $w_{ab}w_{cd}$, which is not the case. Furthermore, an $f$ and a $g$ cannot both survive: if the $f$ comes before the $g$, we get a $d$ before a $c$ in $w_{cd}$, contradicting the equation $w_{cd} = c^n d^{2n}$; the case of $g$ before $f$ is analogous. Without loss of generality we suppose that $q$ occurrences of $f$ in $w$ survive.

Consider the sequence of states of $W$ as $w$ is read, and in particular as each surviving occurrence of $f$ is about to be read. Since $q$ is greater than the number of states, some state is repeated. We write $w = xfufy$, where $x$, $y$ and $u$ are strings over $\Sigma$, the state of $W$ after reading $x$ is equal to the state after reading $xfu$, both occurrences shown of $f$ survive, and $u$ is as long as possible, subject to these conditions. Then $u_{ab}$ is a power of $a$; all other symbols are forbidden, because $u_{ab}$ is a substring of $w_{ab}$ preceded and followed by $ad^2$. Likewise, $u_{cd}$ is a power of $d$. We set $\overline{uf} = \overline{a^i d^j}$, where $n \ge i \ge 1$ and $2n \ge j \ge 2$.

Let $v = xfy$. Then $v \in L$, and no occurrence of $g$ in $v$ survives, for the same reason that no occurrence of $g$ in $w$ survives. Since fewer than $q$ occurrences of $f$ survive in $v$, we have

$$|v| > |v_{ab}v_{cd}| - 2q = |a^{n-i}b^{2n}| + |c^n d^{2n-j}| - 2q = 6n - i - j - 2q.$$

On the other hand, $\overline{v} = \overline{a^{n-i} g^n d^{2n-j}}$, so

$$|\overline{v}| \le 4n - i - j.$$

Putting together the two estimates, we get $|v| > (\frac{3}{2} - \varepsilon)|\overline{v}|$. We also know that $|v| \ge \frac{1}{3}(6n - i - j) \ge N$, so again we have found a long string that is far from being geodesic.

## 3.6. Combable Groups

As we saw in the discussion following Lemma 2.3.2 (lipschitz property), an automatic structure associates to each vertex of the Cayley graph a set of paths leading from the basepoint to the vertex. Paths associated with the same vertex or with neighbouring vertices are within a fixed distance of each other, in the uniform metric of paths. In the language of Section 3.3, the group (and therefore the Cayley graph) can be pseudoisometrically embedded in the space of paths in the Cayley graph that start at the basepoint, and this embedding is a right inverse for the map that associates to each path its endpoint.

The existence of such an embedding is a purely metrical property, and can be extended to spaces other than the Cayley graph of a group. It is also invariant under pseudoisometry. (The image of a path under a pseudoisometry need not be a path, but can be approximated by one; see Theorem 3.6.4.) This makes it a very natural property to consider in the context of contemporary geometry.

By contrast, the condition that the set of accepted strings of an automatic structure is a regular language is not geometric, and it does not extend readily to, say, homogeneous spaces. From the geometric point of view, therefore, it is interesting to investigate what is left of an automatic structure when we drop the regularity condition, while preserving the central property that paths running to nearby points remain near each other. The resulting structure is called a *combing* (see Definition 3.6.1 for a formal definition).

Unfortunately, it is not clear what other nice properties of automatic groups should be part of the definition of a combing. For example, an automatic structure with uniqueness gives a combing by quasigeodesics, by Theorem 3.3.4 (accepted implies quasigeodesic); also, the domains of adjacent paths have approximately the same length, by Lemma 2.3.9 (bounded length difference). Should we insist that our combings have either of these properties, in view of the fact that any automatic group has an automatic structure with uniqueness? Other variants can be conceived, and there are several competing definitions in use (see, for example, [Alo89, Sho90]). Some of them may turn out to be equivalent: the whole situation needs to be clarified by means of proofs of equivalence or counterexamples.

In this book we require that the domains of adjacent paths be of similar length. We choose this definition partly for historical reasons—this was the original definition of a combing, first given explicitly by Epstein and Thurston—and partly because it is strong enough to guarantee that the quadratic isoperimetric inequality (Theorem 2.3.12) and many other results about automatic groups remain true about combable groups.

**Definition 3.6.1 (combing).** Let $X$ be a metric space with a basepoint. A $(k,\varepsilon)$-*broken path* in $X$ is a (possibly discontinuous) map from an interval $J$ to $X$ that is a $(k,\varepsilon)$-pseudomap. We denote by $P_{(k,\varepsilon)}(X)$ the space of $(k,\varepsilon)$-broken paths in $X$ with domain $[0,b]$, for $b \geq 0$, and by $P^*_{(k,\varepsilon)}(X)$ the subspace of broken paths taking 0 to the basepoint. We set $P(X) = \bigcup_{(k,\varepsilon)} P_{(k,\varepsilon)}(X)$ and $P^*(X) = \bigcup_{(k,\varepsilon)} P^*_{(k,\varepsilon)}(X)$.

If $p \in P(X)$ has domain $[0,b]$, we set $p(t) = p(b)$ if $t > b$ (in other words, we extend $p$ to $[0,\infty]$ by making it constant for $t \geq b$). We give $P(X)$ a metric $d_P$, as follows: if $p_1, p_2 \in P(X)$ have domains $[0,b_1]$ and $[0,b_2]$, we set

$$d_P(p_1,p_2) = \max_{0 \leq t \leq \infty} d_u(p_1(t), p_2(t)) + |b_1 - b_2|,$$

where $d_u(p_1,p_2) = \max_{0 \leq t \leq \infty} d(p_1(t), p_2(t))$ is the uniform distance.

The *endpoint map* $P(X) \to X \times X$ associates to a broken path $p \in P(X)$ with domain $[0,b]$ its endpoints $(p(0), p(b))$. If $p \in P^*(X)$ the first coordinate is the basepoint, by definition, so we also refer to the map $p \mapsto p(b)$ on $P^*(X)$ as the endpoint map.

A $(k,\varepsilon)$-*combing* of $X$ is a pseudoisometric embedding of $X$ in $P^*_{(k,\varepsilon)}(X)$ (with the metric $d_P$) that is a right inverse for the endpoint map $P^*(X) \to X$. A space is *combable* if it has a $(k,\varepsilon)$-combing for some $k$ and $\varepsilon$. A group is *combable* if its Cayley graph (with respect to some set of generators) is combable. What this means is that for each $g \in G$ we are given a broken path; this path is almost continuous (for most purposes, we may as well take it to be continuous); as $g$ varies, the path varies in a reasonable way—nearby elements of $G$ give paths which are uniformly near and whose domains are almost the same.

Now suppose we have a group $G$ acting on $X$; we extend the action of $G$ to $P(X)$ in the obvious way. The endpoint map is $G$-equivariant. A $(k,\varepsilon)$-*bicombing* of $X$ is a $G$-invariant pseudoisometric embedding of $X \times X$ in $P_{(k,\varepsilon)}(X)$ that is a right inverse for the endpoint map $P \to X \times X$; a space is *bicombable* if it has a bicombing. We say that $G$ is *bicombable* if its Cayley graph has a bicombing under the left action of $G$. What this means is that we associate to each pair of elements of $G$ a broken path joining them. If we restrict the first element of the pair to be the identity element, we get a combing. This combing determines the bicombing because the set of all broken paths in the bicombing is invariant under left translation by elements of $G$.

As already mentioned, Theorems 2.5.1 (uniqueness) and 2.3.9 (bounded length difference) imply that any automatic group is combable. Likewise, any biautomatic group is bicombable.

**Open Question 3.6.2.** Find a combable group that is not automatic. (It is easy to find a combing that is not an automatic structure: for example, the combing for $\mathbf{Z}$ given by the words $x^{2n}X^n$, for $n \in \mathbf{Z}$.)

**Open Question 3.6.3.** Is a combable group also combable by quasigeodesics? (Again, the example in the previous question is not a combing by quasigeodesics.)

Why do we define combings using broken paths, instead of paths? One reason is that this makes it very easy to prove that the existence of a combing is an invariant of pseudoisometry:

**Theorem 3.6.4 (combable is invariant).** *Let $f : X \to Y$ be a pseudoisometry between metric spaces, taking basepoint to basepoint. If $X$ is combable, so is $Y$. Suppose that $X$ and $Y$ have free $G$-actions, for some group $G$, and that $f$ is equivariant. Then, if $X$ is bicombable, so is $Y$. In particular, the property of a group being combable or bicombable does not depend on the choice of generators.*

*Proof of 3.6.4:* Given $y \in Y$, take $x \in X$ with $f(x) = y$ (see Convention 3.3.1). The combing of $X$ gives a set of broken paths $\alpha$ from the basepoint $*_X$ to $x$. The image of $\alpha$ under $f$ need not be a map, but by choosing a single point in $Y$ for each point in the domain we obtain some broken path from the basepoint $*_Y$ to $y$, which we call $f(\alpha)$. In the bicombable case, we take care to make these choices equivariantly.

The constants of $f(\alpha)$ can be estimated in terms of the constants of $\alpha$ and of $f$, and $f(\alpha)$ depends on $y$ in the desired manner.

The last statement of the theorem follows from Lemma 3.3.3 (change of generators is pseudoisometry). 3.6.4

Another reason for working with pseudopaths is that, in view of Lemma 3.3.2 (subspace) and invariance under pseudoisometries, we can define a combing by specifying it on a subset that is sufficiently widely distributed in the space. In the case of the Cayley graph, this subset might be the set of vertices, that is, the group itself. Thus, we could dispense completely with the edges of the Cayley graph, and work only with the group and the word metric on it.

At the same time, the use of pseudopaths does not augment the class of combable groups in any way, for in the Cayley graph we can always replace the pseudopaths by true paths—and, in fact, by paths that correspond to words on the generators. A path is *wordlike* if it maps integers to vertices and is parametrized by pathlength.

**Lemma 3.6.5 (parametrize by pathlength).** *Any combable or bicombable group can be combed by wordlike paths.*

*Proof of 3.6.5:* By Lemma 3.3.5 (piecewise geodesic), we can replace the broken paths of the given combing by piecewise geodesic paths having the same domain, and having speed less than some bound that depends only on the constants of the combing. However, we cannot reparametrize by pathlength without taking precautions, because this would destroy the uniform distance property.

We may, however, normalize the speed bound to one, and assume that integer times are mapped to vertices. The result is a sequence of edges, each traversed at unit speed, possibly interspersed with periods of integer length where the path is stationary at a vertex. To get rid of the stationary part, we first imagine that some generator is the identity. Then we can go around a loop from a vertex to itself instead of being stuck at a vertex.

If there is no trivial generator, we cannot do this, but instead we use a word $xx^{-1}$, where $x$ is an arbitrary generator. We insert this word at every alternate interval of length one where the path was previously constant, and traverse the corresponding edges in two units of time. The other alternate intervals of length one where the path was previously constant are deleted. 3.6.5

**Theorem 3.6.6 (combable isoperimetric inequality).** *Every combable group satisfies a quadratic isoperimetric inequality. Every combable group has a solvable word problem (but not necessarily in quadratic time).*

*Proof of 3.6.6:* We can assume that the combing is by wordlike paths. By Definition 3.6.1 (combing), paths to neighbouring points have domains of approximately the same length. We can then argue as for automatic groups (Theorem 2.3.12) to show the existence of a quadratic isoperimetric inequality. The solvability of the word problem follows from Theorem 2.2.5. 3.6.6

# Chapter 4

# Abelian and Euclidean Groups

In Section 3.4 we looked at word hyperbolic groups, where the automatic structure derives from the fact that arbitrary geodesic words are a bounded distance apart. This is no longer the case for another interesting class of groups, namely euclidean groups. Euclidean groups are discrete cocompact subgroups of isometries of euclidean space (of any dimension), and by Bieberbach's Theorem they are characterized by having a free abelian subgroup of finite index. Very general considerations—Theorems 4.1.1 (direct product) and 4.1.4 (finite index)—show that all euclidean groups are automatic.

With more work one can show that they can be given a biautomatic structure satisfying three desirable properties simultaneously: prefix closedness, representation by geodesics, and uniqueness (Corollary 4.2.4). This result is due to Epstein and Levy, based on Thurston's idea of using barycentric subdivisions to give the free abelian subgroup of finite index an automatic structure that is invariant under a finite group of automorphisms.

Extending in another direction, we show in Section 4.3 that a free abelian group is ShortLex-automatic with respect to *any* set of generators. This result is due to Holt.

Section 4.4 describes an example, due to Cannon, of a group that is ShortLex-automatic but not strongly geodesically automatic, with respect to the same set of generators.

## 4.1. General Results

The results in this section are due to Epstein and Holt.

Let $P$ be a property which a group may or may not possess. We say that a group is *virtually* $P$ if a subgroup of finite index is $P$. (This definition is only really useful if $P$ is closed under taking finite index subgroups.) For example, a virtually abelian group is a group that has an abelian subgroup of finite index. Bieberbach [Bie12] showed that virtually abelian groups are exactly those that can be realized as discrete groups of isometries of euclidean space $\mathbf{R}^n$, for some $n$. (For an accessible proof, see [Cha86].) We thus call virtually abelian groups *euclidean groups*. We give two independent proofs that such groups are automatic. The first follows from simple general facts: see Theorems 4.1.1 (direct product) and 4.1.4 (finite index). The second proves a stronger result, but is more technical; it is given in Section 4.2 (Euclidean Groups are Biautomatic).

**Theorem 4.1.1 (direct product).** *The direct product of two automatic groups is automatic. The direct product of two biautomatic groups is biautomatic.*

*Proof of 4.1.1:* Let $G_1$ and $G_2$ be the groups, with automatic structures $(A_1, L_1)$ and $(A_2, L_2)$. By Theorem 2.5.1 (uniqueness), we can assume that $L_1$ consists of unique representatives for the elements of $G_1$. We show that $(A_1 \cup A_2, L_1L_2)$ is an automatic structure for $G = G_1 \times G_2$, where the union is disjoint and $G_1$ and $G_2$ are seen as subgroups of the direct product.

Clearly $L_1L_2$ maps onto $G$. Let $u_1, v_1 \in L_1$ and $u_2, v_2 \in L_2$, and let $u_1u_2$ and $v_1v_2$ represent elements of $G$ which are within a distance of one in the Cayley graph of $G$. Then $u_1$ and $v_1$ represent elements of $G_1$ which are at most one apart in the Cayley graph of $G_1$, and likewise for $u_2$ and $v_2$. By Theorem 2.3.5 (characterizing synchronous), the uniform distance between the paths $\widehat{u_1}$ and $\widehat{v_1}$ in the Cayley graph of $G_1$ is bounded by a constant $c$, and likewise for the distance between $\widehat{u_2}$ and $\widehat{v_2}$. By the uniqueness property of $L_1$ and Lemma 2.3.9 (bounded length difference), the difference in length between $u_1$ and $v_1$ is likewise bounded by a constant $k$. Therefore the uniform distance between $\widehat{u_1u_2}$ and $\widehat{v_1v_2}$ is bounded by $c + k$.

The statement about biautomatic structures follows in the same way, from the observation that $(L_1L_2)^{-1} = L_2^{-1}L_1^{-1}$ (for the existence of a biautomatic structures with uniqueness, see remark following Lemma 2.5.5 (characterizing biautomatic)). 4.1.1

**Open Question 4.1.2.** Here is a question which demonstrates how little we know about automatic groups. If a direct product of two groups is automatic, are the factors automatic? More generally, if $H$ is a retract of an automatic group $G$ (that is, $H$ is a subgroup of $G$ and there is a homomorphism $G \to H$ that equals the identity on $H$), is $H$ automatic?

Theorem 4.1.1 immediately shows that any finitely generated abelian group is automatic, since such a group can be written as the product of a finite group and finitely many copies of the rank-one free abelian group $\mathbf{Z}$. In fact we can conclude that any finitely generated abelian group is biautomatic, either from

Theorem 4.1.1 or from the remark following Lemma 2.5.5 (characterizing biautomatic), to the effect that any automatic structure on an abelian group is biautomatic.

The simplest automatic structure for $\mathbf{Z}$ is given by the regular expression $x^* \vee X^*$, where $x$ is a generator. For a free abelian group on two generators $x$ and $y$, the language constructed in the proof of Theorem 4.1.1 has regular expression $(x^* \vee X^*)(y^* \vee Y^*)$, so it coincides with the ShortLex language on the ordered alphabet $(x, X, y, Y)$. If we use the same generators in the order $(x, y, Y, X)$, the corresponding ShortLex language is even symmetric, being given by $(x^*y^*) \vee (x^*Y^*) \vee (y^*X^*) \vee (Y^*X^*)$. This idea applies to abelian groups of any rank:

**Theorem 4.1.3 (abelian has symmetric automation).** *Any finitely generated abelian group $G$ has a symmetric automatic structure.*

*Proof of 4.1.3:* We write $G$ as the product of a finite group $H$ and a free abelian group $F$ on generators $x_1, \dots, x_k$. Let $A$ be the ordered alphabet $(x_1, \dots, x_k, X_k, \dots, X_1)$. Then the language $L = \mathsf{ShortLex}(F, A)$ gives a symmetric automation for $F$, and the language $HLH$ over $H \cup A$ gives a symmetric automation for $G$. $\boxed{4.1.3}$

To extend the results above to euclidean groups, we need the following fact:

**Theorem 4.1.4 (finite index).** *Let $G$ be a group and $H$ a subgroup of finite index. $G$ is automatic if and only if $H$ is. If $G$ is biautomatic, so is $H$.*

**Open Question 4.1.5.** It is not known if one can deduce that $G$ is biautomatic from the knowledge that $H$ is biautomatic. A counterexample would also provide a group which is automatic but not biautomatic.

*Proof of 4.1.4:* Assume first that $H$ is automatic, with automatic structure $(A, L)$, and let $P$ be a set of representatives of the left cosets of $H$ in $G$. We show that $(A \cup P, LP)$ is an automatic structure for $G$. Clearly $LP$ is regular and maps onto $G$. Now take two strings $u_1p_1$ and $u_2p_2$, where $u_1, u_2 \in L$ and $p_1, p_2 \in P$, and suppose that their images in the Cayley graph of $G$ are a distance at most one apart, that is, $\overline{u_1p_1x} = \overline{u_2p_2}$, where $x \in L \cup P \cup \{\varepsilon\}$. Then we have $\overline{u_1(p_1xp_2^{-1})} = \overline{u_2}$, so $p_1xp_2^{-1} \in H$. This implies that $\overline{u_1}$ and $\overline{u_2}$ are a bounded distance apart in the Cayley graph $\Gamma(H, A)$ of $H$, the bound being given by the maximum of the word lengths of $p_1xp_2^{-1} \in H$ as $x, p_1, p_2$ take on all their (finitely many) possible values. By Theorem 2.3.5 (characterizing synchronous), the uniform distance between $\widehat{u_1}$ and $\widehat{u_2}$ in $H$ is also bounded, and so is the uniform distance between $\widehat{u_1p_1}$ and $\widehat{u_2p_2}$. Again by Theorem 2.3.5, $LP$ defines an automatic structure.

To deal with the converse direction, assume that $G$ is automatic with automatic structure $(A, L)$. By Theorem 2.4.1 (changing generators), we can assume that $A$ contains a set $P$ of representatives of the left cosets of $H$ in $G$. We also assume that $P$ contains the identity element $e$. Any element of $G$ can be written uniquely as a product of an element of $H$ followed by an element of $P$. For each $p \in P$ and $x \in A$ we let $\sigma(p, x)$ be the element of $H$ such that $px = \sigma(p, x)q$, with $q \in P$. The elements $\sigma(p, x)$ form a set $\Sigma$ of generators for $H$, called the *Schreier generators*. This is easy to see: Given a string $x_1 \ldots x_r \in A^*$, we first replace $x_1 = ex_1$ by $\sigma(e, x_1)p_1$, for some $p_1 \in P$, then we replace $p_1x_2$ by $\sigma(p_1, x_2)p_2$, and so on until we have a string in $\Sigma^* P$, of length $r + 1$. If $x_1 \ldots x_r$ represents an element of $H$, the new string must end with the identity element, which can be omitted. In particular, distances in the Cayley graph $\Gamma(H, \Sigma)$ are no greater than in $\Gamma(H, A)$.

Let $L'$ be the language over $\Sigma$ consisting of strings obtained by the process above from strings of $L$ representing elements of $H$. We show that $(\Sigma, L')$ is an automatic structure for $H$. Let $W$ be a normalized finite state automaton accepting $L$, and let $S$ be the set of its live states. We will construct a partial deterministic automaton over $\Sigma$ which accepts $L'$. We first construct an automaton over $A$, with state set $S \times P$. The start state is $(s_0, e)$, where $s_0$ is the start state of $S$, and the accept states are $(s, e)$, where $s$ is an accept state of $W$. The transition function is defined by $(s, p)x = (sx, q)$, where $px = \sigma(p, x)q$. We now change the arrows' labels as follows: if an arrow labelled $x$ has source $(s, p)$, we replace the label by $\sigma(p, x)$. This gives a partial automaton over $\Sigma$, whose language of accepted words is easily seen to be $L'$.

Next we must show that if $u_1, u_2 \in L'$ represent elements at most one generator apart in $H$, the corresponding paths are at a bounded uniform distance apart. Suppose $u_1$ and $u_2$ come from strings $v_1$ and $v_2$ in $L$. Then $\overline{v_1} = \overline{u_1}$ and $\overline{v_2} = \overline{u_2}$, so $v_1$ and $v_2$ have endpoints a bounded distance apart in the Cayley graph $\Gamma(G, A)$, the bound being given by the maximum word length of any element of $\Sigma$ in terms of $A$. Thus the paths $\widehat{v_1}$ and $\widehat{v_2}$ are a bounded uniform distance apart in $\Gamma(G, A)$. By construction, the prefixes $u_1(t)$ and $v_1(t)$ differ by at most an element of $P$, for all $t \geq 0$, and similarly for $u_2(t)$ and $v_2(t)$; therefore the paths $\widehat{u_1}$ and $\widehat{u_2}$ are also a bounded uniform distance apart in $\Gamma(H, A)$. But we already know that distances in $\Gamma(H, \Sigma)$ are no greater than in $\Gamma(H, A)$, so that $\widehat{u_1}$ and $\widehat{u_2}$ are a bounded uniform distance apart in $\Gamma(H, A)$, which shows that $(\Sigma, L')$ gives an automatic structure.

To show that $G$ biautomatic implies $H$ biautomatic, we follow the same reasoning, using Lemma 2.5.5 (characterizing biautomatic) instead of Theorem 2.3.5 (characterizing synchronous). 4.1.4

**Corollary 4.1.6 (euclidean implies automatic).** *Discrete groups of euclidean isometries, or equivalently finitely generated virtually abelian groups, are automatic.* 4.1.6

In the next section we will prove that such groups are in fact biautomatic, and that the structure can be made to satisfy several nice properties.

## 4.2. Euclidean Groups are Biautomatic

In this section we show that a euclidean group has a biautomatic structure that is prefix closed and has unique, geodesic representatives. We can do this by choosing the generators carefully. In the next section we will prove that abelian groups are ShortLex-automatic for any set of generators; this also gives a biautomatic structure with the same nice properties, but the proof does not extend to the euclidean case.

We start by giving the free abelian subgroup of finite index an automatic structure that is invariant under a finite group of automorphisms (these will be the inner automorphisms in the euclidean group). The idea of doing this is due to Thurston, who also suggested the use of barycentric subdivisions. Theorem 4.2.1 itself was proved by Epstein and Levy.

**Theorem 4.2.1 (automatic structure with invariance).** *Let $H$ be a finite group acting by automorphisms on a free abelian group $G$ of finite rank. There is a finite set $A$ of distinct generators for $G$, invariant under $H$ and under inversion, and a regular language $L$ over $A$, also invariant under $H$, such that $(A, L)$ defines a prefix closed biautomatic structure for $G$. Moreover, $L$ consists of unique geodesic representatives for the elements of $G$.*

The word "distinct" is necessary so we can define an action of $H$ on $A$; if different elements of $A$ map into the same group element, this action is not well-defined. The $H$-invariance of $L$ means that $L$ is invariant under the extension to $A^*$ of this action of $H$ on $A$.

The extension of the involution $x \mapsto x^{-1}$ to $A^*$ gives rise to an automorphism of $G$, the central reflection map, which in additive notation is written $g \mapsto -g$. This automorphism commutes with all automorphisms of $G$, so we can assume that it lies in $H$, by adjoining it to $H$ if necessary. It might seem at first that the automation given by the theorem in this case is symmetric, that is, $L = L^{-1}$; but this is not the case because the strings $x^{-1}y^{-1}$ and $(xy)^{-1}$ are distinct, although they represent the same element in $G$.

*Proof of 4.2.1:* In order to make geometric language natural, we use additive notation for elements of $G$ and tensor the group with $\mathbf{R}$, obtaining an $n$-dimensional vector space $V$ over $\mathbf{R}$, where $n$ is the rank of $G$. We see $G$ as a lattice in $V$. We may as well assume that $H$ contains the map $x \mapsto -x$, as discussed in the previous paragraph.

We start by choosing an $H$-invariant convex polyhedron $K$ with vertices in $G$. For example, we can take $K$ to be the convex hull of the union of $H$-images of an arbitrary set of generators for $G$. Clearly, $H$ contains the origin. By multiplying $K$ by an appropriate integer, we can ensure that the barycentre $x_F$ of each face $F$ of $K$ belongs to $G$. (For us, a face can be of any dimension less than $n$, and the barycentre of a face is simply the average of its vertices.) In fact, we will assume that $x_F/N$ belongs to $G$, where $N$ is a fixed integer to be determined later.

We then form the barycentric subdivision of the boundary of $K$; this is an $(n-1)$-dimensional simplicial complex $\mathcal{K}$, whose simplices are of the form $(x_{F_0}, x_{F_1}, \ldots, x_{F_k})$, where $F_0 \subset \cdots \subset F_k$ is any strictly increasing sequence of faces of $K$. Let $X$ be the set of vertices of $\mathcal{K}$; we give $X$ a partial order by saying that $x_F \leq x_{F'}$ if $F \subset F'$. We always list the vertices of a simplex of $\mathcal{K}$ in increasing order. Any element of $H$ gives rise to a permutation of the $k$-simplices of $\mathcal{K}$, for $0 < k < n$, and if an element of $H$ preserves a simplex of $\mathcal{K}$, it acts as the identity there.

Let $\sigma = (x_0, \cdots, x_k)$ be a simplex. (We are distinguishing between $\sigma$, which is an ordered set of affinely independent points, and its convex hull, which we denote by $\langle\sigma\rangle$.) The set of points $v \in V$ of the form

**4.2.2.** $$v = \lambda_0 x_0 + \cdots + \lambda_k x_k,$$

where $\lambda_i > 0$ for $0 \leq i \leq k$, is called the (open) cone over $\sigma$. The cones over all the simplices of $\mathcal{K}$ (including the cone over $\varnothing$, which is just the origin) form a partition of $V$; we will denote by $\sigma(v)$ the simplex in whose cone a given point $v$ lies. (See Figure 4.1.)

If $v$ has the form 4.2.2, we set $\lambda_{x_i}(v) = \lambda_i$ for $0 \leq i \leq k$, and $\lambda_x(v) = 0$ for $x \in X \setminus \{x_0, \ldots, x_k\}$. Then each $\lambda_x$ is a lipschitz function on $V$, and for $h \in H$ we have $\lambda_x \circ h = \lambda_{h^{-1}(x)}$. Furthermore, we have $\|v\|_K = \sum_{x \in X} \lambda_x(v)$, where $\|\cdot\|_K$ is the norm on $V$ whose unit ball is $K$.

Now let our set of generators be $A = G \cap K$, which is $H$-invariant by construction. By definition, $\|a\|_K \leq 1$ for $a \in A$, so any word over $A$ representing $g \in G$ must have length at least $\|g\|_K$. Conversely, we will construct for each $g \in G$ a word $u(g)$ of length $\lceil \|g\|_K \rceil$, where $\lceil \|g\|_K \rceil$ is the smallest integer greater than or equal to $\|g\|_K$. This word will therefore be a geodesic.

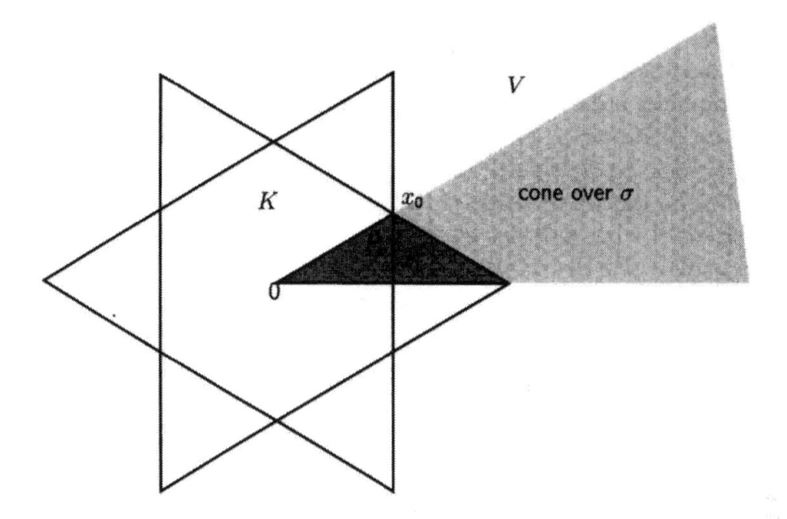


**Figure 4.1. A star of David.** If $G$ has rank two and $H$ is the dihedral group of order 12, we can choose $K$ as a regular hexagon. Then $P$ is a six-cornered star. The cone over the simplex $\sigma = \langle x_0, x_1 \rangle$ is shown in light gray, and the set $P_\sigma$ in dark gray.

We start building the word with the string

$$w(g) = \prod_{x \in X} x^{\lfloor \lambda_x(g) \rfloor},$$

where $\lfloor \lambda_x \rfloor$ is the integral part of $\lambda_x$. The product is taken in order of increasing $x$; there is no ambiguity because the $\lambda_x$ are non-zero only for a sequence $x_0 < \cdots < x_k$ of values of $x$. Let $P$ be the set of points $v \in V$ such that $\lambda_x(v) < 1$ for all $x \in X$. We can also see $P$ as the union of open parallelepipeds $P_\sigma$, containing 0 and with sides equal to the vectors $x_0, \ldots, x_k$, for all choices of $\sigma = (x_0, \ldots, x_k) \in \mathcal{K}$. It is easy to check that if $g$ lies in the open cone on $\sigma$, $g - w(g)$ lies in $P_\tau$ for some simplex $\tau \subset \sigma$, and that it has norm $\|g\|_K - |w(g)|$, so that we have reduced the problem to finding geodesic representatives for points in $G \cap P$.

So let $p \in G \cap P$. If $p = 0$, we use the empty string for our geodesic representative $u(p)$. Otherwise, let $\sigma = \sigma(p) = (x_0, \ldots, x_k) \in \mathcal{K}$ be the unique open simplex whose cone contains $p$. Let $l$ be the greatest integer strictly less than $\|p\|_K$. If $l = 0$ we have $p \in K$, and we set $u(p) = p$; otherwise $1 \leq l < \|p\|_K < k + 1 \leq n + 1$.

We choose $y \in G \cap \langle \sigma \rangle$ such that $z = p - ly \in K$ and $ly \in P$; we will see in a moment why such a $y$ exists. In general there are several candidates for $y$. We pick $p/\|p\|_K$, if that happens to belong to $G$; otherwise we pick the candidate whose barycentric coordinates with respect to $\sigma$ are least in lexicographical order. The important thing is that we have a well-defined procedure to choose $y$, and that the choice is invariant under $H$. We then set $u(p) = y^l z$; by construction, $|u(p)| = \lceil \|p\|_K \rceil$, so $u(p)$ is geodesic.

Let $L$ be the set of words of the form $w(g)u(g - \overline{w(g)})$ as $g$ varies over $G$. The set of $w(g)$ is obviously a regular language. Since $P \cap G$ is finite, $L$ is also regular. By the construction, it is obvious that $L$ is prefix closed, and we have gone to great pains to ensure that each element of $L$ is a geodesic. Moreover $L$ gives us an automatic structure because the functions $\lambda_i$ which occur in Equation 4.2.2 are lipschitz, which ensures that $w(g)$ is a lipschitz function of $g$, and the other choices vary over a fixed finite set of elements of $G$. (We use here Theorem 3.3.6: since $G$ acts on $V$ properly discontinuously with compact quotient, the word metric on $G$ is equivalent to the metric given by the norm on $V$.) On an abelian group, any automatic structure is biautomatic. Since the construction of $w$ and $u$ is $H$-equivariant, the language $L$ is $H$-equivariant. To complete the proof of the theorem, we therefore need only prove the following lemma.

**Lemma 4.2.3.** *There exists $y \in G \cap \langle\sigma\rangle$ such that $p - ly \in K$ and $ly \in P$.*

*Proof of 4.2.3:* We work in the $(k+1)$-dimensional vector subspace of $V$ spanned by $x_0, \ldots, x_k$, and represent points by coordinates with respect to this basis. Because $K$ is convex and invariant under central reflection, the norm $\|\cdot\|_K$ is dominated by the norm $\|\cdot\|$ given by the sum of absolute values of coordinates; in other words, a point with coordinates $(a_0, \ldots, a_k)$ with $\sum_{i=0}^k |a_i| \leq 1$ certainly lies in $K$.

We now use the fact that we multiplied $K$ by some integer $N$ at the beginning of the proof of Theorem 4.2.1. This implies that any point whose coordinates are integral multiples of $1/N$ is in $G$. Let $Y_l$ be the set of points with coordinates $y_i = m_i/N$, where the $m_i$ are integers with $0 \leq lm_i < N$ and $\sum y_i = 1$. Then any point $y \in Y_l$ automatically satisfies all the desired conditions, except possibly for $p - ly \in K$ (the condition $ly \in P$ is equivalent to $y_i < 1/l$ for every $i$).

We project $p$ with center at the origin, obtaining a point $p' = p/\|p\|_K \in \langle\sigma\rangle$ with coordinates $(p'_0, \ldots, p'_k)$ such that $\sum_{i=0}^k p'_i = 1$. To simplify the notation, we assume that $p'_0$ is the largest of the $k+1$ coordinates. As a candidate for $y$, we take the point $(y_0, \ldots, y_k)$, where each $y_i$, for $1 \leq i \leq k$, is obtained by rounding $p'_i$ down, and $y_0$ is rounded up to preserve the sum of the coordinates. Then $0 \leq p'_i - y_i < 1/N$ for $1 \leq i \leq k$ and $-k/N \leq p'_0 - y_0 \leq 0$. A simple calculation shows that $y + 2k(p' - y)$ is in $\langle\sigma\rangle$ for $N$ large (but depending only on $k$; in particular, the different possibilities for $k$ can all be covered at once).

Since $y + 2k(p' - y) \in \langle\sigma\rangle \subset K$, the convexity of $K$ implies, successively, that $y + (l+1)(p' - y) \in K$ (since $l + 1 \leq 2k$), $l(p' - y) \in K$ (since $-y \in K$ by symmetry and $l \leq k$), and finally $p - ly \in K$ (because $p - ly$ is in the segment joining $l(p' - y)$ and $y + (l+1)(p' - y) = (l+1)p' - ly$).

The point $y$ found above fails to be in $Y_l$ if $y_0 \geq 1/l$. But if that is so we have $p'_0 \geq 1/l - k/N$, so the strategy in this case is to bound $p - lp'$ and choose a different $y$ that does lie in $Y_l$. We choose $y$ by rounding down all but the smallest coordinate of $p'$, and rounding up the smallest coordinate to make up for the difference. If $p'_s$ is the smallest coordinate, then $p'_s \leq 1/(k+1) < l$. Therefore, if $N$ is large enough, $y \in Y_l$. Then $\|p' - y\| < 2k/N$. On the other hand, we have

$$\frac{1}{\|p\|} > \frac{p_0}{\|p\|} = p'_0 \geq \frac{1}{l} - \frac{k}{N},$$

so that $\|p - lp'\| = \|p\| - l < kl^2/(N - kl)$. Adding the two estimates, we can find $N$ such that $\|p - ly\| \leq 1$. As we observed at the beginning of the proof, this implies that $p - ly \in K$. 4.2.3

4.2.1

**Corollary 4.2.4 (euclidean implies biautomatic).** *Any euclidean group has a prefix closed biautomatic structure consisting of unique geodesic representatives.*

*Proof of 4.2.4:* Let $G$ be the euclidean group and $G'$ a finitely generated free abelian group that has finite index in $G$. We can assume without loss of generality that $G'$ is normal. Let $P$ be a set of coset representatives for the cosets of $G'$ in $G$, and assume that the identity $e$ is in $P$. We set $P' = P \backslash \{e\}$. Let $(A, L)$ be the automatic structure obtained by applying Theorem 4.2.1 (automatic structure with invariance) to $G'$, under the action of $H = G/G'$. As in the first part of the proof of Theorem 4.1.4 (finite index) shows that, we can easily check that $(A \cup P', L \cup LP')$ is an automatic structure for $G$. The invariance of $L$ under conjugation by elements of $H$ enables us to establish that the structure is biautomatic. The uniqueness property and prefix closure are immediate.

It remains to be shown that $L$ consists of geodesics. For this we form the Schreier generators $\sigma(p, q) \in G'$, for $p, q \in P$; recall from the proof of Theorem 4.1.4 (finite index) that $\sigma(p, q)$ is defined by $pq = \sigma(p, q)r$, where $r \in P$. We can assume that we chose the convex set $K$ at the beginning of the proof of Theorem 4.2.1 so large that all the $\sigma(p, q)$ are contained in it.

Given a word over $A \cup P'$, we start by moving all generators that belong to $P'$ to the right past those belonging to $A$, by conjugation; this is possible because $pxp^{-1} \in A$ for $x \in A$ and $p \in P'$. We take the resulting string over $P'$ and write it as a string over the Schreier generators, followed by at most one generator in $P'$. The result is a word of the form $wz$, where $w \in A^*$ and $z \in Z$. We now replace $w$ by the unique representative of $\overline{w}$ in $L$. The resulting word is no longer than the original word, and it lies in $L \cup LP'$.

4.2.4

## 4.3. Abelian Groups and ShortLex

The material in this section is due to Holt.

**Theorem 4.3.1 (abelian implies ShortLex-automatic).** *Let $G$ be a finitely generated abelian group. Then $G$ is ShortLex-automatic with respect to any ordered set of semigroup generators.*

*Proof of 4.3.1:* Let $A = \{x_1, \ldots, x_n\}$ be an ordered set of semigroup generators for $G$. Since $G$ is abelian, we can put every string in *normal form* $x_1^{a_1} \ldots x_n^{a_n}$, where the $a_i$ are non-negative integers; we identify $x_1^{a_1} \ldots x_n^{a_n}$ with the $n$-tuple $(a_1, \ldots, a_n)$. Elements of ShortLex$(G, A)$ are already in normal form. In this proof all strings are assumed to be in normal form, unless we say otherwise.

Given two $n$-tuples $v = (v_1, \ldots, v_n)$ and $w = (w_1, \ldots, w_n)$, we say that $v$ is *contained in* $w$, and write $v \prec w$, if $v_i \leq w_i$ for all $1 \leq i \leq n$. Then the *complement* of $v$ with respect to $w$ is $(w_1 - v_1, \ldots, w_n - v_n)$.

If $w$ and $w'$ are distinct strings representing the same group element, we say that we have a *relation* $w \sim w'$. A *minimal relation* is one that is minimal for the partial order $\prec$ on pairs of strings (where $(v \sim v') \prec (w \sim w')$ if $v \prec w$ and $v' \prec w'$). Notice that if $w \sim w'$, we cannot have both $w, w' \in$ ShortLex$(G, A)$, since ShortLex$(G, A)$ consists of unique representatives.

Let $w$ and $w'$ be words in ShortLex$(G, A)$ such that $\overline{wx} = \overline{w'}$ for some generator $x \in A$. We interpret $wx$ as a string in normal form. We claim that $wx$ and $w'$ differ only by a minimal relation, if at all. For if $wx \neq w'$, let $v \sim v'$ be a minimal relation contained in $wx \sim w'$. Let $u$ be the complement of $v$ in $wx$, and $u'$ the complement of $v'$ in $w'$; we have $\overline{u} = \overline{u'}$. At least one of $u$ and $v$ is contained in $w$. But $v \prec w$ would imply $v, v' \in$ ShortLex$(G, A)$ (since ShortLex$(G, A)$ is closed under $\prec$), which is impossible. Thus $u \prec w$. But then $u, u' \in$ ShortLex$(G, A)$, so $u = u'$. Thus $wx$ and $w'$ differ only by the minimal relation $v \sim v'$.

If there is a bound on the size of minimal relations, the previous paragraph implies that ShortLex$(G, A)$ satisfies the conditions of Theorem 2.3.5 (characterizing synchronous), and the theorem follows. That there is such a bound is a consequence of the next lemma, where we look at relations as $(2n)$-tuples of non-negative integers, with the order $\prec$. [4.3.1]

**Lemma 4.3.2.** *Given a set $Z$ of $N$-tuples of non-negative integers, the subset $Y \subset Z$ of $N$-tuples which are minimal in $Z$ for the order $\prec$ is finite.*

*Proof of 4.3.2:* We prove this by induction on $N$. If $Y$ is infinite, consider an enumeration $y^{(1)}, y^{(2)}, \ldots$ of $Y$, with $y^{(i)} = (y_1^{(i)}, \ldots, y_N^{(i)})$. For any $s$, we have $y_i^{(s)} < y_i^{(1)}$ for some $i$ (depending on $s$), otherwise $y^{(s)}$ would not be minimal. This implies that, for some $i$ and some $y$, the set $Y' \subset Y$ of $N$-tuples whose $i$-th coordinate equals $y$ is infinite. But $N$-tuples in $Y'$ are minimal in the set $Z'$ of $N$-tuples in $Z$ whose $i$-th coordinate is $y$, and $Z'$ can be seen as a set of $(N-1)$-tuples. By the induction step, $Y'$ is finite, and we get a contradiction. [4.3.2]

## 4.4. A Euclidean Counterexample

We discuss now an example of a euclidean group that is ShortLex-automatic with respect to a certain set of generators, but is very far from being strongly geodesically automatic for the same set of generators. It is also not ShortLex-automatic with respect to the same generators in reverse order.

**Example 4.4.1 (order matters).** Let $G$ be the wreath product of $\mathbf{Z}$ with $\mathbf{Z}_2$, which can be described as the group generated by translations plus the diagonal flip $(a,b) \mapsto (b,a)$ in $\mathbf{Z}^2 = \mathbf{Z} \times \mathbf{Z}$. A group presentation for $G$ is

$$G = \langle \{x, y, z\} / \{x^2, zyz^{-1}y^{-1}, yxz^{-1}x^{-1}\} \rangle,$$

where $x$ is the flip, and $y$ and $z$ are unit translations along the two coordinate axes in $\mathbf{Z}^2$. $G$ has $\mathbf{Z}^2$ as an index-two subgroup, and so it is biautomatic by Corollary 4.2.4.

We show that $G$ is not ShortLex-automatic with respect to the set of generators $A = (x, y, Y, z, Z)$, in the order $x < y < Y < z < Z$. In fact, ShortLex$(G, A)$ is the language given by the regular expression

$$(x \vee \varepsilon)(y^* \vee Y^*)(z^* \vee Z^*),$$

since the projection $G \to G/\mathbf{Z}^2 = \mathbf{Z}_2$ corresponds to counting occurrences of $x$ modulo 2. Thus the strings $u_n = xy^n z^n$ and $v_n = y^n z^n$ are ShortLex-minimal; their distance in the Cayley graph is one, because $\overline{v_n} = \overline{u_n x}$. But $\overline{u_n(n)^{-1} v_n(n)} = Y^{n-1} x y^n$, which reduces to $xy^n Z^{n-1}$, a word of length $2n$. Thus the uniform distance between $\widehat{u_n}$ and $\widehat{v_n}$ is at least $2n$, and by

Lemma 2.3.2 (lipschitz property) the language ShortLex($G, A$) cannot be part of an automatic structure.

On the other hand, $G$ is ShortLex-automatic with respect to the same generators in the opposite order, namely $(Z, z, Y, y, x)$. The ShortLex language in this case is

$$(z^* \vee Z^*)(y^* \vee Y^*)(x \vee \varepsilon).$$

The two ShortLex languages are inverse to each other, so this gives an example of an automatic structure that is not biautomatic. This was brought to our attention by John Sullivan.

**Example 4.4.2 (no geodesic automaton).** We consider again the group $G$ of Example 4.4.1, but this time we add $a = y^2$ and $b = yz$ to the set of generators. Then the set of geodesic strings over $\{x, y, Y, z, Z, a, A, b, B\}$ is not a regular language; this was shown by Cannon. Take the word $a^ixa^jx$, where $i$ and $j$ are non-negative integers. Any word representing $g = a^ixa^jx$ must have an even number of occurrences of $x$, because none of the relators allows us to replace an odd number of $x$'s with an even number. Also $g = y^{2i}z^{2j}$; by reducing any equivalent word to a form containing only $y$, $Y$, $z$ and $Z$, we see that there must be at least $i+j$ occurrences of symbols other than $x$. If the equivalent word does not involve $x$, we are in the free abelian case, and there must be at least $i+j$ symbols if $i \geq j$ (as in $b^{2j}a^{i-j}$), or at least $2j$ symbols if $i < j$ (as in $b^{2i}z^{2(j-i)}$). The upshot of all this is that $a^ixa^jx$ is geodesic if and only if $j > i+1$. But a finite state automaton is not capable of checking this condition.

Thurston suggested that, with a certain ordering of this enlarged set of generators, this example might be ShortLex-automatic, giving a counterexample to the conjecture that, for any ShortLex-automatic group, the geodesics form a regular language. This was tried using the *automata* programs developed at Warwick by Epstein, Holt and Sarah Rees, but initial experiments failed to find any such ordering. These programs are capable of proving that the ShortLex language associated to a given ordering of generators gives an automatic structure; however they are not capable of proving the ShortLex language is not automatic. There are $9! = 362880$ different orderings, and each one should be allowed to run for a few hours, to give it a fair chance. This meant that a brute force approach would take a few centuries, without any certainty of reaching an answer.

After some thought and discussion, Holt and Paterson decided to test the ordering

$$b < B < a < A < x < y < Y < z < Z,$$

for which the *automata* program produced a finite state automaton with 41 states recognizing the ShortLex language. The program also proved that the group is ShortLex-automatic with this ordering of the generators. Thus we have a ShortLex-automatic structure that is not strongly geodesically automatic.

# Chapter 5

# Finding the Automatic Structure: Theory

The main objective of this chapter is to present a collection of axioms for an automatic structure, which are all expressed as regular predicates—see Definition 1.4. In particular, the axioms refer to properties of a finite number of finite state automata. The axioms are satisfied if and only if the finite state automata form an automatic structure for some automatic group. Thus, given a collection of automata, we can check algorithmically whether or not they form an automatic structure on a group.

The idea that it is possible to give axioms for an automatic group, which are checkable because they are regular predicates, is due to Thurston. The actual axioms and the associated proofs are due to Epstein. Mistakes were corrected and improvements suggested by Holt, Uri Zwick and Chuya Hayashi.

We then discuss how one might search for an automatic structure on a group, given a presentation by a finite number of generators and relators. Section 5.2 presents a procedure, due to Epstein, Holt and Thurston, that can do this in principle, although in practice it would take far too long.

The most that one can hope for such a procedure is that it will eventually find an answer when the group is automatic, and will go on forever if it isn't. The reason that one cannot realistically hope for more is that there is no algorithm which takes as its input a finite presentation of a group and gives as output the answer Yes or No, depending on whether the group is trivial or not. If an automatic structure is found for the trivial group, one knows that the group is trivial. Therefore there are bound to be finite presentations for which our procedure is non-terminating (or terminates with "Don't know").

Likewise, even if we know that a group is automatic, we cannot expect to predict how long it will take for an algorithm to find the automatic structure from a given presentation, or how much storage will be needed during the computation. The reason is again already apparent for the trivial group: one can find arbitrarily complex presentations for the trivial group, and of course our algorithm will terminate when applied to them. But one can prove that there is no recursive function (i.e., a computable function in the intuitive sense) that provides an upper bound for the time taken by such an algorithm on the trivial group, as a function of the complexity of the presentation.

## 5.1. Axiom Checking

Suppose we are presented with an alphabet $A$, a finite state automaton $W$ over $A$, accepting a language $L$, and finite state automata $M_x$ over $(A, A)$ for $x \in A \cup \{\varepsilon\}$, accepting languages $L_x$. Do these data form an automatic structure for some group? In this section we present a set of axioms which give necessary and sufficient conditions for this to be the case. All the axioms are fully quantified regular predicates. Therefore, given the finite state automata and the respective languages, we can verify algorithmically whether or not the axioms are satisfied.

The first axiom states that $L$ is non-empty.

**Axiom 1 (non-empty).** $(\exists w)(w \in L)$.

The next axiom limits the "natural domain" for the $M_x$ to pairs of elements in $L$. Formally, this axiom must be considered as a conjunction of several statements, one for each $x \in A \cup \{\varepsilon\}$, because we are not allowed to quantify over the set of generators.

**Axiom 2.** For each $x \in A \cup \{\varepsilon\}$,

$$(\forall w_1, w_2)((w_1, w_2) \in L_x \Longrightarrow (w_1 \in L \wedge w_2 \in L)).$$

The next set of axioms ensures that the condition $(w_1, w_2) \in L_\varepsilon$ is an equivalence relation on $L$.

**Axiom 3.** $(\forall w)(w \in L \Longrightarrow (w, w) \in L_\varepsilon)$.

**Axiom 4.** $(\forall w_1, w_2)((w_1, w_2) \in L_\varepsilon \Longrightarrow (w_2, w_1) \in L_\varepsilon)$.

**Axiom 5.** $(\forall w_1, w_2, w_3)(((w_1, w_2) \in L_\varepsilon \wedge (w_2, w_3) \in L_\varepsilon) \Longrightarrow (w_1, w_3) \in L_\varepsilon)$.

Once these axioms are verified, we can form $X$, the set of equivalence classes under the equivalence relation defined by $M_\varepsilon$. The equivalence class of $w$ is denoted by $[w]$.

The next six axioms will imply that the relations on $L$ defined by the $L_x$ give rise to permutations of $X$. Each of these axioms is again implicitly assumed to be the conjunction of several axioms, one for each generator.

The first requirement is that the relations should involve all elements in the domain.

**Axiom 6.** For each $x \in A$,

$$(\forall w_1)(w_1 \in L \Longrightarrow (\exists w_2)((w_1, w_2) \in L_x)).$$

The next axiom implies that the relation gives a well-defined map $L \to X$.

**Axiom 7.** For each $x \in A$,

$$(\forall w_1, w_2, w_3)(((w_1, w_2) \in L_x \wedge (w_1, w_3) \in L_x) \Longrightarrow (w_2, w_3) \in L_\varepsilon).$$

The next axiom implies that the map $L \to X$ thus defined factors through $X$, inducing a map $\sigma_x : X \to X$:

**Axiom 8.** For each $x \in A$,

$$(\forall w_1, w_2, w_3)(((w_1, w_2) \in L_\varepsilon \wedge (w_1, w_3) \in L_x) \Longrightarrow (w_2, w_3) \in L_x).$$

Dually, the relation needs to involve all elements in the range, its inverse must give a well-defined map $L \to X$, and this map must factor through $X$:

**Axiom 9.** For each $x \in A$,

$$(\forall w_2)(w_2 \in L \Longrightarrow (\exists w_1)((w_1, w_2) \in L_x)).$$

**Axiom 10.** For each $x \in A$,

$$(\forall w_1, w_2, w_3)(((w_1, w_2) \in L_x \wedge (w_3, w_2) \in L_x) \Longrightarrow (w_1, w_3) \in L_\varepsilon).$$

**Axiom 11.** For each $x \in A$,

$$(\forall w_1, w_2, w_3)(((w_1, w_2) \in L_\varepsilon \wedge (w_3, w_1) \in L_x) \Longrightarrow (w_3, w_2) \in L_x).$$

Putting all these axioms together, we see that $\sigma_x$ is a permutation and that $\sigma_x[w_1] = [w_2]$ if and only if $(w_1, w_2) \in L_x$. The permutations are seen as acting on $X$ on the right, so that $\sigma_y\sigma_x$ means apply $\sigma_y$ first and then $\sigma_x$. We define $H$ to be the group of permutations generated by $\sigma_x$, for $x \in A$, and we extend the map $x \mapsto \sigma_x$ to a semigroup homomorphism $A^* \to H$, the image of a word $w$ over $A$ being denoted by $\sigma_w$. We wish to identify $H$ with $X$, in such a way that the diagram

**5.1.1.**

$$\begin{array}{ccc} L & \rightarrow & X \\ \cap & & \uparrow \\ A^* & \rightarrow & H \end{array}$$

commutes. Now when a group $H$ acts on a set $X$, we can identify $X$ with $H$ if the action is free (no point of $X$ is fixed by a non-trivial element) and transitive (any point of $X$ can be taken to any other). In that case, the actual identification consists of fixing a basepoint and then mapping $g \in H$ to the image of the basepoint under $g$.

The commutativity of Diagram 5.1.1 does not follow from the axioms listed so far, because the $x$'s are not yet related in any way to the $\sigma_x$'s. This is easiest to see if $L$ is prefix closed, so $\varepsilon \in L$: then commutativity would force the choice of $[\varepsilon]$ as a basepoint, and we see by induction on the length of $u$ that we would have to have $[\varepsilon]\sigma_{ux} = [u]\sigma_x = [ux]$ whenever $ux \in L$. As things stand, there is no reason why this should be true: without affecting the truth of any of the axioms so far, we could relabel the languages $L_x$, and hence the permutations $\sigma_x$, thus destroying the relationship $[u]\sigma_x = [ux]$ even if it happened to hold to begin with.

To express this relationship in the case when $L$ is not prefix closed, we must extend the equivalence currently defined on $L$ to the prefix closure of $L$. Let $k$ be the number of states of the word acceptor $W$. For each prefix $u$ of an element of $L$, there is a string $w$ of length less that $k$ and such that $uw \in L$. We would like to define $[u]$ as the result of applying $\sigma_w^{-1}$ to $[uw]$. The next axiom tells us that this procedure gives a consistent answer.

**Axiom 12 (extending to prefixes).** For each $w, w' \in A^*$ such that the lengths of $w$ and $w'$ are no more than the number $k$ of states of $W$,

$$(\forall u)(uw \in L \wedge uw' \in L \Longrightarrow [uw]\sigma_w^{-1} = [uw']\sigma_{w'}^{-1}).$$

As usual, if we were to be completely formal, this axiom should be written out as a conjunction of many statements, one for each pair $(w, w')$. The number of statements is likely to be extremely large, but it depends only on $k$ and the size of $A$.

If $u$ is a prefix of an element of $L$, we will write $[u]$ for the element of $X$ assigned unambiguously to $u$ as a result of the previous axiom. If $u \in L$, this agrees with the old definition of $[u]$, since $w$ can then be taken to be the nullstring. Axiom 12 clearly implies that $[u]\sigma_x = [ux]$ for each $ux \in L$, if $L$ is prefix-closed. More generally, for arbitrary $L$ and $ux$ a prefix of a string in $L$, we still have $[u]\sigma_x = [ux]$, as can be seen by taking a string $w$ of length less than $k$ such that $uxw \in L$, and observing that

$$[u]\sigma_x\sigma_w = [u]\sigma_{xw} = [uxw] = [ux]\sigma_w$$

by the definition of $[u]$ and $[ux]$. By induction, we obtain the following lemma:

**Lemma 5.1.2 (transitivity).** *Assume Axioms 1–12 are satisfied. If $u$ and $w$ are strings over $A$ such that $uw$ is a prefix of $L$, then $[uw] = [u]\sigma_w$. In* *particular, the action of $H$ on $X$ is transitive, and Diagram 5.1.1 is commutative if the identity in $H$ is mapped to $[\varepsilon]$.* 5.1.2

The above axioms are not yet sufficient to characterize an automation of a group, because they do not guarantee that the action of $H$ on $X$ is free. To see this, let $P$ be a finite group and let $K$ be a subgroup whose conjugates have a trivial intersection. (For example, take $P$ to be the symmetric group on three letters, and take for $K$ a subgroup of order two.) Let $A$ be the set of all non-identity elements of $P$. Take for $W$ a finite state automaton which accepts all strings over $A$, and define $M_x$, for $x \in A \cup \{\varepsilon\}$, as follows. The state set is $P \times P$. An element $(x, y)$ of the padded alphabet associated to $(A, A)$ sends a state $(g_1, g_2)$ to $(g_1 x, g_2 y)$, where an end-of-string symbol $x$ or $y$ is interpreted as the identity element. An accept state of $M_x$ is a pair $(g_1, g_2)$, with $Kg_1 x = Kg_2$. It is immediate that the axioms are satisfied, that $X$ is the set of cosets of the form $Kg$, and that the group $H$ generated by the $\sigma_x$ is equal to $P$. (In the example of the symmetric group, $H$ is all permutations of the three cosets.)

It turns out that, in order to guarantee that $H$ acts freely, we need only assume that any element of $H$ that is short and fixes some element of $X$ is the identity element.

**Axiom 13 (trivial stabilizer).** Let $k$ be greater than the number of states in the finite automata, and greater than the length of some accepted string representing the identity. For each string $w$ over $A$ of length at most $4k$,

$$(\exists u \in L)([u]\sigma_w = [u]) \Longrightarrow (\forall u \in L)([u]\sigma_w = [u]).$$

Again, this should be seen as the conjunction of a large but precisely determined number of regular predicates, one for each $w$. Notice also that an accepted string representing the identity can be found constructively: we start by using Axiom 1 (non-empty) to find some $u \in L$, then use the $M_x$ to find accepted representatives for successively shorter prefixes of $u$, just as in the proof of Theorem 2.3.10 (quadratic algorithm).

**Lemma 5.1.3 (free action).** *If Axioms 1–13 are satisfied, the stabilizer of the basepoint (and hence of any point) of $X$ is trivial.*

*Proof of 5.1.3:* Let $w = x_1^{\varepsilon_1} x_2^{\varepsilon_2} \dots x_n^{\varepsilon_n}$ be a word over $A$, where $x_i \in A$ and $\varepsilon_i = \pm 1$ for $i = 1, \dots, n$ (see Definition 2.1.7 for the notation $x^{-1}$). Assume that $[\varepsilon]\sigma_w = [\varepsilon]$; we must show that $\sigma_w$ is the identity in $H$.

As usual, $w(t)$ will denote the prefix of $w$ of length $t$, for $0 \le t \le |w| = n$. For each $t \le |w|$, let $u_t$ be a shortest element of $L$ such that $[\varepsilon]\sigma_{w(t)} = [u_t]$. Now $[u_0] = [u_n] = [\varepsilon]$, so we can take $u_0 = u_n$ and its length will be less than $k$, by the definition of $k$ in Axiom 13. By Lemma 5.1.2 (transitivity), $[\varepsilon]\sigma_{u_0} = [u_0]$ and so $\sigma_{u_0}$ fixes $[\varepsilon]$. By Axiom 13 (trivial stabilizer), $\sigma_{u_0} = \sigma_{u_n}$ is equal to the identity in $H$.

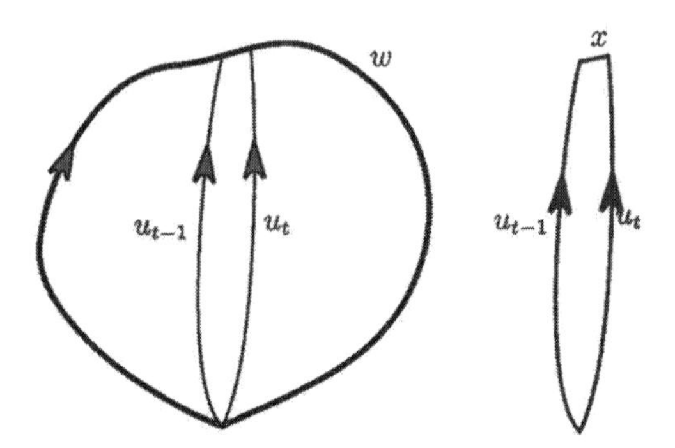


**Figure 5.1. The first factorization of a group element fixing $[\varepsilon]$.** We illustrate the proof of Lemma 5.1.3 (free action). The element to be factorized is represented by the word $w$. The words $u_{t-1}$ and $u_t$ are certain elements of $L$, the language of the finite state automaton $W$. The picture shows how to write $\sigma_w$ as a product of elements as in Equation 5.1.4.

Therefore

**5.1.4.**
$$\sigma_w = \prod_{t=1}^{n} \sigma_{u_{t-1}} \sigma_{x_t}^{\varepsilon_t} \sigma_{u_t}^{-1},$$

as illustrated in Figure 5.1. Setting $v_t = u_{t-1} x_t^{\varepsilon_t} u_t^{-1}$, we have now

$$\begin{aligned} [\varepsilon]\sigma_{v_t} &= [\varepsilon]\sigma_{u_{t-1}}\sigma_{x_t}^{\varepsilon_t}\sigma_{u_t}^{-1} = [u_{t-1}]\sigma_{x_t}^{\varepsilon_t}\sigma_{u_t}^{-1} = [\varepsilon]\sigma_{w(t-1)}\sigma_{x_t}^{\varepsilon_t}\sigma_{u_t}^{-1} \\ &= [\varepsilon]\sigma_{w(t)}\sigma_{u_t}^{-1} = [u_t]\sigma_{u_t}^{-1} = [\varepsilon]. \end{aligned}$$

The above equations reduce our task to considering the following special case. We are given $u$ and $u'$ in $L$, with $[u]\sigma_x = [u']$ for some $x \in A$, and $[\varepsilon]\sigma_v = [\varepsilon]$ for $v = uxu'^{-1}$. We need to show that $\sigma_v$ is the identity, and this will complete the proof of the lemma.

The pair $(u, u')$ is accepted by $M_x$, by the definition of $\sigma_x$. For a fixed $t$ with $0 \le t \le \max(|u|, |u'|)$, we look at the prefixes $u(t)$ and $u'(t)$ of length $t$ (recall that $u(t) = u$ if $t \ge |u|$). Since $k$ is greater than the number of states of any of our automata, there are strings $w_t$ and $w_t'$ of length less than $k$ and such that $(u(t)w_t, u'(t)w_t')$ is accepted by $M_x$. We set $\tau_t = \sigma_{w_t}\sigma_x\sigma_{w_t'}^{-1}$ (Figure 5.2). Note that $\tau_t = \sigma_x$ if $t = \max(|u|, |u'|)$.

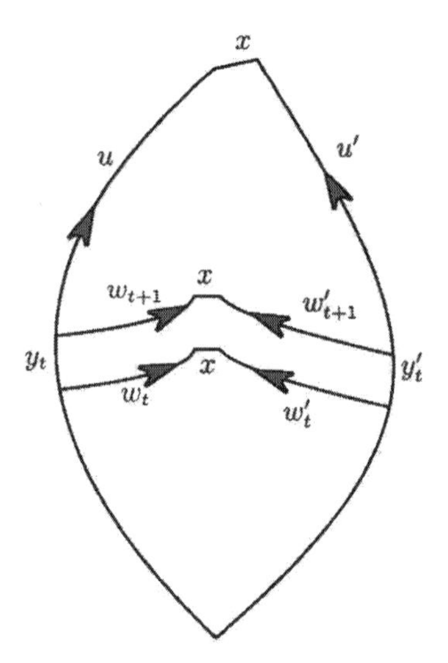


**Figure 5.2. The factorization of $v = uxu'^{-1}$.** This illustrates how the element $\sigma_v \in H$ can be factorized into elements of the form $\sigma_{y_t}\tau_{t+1}\sigma_{y'_t}^{-1}\tau_t^{-1}$ in the last part of the proof of Lemma 5.1.3 (free action).

We have

$$[u(t)]\tau_t = [u(t)w_t]\sigma_x\sigma_{w'_t}^{-1} = [u'(t)w'_t]\sigma_{w'_t}^{-1} = [u'(t)].$$

For $t = 0$ this shows that $\tau_0$ fixes $[\varepsilon]$. From Axiom 13 (trivial stabilizer), it follows that $\tau_0$ is the identity in $H$.

Let $y_t$ and $y'_t$ be the elements of $A$ at position $t+1$ in $u$ and $u'$. We have

$$\begin{aligned}[u(t)]\tau_t\sigma_{y'_t}\tau_{t+1}^{-1}\sigma_{y_t}^{-1} &= [u'(t)]\sigma_{y'_t}\tau_{t+1}^{-1}\sigma_{y_t}^{-1} = [u'(t+1)]\tau_{t+1}^{-1}\sigma_{y_t}^{-1} \\ &= [u(t+1)]\sigma_{y_t}^{-1} = [u(t)].\end{aligned}$$

It follows from Axiom 13 (trivial stabilizer) that $\tau_t\sigma_{y'_t}\tau_{t+1}^{-1}\sigma_{y_t}^{-1}$, which corresponds to the action of a word of length at most $4k$, is equal to the identity in $H$. Thus for each $t$, $\tau_t\sigma_{y'_t}\tau_{t+1}^{-1} = \sigma_{y_t}$; if we multiply these equations for $t = 1, \ldots,$ most of the $\tau$'s cancel on the left side of the equality sign, and we find that $\sigma_{u'}\tau_x^{-1} = \sigma_u$, so that $\sigma_v$ is the identity in $H$, as we needed to show. $\boxed{5.1.3}$

If all the axioms above are satisfied, we conclude that $H$ acts freely and transitively on $X$, and therefore there is a canonical identification between $H$ and $X$ that makes Diagram 5.1.1 commute. Therefore the generators and automata we started with give an automatic structure for $H$. Moreover, the words of the form $w_t x w_t'^{-1} y_t' w_{t+1}' x^{-1} w_{t+1}^{-1} y_t^{-1}$, where these symbols have the meaning assigned to them in the proof of Lemma 5.1.3, form a set of relators for $H$. In particular, $H$ has a presentation with relators of length at most $4k$, where $k$ is as in Axiom 13. (Thus we recover the result that automatic groups are finitely presented, first stated as part of Theorem 2.3.12 (quadratic isoperimetric inequality). In fact, the construction here is very similar to the proof of that theorem.)

The following theorem summarizes the results of this section:

**Theorem 5.1.5 (axioms true).** *Let $A$ be a finite alphabet. Let $W$ be a finite state automaton over $A$ and let $M_x$, for $x \in A \cup \{\varepsilon\}$, be finite state automata over $(A, A)$. A necessary and sufficient condition for these data to form an automatic structure for a group is that Axioms 1–13 above should be satisfied. Moreover, the discussion above constructs the group, in terms of a finite number of generators and relators, from the finite state automata.*

*Conversely, let $G$ be a group with an automatic structure. Then the structure's finite state automata satisfy Axioms 1–13, and the group constructed from them by the process above is isomorphic to $G$.* 5.1.5

## 5.2. A Naive Algorithm

In this section we present a procedure for constructing an automatic structure from a finite group presentation. The procedure will terminate if the group has an automatic structure; otherwise it will go on forever. There is no computable upper bound on the time it might take for an automatic structure to be found: no matter how long the procedure has been running, we will not know whether the group is automatic until the procedure terminates, if it terminates. We cannot hope for an algorithm that answers Yes or No depending on whether or not the group is automatic, because the problem of whether or not a group presentation yields the trivial group is undecidable. More precisely, suppose we had such an algorithm and applied it to a group presentation. If it gave the answer No, we would know the group is not the trivial group. If the answer were Yes, we could apply Theorem 2.5.1 to get a structure with uniqueness, and again we would know whether or not the group is trivial.

Let $G$ be a group given by a finite presentation, with $A$ the set of (group) generators. Let $A' = A \cup A^{-1}$ be the set of generators and their inverses (see Convention 2.1.2); we will look for automatic structures over $A'$. The first problem is to find the word acceptor $W$. This is hard, and an efficient algorithm needs some intelligent guesswork, based on the nature of the group.

We will instead adopt an idiot's approach, and apply the same procedure to every finite state automaton over $A'$ "simultaneously." (This is done using the old diagonal trick: given a countable list of procedures, we can carry them out "simultaneously" by doing step 1 of the procedure 1, then step 1 of procedure 2, then step 2 of procedure 1, and so on. At any moment, we will have performed the first $i$ steps of the $(n-i)$-th procedure. Any procedure that terminates if carried out separately also terminates under this scheme.) This is not practical, of course, but it does allow us to pretend that $W$ is fixed, which makes our task somewhat more feasible.

We therefore assume that we know an automaton $W$ that is part of an automatic structure for $G$. By Theorem 2.3.4, for any sufficiently large finite neighbourhood $N$ of the identity element in $G$, the standard automata with respect to $(W, N)$ give an automatic structure on $G$. Of course at the beginning we do not know how big to take $N$, or what the Cayley graph looks like near the identity: see the discussion following Definition 2.3.3 (standard automata). The best we can do is to build the Cayley graph little by little, using the relators to combine vertices and create loops, and eventually we will have a faithful picture of a sufficiently large $N$. This procedure is known as the *Todd–Coxeter enumeration process*, and there are several versions of it. The version we give here is not the most efficient, but it suits our theoretical purposes.

In the following material we will assume that all graphs have a preferred vertex, or basepoint, denoted by $*$. We also assume that the basepoint of a Cayley graph is the identity element.

**Definition 5.2.1 (partial Cayley graph).** A *partial Cayley graph* over $A'$ is a finite, connected, directed graph $T$ with edges labelled by the elements of $A'$, and such that, if there is an edge labelled $x$ from $v_1$ to $v_2$, there is also an edge labelled $x^{-1}$ from $v_2$ to $v_1$. A partial Cayley graph is *unambiguous* if no two directed edges with the same vertex as source have the same label.

Let $v$ be a vertex of an unambiguous partial Cayley graph $T$, and let $w$ be a string over $A'$. Then $w$ may possibly define a path $\alpha(w,v)$ in $T$ starting at $v$, or, at a certain point, one may arrive at a vertex of $T$ from which the next letter in $w$ is not defined. If the complete path in $T$ is defined, we set $p(w,v)$ equal to the vertex of $T$ at which the path ends; otherwise, $\alpha(w,v)$ and $p(w,v)$ are not defined. The path $\alpha(w,v)$ is a *loop* if $p(w,v)=v$, and a *simple loop* if, in addition, there are no other repeated vertices in $\alpha(w,v)$.

An unambiguous partial Cayley graph $T$ is said to be *partially homogeneous* if, for any simple loop $\alpha(w,v)$ and any vertex $v'$ in $T$, the path $\alpha(w,v')$ is either undefined or also a loop. In the second case, it follows that $\alpha(w,v')$ is a simple loop.

Any subgraph of the Cayley graph $\Gamma(G, A)$ of a group is an unambiguous, partially homogeneous partial Cayley graph. Any partial Cayley graph $T$ can be made unambiguous and partially homogeneous in a finite number of steps of two types: scissor moves and loop closings. A scissor move consists of identifying two distinct edges which have the same label and same source, and also identifying with each other their target vertices. Since the total number of vertices is finite, $T$ becomes unambiguous after enough scissor moves. Then, if there are vertices $v$ and $v'$ and a word $w$ such that $\alpha(w, v)$ is a simple loop, but $\alpha(w, v')$ is defined and is not a loop, we identify $p(w, v')$ with $v'$. This loop-closing process can make $T$ ambiguous again, so we must follow it with more scissor moves. After a finite number of loop closings, $T$ will be partially homogeneous and unambiguous.

A *homomorphism* $\phi : T' \to T''$ between labelled graphs over $A$ is a map which sends vertices to vertices, basepoints to basepoints and edges to edges, preserving labels and directions. If $G$ and $H$ are groups over the same set of generators $A$ (that is, quotients of the free group $F(A)$), group homomorphisms $G \to H$ are in one-to-one correspondence with homomorphisms of Cayley graphs $\Gamma(G, A) \to \Gamma(H, A)$.

A partial Cayley graph $T$ over $A$ has an associated finitely presented group $G(T)$, defined as follows: the set of generators is $A$, and the relators are the words formed by reading off the labels of simple loops in $T$. There is a natural homomorphism $p : T \to \Gamma(G(T), A)$, which is unique with the following universal property: If $H$ is another group over $A$, any homomorphism $T \to \Gamma(H, A)$ factors through $p$, and therefore gives rise to a group homomorphism $G(T) \to H$. In particular, a homomorphism $T \to T'$ between partial Cayley graphs gives rise to a homomorphism $G(T) \to G(T')$. If $T'$ is obtained from $T$ by a scissor move or a loop closing, the homomorphism $T \to \Gamma(G(T))$ factors through $T'$, and therefore the universal property implies that $G(T)$ and $G(T')$ can be naturally identified.

**Algorithm 5.2.2 (Todd–Coxeter).** Let $G = \langle A/R \rangle$ be a finitely presented group. We construct a sequence of unambiguous, partially homogeneous partial Cayley graphs $T_i$, for $i \geq 0$, as follows:

We start by taking a basepoint $*$ and attaching to it one loop for each relator in $R$. Each loop starts and ends at $*$ and successive edges describe the word in $R$. We make the resulting graph unambiguous and partially homogeneous, as described above, and call the result $T_0$.

Assume we have constructed $T_i$. If all arrows are defined, $G$ must be finite and $T_i$ is the complete Cayley graph of $G$. Otherwise, we define all missing arrows on vertices of $T_i$, making them point to newly created vertices. We then make the resulting graph unambiguous and partially homogeneous. We call the result $T_{i+1}$.

By definition, we have $G(T_0) = G$. The discussion above implies that the induction step does not modify $G(T_i)$, so all these partial Cayley graphs have $G$ as their associated group. In particular, there are maps $\phi_i : T_i \to \Gamma(G, A)$.

**Proposition 5.2.3 (reconstructed neighbourhood).** *Let $G = \langle A/R \rangle$ be a finitely presented group, and let $T_i$, for $i \geq 0$, be the corresponding partial Cayley graphs given by the Todd–Coxeter algorithm. If $U$ is the closed metric ball of radius $n$ around the identity in the Cayley graph $\Gamma(G, A)$, we have $U \subset T_i$ for $i$ large enough. More precisely, there is a homomorphism of graphs $f : U \to T_i$ such that $\phi_i \circ f$ is the identity (so that $f$ is an isomorphism between $U$ and $f(U)$).*

*Proof of 5.2.3:* The obvious way to try to define $f$ is this: if $g \in U$ is represented by a word $w$ of length at most $n$, we would like to set $f(g) = p(w, *)$ in $T_i$. This is defined if $i \geq n$, but different representatives $w$ may give different results. To ensure that the result does not depend on $w$, we must check that every word $v$ of length at most $2n$ and representing the identity gives rise to a closed loop $\alpha(v, *)$ in $T_i$. There are finitely many such words, each of the form $v = \prod_j w_j r_j^{\pm 1} w_j^{-1}$ for appropriate relators $r_j \in R$ and words $w_j$ (see Definition 2.1.8). Let $N$ be the maximum value of $|w_j| + |r_j|$, for all $j$ and all words $v$. Then $\alpha(r_j^{\pm 1}, p(w_j, *))$ is defined in $T_i$, for $i \geq N$, and is in fact a loop, because $T_i$ is partially homogeneous and $\alpha(r_j, *)$ is a loop (already in $T_0$). Thus $\alpha(w_j r_j^{\pm 1} w_j^{-1}, *)$ is a loop for every $j$, and $\alpha(v, *)$ is a loop for every $v$. 5.2.3

In general, we cannot predict how large $N$ has to be in terms of the presentation, because the problem of whether or not a group presentation yields the trivial group is undecidable (see the first paragraph of this section).

Given an unambiguous, partially homogeneous partial Cayley graph $T$, and a finite state automaton $W$, we define the standard automata $M_x(W, T)$ associated to $(W, T)$ by adapting the prescription given in Definition 2.3.3. The difference is that here the third component of the state set is a vertex of $T$, rather than a vertex in a finite neighbourhood of the Cayley graph of a group. The interpretation of the formula $h = y_1^{-1} g y_2$, where now $g \in T$ and $y_1, y_2 \in A'$, also requires some elaboration. Suppose that $g = p(w, *)$, where $w$ is a word over $A$; we set $h = p(y_1^{-1} w y_2, *)$ if this element is defined. Since $T$ is partially homogeneous, this definition depends only on $g$, not on the representative $w$. If there is no $w$ for which $p(y_1^{-1} w y_2, *)$ is defined, we go into the failure state.

Suppose that $T'$ and $T''$ are unambiguous, partially homogeneous partial Cayley graphs and that there is a homomorphism $\phi : T' \to T''$ between the two. If $(s_1, s_2, g)(x, y) = (s_1x, s_2y, h)$ in the standard automaton of $(W, T')$, then $(s_1, s_2, \phi(g))(x, y) = (s_1x, s_2y, \phi(h))$ in the standard automaton of $(W, T'')$. It follows that the language accepted by $M_x(W, T')$ is a subset of the language accepted by $M_x(W, T'')$. Applying this to the graphs $T_i$ and $U$ of Proposition 5.2.3, we see that

$$L(M_x(W,U)) \subset L(M_x(W,T_i)) \subset L(M_x(W,U')),$$

where $U'$ is a ball in $\Gamma(G, A)$ containing the image of $T_i$. Now if $G$ is automatic, the language $L(M_x(W, U))$ equals the language $L_x$ of a multiplier automaton for $G$ if $U$ is large enough, by Theorem 2.3.4 (standard automata theorem); in particular, $L(M_x(W, U))$ stops growing with $U$. Thus $L(M_x(W, T_i)) = L_x$ for $i$ large enough.

We have shown that, if we keep cranking the wheels of the Todd–Coxeter algorithm, we will eventually find an automatic structure for $G$. The question is, How do we know when we can stop? We must check that the axioms given in Section 5.1 (Axiom Checking) are satisfied by $W$ and the standard automata $M_x(W, T_i)$. If so, we will have a group $H$ derived from these axioms as explained in that section, and the only thing left to check is that $H$ is the same group $G$ we started from. Clearly there is a well-defined homomorphism from $H$ to $G$, because if we have two representatives $w_1$ and $w_2$ of the same element in $H$, the pair $(w_1, w_2)$ is accepted by $M_0(W, T_i)$ by definition, and therefore they represent the same element of $G$. The homomorphism is surjective because all elements of $G$ are represented by strings over $A$. To check that it is injective, we just need to test whether the relators of $G$ hold in $H$. More precisely, for each relator $w \in R$, we use the candidate multiplier automata to test whether $[u]\sigma_w = [u]$ for some fixed $u \in L$.

We summarize the results of this section:

**Theorem 5.2.4.** *Let $G = \langle A/R \rangle$ be a finitely presented group and $W$ a finite state automaton over $A$. Consider the following procedure: For each $i \geq 0$, use Todd–Coxeter (Algorithm 5.2.2) to construct the partial Cayley graph $T_i$; then form the associated standard automata $M_x(W, T_i)$, for $x \in A \cup \{\varepsilon\}$; then test whether $W$ and the $M_x(W, T_i)$ form an automatic structure for $G$, by checking Axioms 1–13 of Section 5.1 and the equality $[\varepsilon]\sigma_w = [\varepsilon]$ for each $w \in R$, stopping when the answer is positive. This procedure terminates if $G$ is an automatic group. If $G$ is not automatic, it will never terminate, and there cannot be any general way of knowing that it will not terminate.* 5.2.4

# Chapter 6

# Finding the Automatic Structure: Practical Methods

In the previous chapter we described a completely general, albeit impractical, procedure to find a group's automatic structure. In this chapter we trade generality for expediency, and limit ourselves to searching for a ShortLex-automatic (see page 56). This is a much less daunting task than to search among all automatic structures, for at least two reasons: First, ShortLex is a prefix closed language with uniqueness. This allows us to greatly simplify the axioms given in Section 5.1. We will essentially reprove the results of Section 5.1 in this context; the repetition is justified by the greater simplicity of the proof of the specialized result. A second reason is that we can use the Knuth–Bendix procedure [BL82] to find reduction rules that encode, in a much more efficient way, the same information given by the Cayley graph of the group.

Unfortunately our strategy may fail because the ShortLex language may not be regular, even if the group is automatic: Section 3.5 discusses a group that is ShortLex-automatic with respect to one set of generators, but not with respect to another. This is in contrast with the result in Theorem 2.4.1 (changing generators). Nonetheless, there are large classes of groups that are ShortLex-automatic for any set of generators. In fact, Gromov [Gro87] has indicated that in some sense most finite presentations define groups which are word hyperbolic and therefore ShortLex-automatic, by Theorem 3.4.5 (word hyperbolic implies automatic) and Corollary 2.5.2 (geodesic hierarchy).

Moreover, Holt, Epstein and Sarah Rees have developed a suite of computer programs [EHR91] that implement the algorithms described here, and these programs have successfully found automatic structures for a number of examples.

The algorithm in Section 6.2 is a minor extension by Holt and Epstein of the standard Knuth–Bendix procedure. Although the idea of using Knuth–

Bendix in this context is due to Holt, it was foreshadowed in the work of Gilman [Gil84b, Gil87], where there is also clear evidence of the notion of automatic structures on groups, before the concept was made explicit.

## 6.1. Semigroups and Specialized Axioms

In Section 5.1 we found necessary and sufficient conditions for a set of candidate automata over an alphabet to form an automatic structure of the group defined by a given presentation. Here we find similar—but much simpler—conditions for the automata to form an automatic structure *with prefix closure and uniqueness.* To state these conditions, it is convenient to introduce the notion of a semigroup presentation, and show how it relates to group presentations.

There is no semigroup analogue to the idea of a normal subgroup, but there is still a sensible way to discuss the quotient of a semigroup:

**Definition 6.1.1 (quotient).** Let $S$ be a semigroup and $R$ a symmetric relation (that is, a set of unordered pairs) on $S$. The *quotient* of $S$ by $R$ is the set of equivalence classes of $S$ under the equivalence relation $\sim$ generated by the equivalences $w_1uw_2 \sim w_1vw_2$ for all $u, v, w_1, w_2 \in S$ with $(u, v) \in R$. The quotient $Q$ has an obvious semigroup structure inherited from $S$, and the quotient map $\pi : S \to Q$ is a homomorphism with the property that $\pi(u) = \pi(v)$ whenever $(u, v) \in R$. In addition, any semigroup homomorphism $\psi : S \to S'$ such that $\psi(u) = \psi(v)$ for all $(u, v) \in R$ must factor through $\pi$, that is, there exists a semigroup homomorphism $\psi' : Q \to S'$ such that $\psi = \psi' \circ \pi$. This factorization is clearly unique.

If $A$ is an alphabet set, the *free semigroup* on $A$ is simply the set $A^*$ of strings over $A$, with multiplication by concatenation, and the nullstring for identity. If $R$ is any set of pairs of strings over $A$ and $Q$ is the quotient of $A^*$ by $R$, we say that $A$ and $R$ together form a *semigroup presentation* for $Q$, and we write $Q = \langle A/R \rangle$. The pairs $(u, v) \in R$ are the *relations* for the presentation. The image $\pi(u)$ of a string $u \in A^*$ in $\langle A/R \rangle$ is generally denoted by $\overline{u}$.

A group presentation $\langle A/R \rangle$ (see Definition 2.1.8) can be interpreted as a semigroup presentation $\langle A'/R' \rangle$, where $A'$ is obtained from $A$ by adding formal inverses, and each relation in $R'$ is a pair $(\varepsilon, w)$, where $w \in R$ is a word over $A$ (and therefore a string over $A'$). We must also adjoin to $R'$ relations of the form $(\varepsilon, xx^{-1})$ and $(\varepsilon, x^{-1}x)$, for all $x \in A$.

Conversely, the semigroup defined by the presentation $\langle A/R \rangle$, where $A$ is closed under inversion, is a group if $R$ contains all the pairs $(\varepsilon, xx^{-1})$, for $x \in A$. We get a group presentation for this group by taking as relators all words of the form $uv^{-1}$, for all pairs $(u, v) \in R$.

**Theorem 6.1.2 (specialized axioms).** *Let $G$ be a finitely presented group with presentation $\langle A/R \rangle$, and let $\langle A'/R' \rangle$ be the associated semigroup presentation for $G$. Suppose we are given a finite state automaton $W$ over $A'$ and, for each $x \in A'$, a finite state automaton $M_x$ over $(A', A')$. These data form an automatic structure for $G$, with $L(W)$ prefix closed and consisting of unique representatives for the elements of $G$, if and only if the following conditions hold:*

(1) *If $(v, w) \in L(M_x)$ for some $x \in A'$, then $v, w \in L(W)$.*

(2) *If $(v, w) \in L(M_x)$, then $\overline{vx} = \overline{w} \in G$.*

(3) *$L(W)$ is non-empty.*

(4) *Let $v \in A^*$ and $x \in A$. If $vx \in L(W)$, then $(v, vx) \in L(M_x)$.*

(5) *Let $(\varepsilon, u) \in R'$ be a relation, where $u = x_1 \ldots x_n$. Then two accepted strings $w_0, w_n \in L(W)$ are equal if and only if there exist accepted strings $w_1, \ldots, w_{n-1} \in L(W)$ with $(w_{i-1}, w_i) \in L(M_{x_i})$ for $1 \le i \le n$.*

*Proof of 6.1.2:* If the given automata form an automatic structure for the group, (1)–(3) follow immediately from Definition 2.3.1 (automatic group). If, in addition, $L(W)$ is assumed to be prefix closed, $v$ is accepted by $W$ if $vx$ is. The pair $(v, vx)$ is then accepted by $M_x$, by Definition 2.3.1. This proves (4). To prove (5), assume, in addition, that $G$ has unique representatives in $L(W)$. Let $v_i$ be the unique representative in $L(W)$ of $\overline{w_0 x_1 \ldots x_i} \in G$ for $1 \le i \le n$. Then $(v_{i-1}, v_i) \in L(M_{x_i})$ for $1 \le i \le n$. Since $u \in R'$, $w_0$ and $v_n$ represent the same element of $G$, and they are therefore equal. If $w_0 = w_n$, we get the "only if" part by defining $w_i = v_i$ for $1 \le i < n$. If, conversely, the $w_i$ are given, we have $w_i = v_i$ by induction on $i$. In particular, $w_n = v_n = w_0$, and the "if" part follows.

Now suppose, conversely, that conditions (1)–(5) are satisfied. If $vx \in L(W)$, the pair $(v, vx)$ is accepted by $M_x$ according to (4), so $v \in L(W)$ by (1). Thus $L(W)$ is prefix closed. In particular, $\varepsilon \in L(W)$ because of (3).

We want to define an action of $G$ on $L(W)$ as we did in Section 5.1. Let $x \in A$ and $w \in L(W)$. By assumption, the set $R'$ contains the pairs $(\varepsilon, xx^{-1})$ and $(\varepsilon, x^{-1}x)$, so applying (5) with $w_0 = w_n = w$ and $u = xx^{-1}$ gives a string $v \in L(W)$ such that $(w, v)$ is accepted by $M_x$ and $(v, w)$ is accepted by $M_{x^{-1}}$. Applying (5) again to $v$ and $x^{-1}x$, we see that $v$ is uniquely determined. Thus we obtain a map $L(W) \to L(W)$, which we denote by $\sigma_x$ and write to the right of its operand: $w\sigma_x = v$. Clearly, $\sigma_x$ is a bijection, with inverse $\sigma_{x^{-1}}$. We extend the map $x \mapsto \sigma_x$ to an action of $F(A)$ on $L(W)$ (that is, a homomorphism from $F(A)$ into the permutation group of $L(W)$), but setting $\sigma_{x_1 \ldots x_n} = \sigma_{x_1} \ldots \sigma_{x_n}$.

By condition (2) we have $\overline{w\sigma_x} = \overline{wx}$ for $w \in L(W)$. Condition (5) says that $\sigma_u$ is the identity if $u \in R'$. It follows that $\sigma_u = \sigma_{u'}$ if $\overline{u} = \overline{u'}$, so we get an action $g \mapsto \sigma_g$ of $G$ on $L(W)$, satisfying $\overline{w\sigma_g} = \overline{w}g$ for $w \in L(W)$ and $g \in G$. In particular, $\overline{\varepsilon\sigma_g} = g$, so the map $g \mapsto \varepsilon\sigma_g$ is a right inverse for the map $\pi : L(W) \to G$. It is also a left inverse, that is, $\varepsilon\sigma_{\overline{w}} = w$, by condition (4). Thus $\pi : L(W) \to G$ is a bijection, and $G$ has unique representatives in $L(W)$.

To complete the proof, we must verify the conditions of Definition 2.3.1. Suppose that $w$ and $v$ are elements of $L(W)$ and that $\overline{wx} = \overline{v}$. Then $w\sigma_x$ and $v$ are two elements of $L(W)$ representing the same element of $G$. Therefore they are equal, and, by the definition of $\sigma_x$, the pair $(u, v)$ is accepted by $M_x$. 6.1.2

Notice that conditions (1)–(5) in the statement of Theorem 6.1.2, like the axioms in Section 5.1, are fully quantified regular predicates. Formally, condition (5) should be considered as the conjunction of several predicates, one for each element of $R'$. We remark that, in our programs, the method by which the automata are constructed ensures that all the conditions except possibly the last are always satisfied. So if we can find some way of guessing the automata, we only have one condition to check, a situation much more manageable that the one in Section 5.2 (A Naive Algorithm).

## 6.2. The Knuth–Bendix Procedure

We are looking for a procedure to produce a word acceptor automaton for the ShortLex language of a group. It is not immediate to think of a way of doing this effectively. Methods based on the cone type or the $n$-level of an element (see pages 66 and 70) are not easy to work with, because the potential number of states is huge. In this section and the next we describe an adaptation of a powerful general procedure, due to D. E. Knuth and P. B. Bendix [KB70], which has proved very fruitful in tackling this problem. Comments by Uri Zwick have greatly improved this section.

The Knuth–Bendix procedure tries to manufacture a set of rules to reduce algebraic expressions to canonical form—in our case, to reduce a string over a set of generators to the unique ShortLex string that represents the same group element. The general procedure is very powerful, and is applicable to a large variety of situations. By restricting its generality, we can simplify it considerably. We can also adapt it to make it especially suitable for finding the automatic structure.

Rather than presenting the general Knuth–Bendix procedure, we will concentrate on the version most suitable to our situation. We make one concession to generality: all the results in this section are valid for semigroups as well as groups, so we will not assume in this section that we are dealing with groups, unless this is stated explicitly.

Given an ordered alphabet $A$, a *reduction rule* over $A$, or simply *rule*, is a formula of the form $\lambda \Rightarrow \rho$, where $\lambda$ and $\rho$ are strings over $A$ and $\lambda > \rho$ in the ShortLex order. We call $\lambda$ and $\rho$ the *left-* and *right-hand sides* of the rule.

Given a set $E$ of rules over $A$, the idea is that we can repeatedly "simplify" a string by replacing a portion of it that matches the left-hand side of a rule by the corresponding right-hand side.

More precisely, let $E$ be a set of rules over $A$. Let $u$ and $v$ be strings over $A$. We write $u \to v$, or $u \to_E v$ if we want to be explicit, if there exist strings $w_1$ and $w_2$ over $A$ and a rule $\lambda \Rightarrow \rho$ in $E$ such that

$$u = w_1 \lambda w_2 \qquad \text{and} \qquad v = w_1 \rho w_2.$$

We say that $v$ is obtained from $u$ by applying the rule $\lambda \Rightarrow \rho$. Notice that this implies $u > v$ in the ShortLex order.

Let $\to^+$, or $\to_E^+$, denote the transitive closure of $\to$; thus $u \to^+ v$ if and only if $v$ can be obtained from $u$ by repeated application of rules in $E$. In this case we say that $u$ *reduces to* $v$ and that $v$ is a *reduction* of $u$. Let $\to^*$ denote the transitive, reflexive closure of $\to$, and let $\leftrightarrow^*$ denote the symmetric, reflexive, transitive closure of $\to$.

Notice that $\leftrightarrow^*$ is exactly the equivalence relation from the definition of a quotient semigroup (Definition 6.1.1), so the set of $\leftrightarrow^*$ equivalence classes can be identified with the semigroup $\langle A/E \rangle$, where $E$ is interpreted as a set of unordered pairs $\{\lambda, \rho\}$.

Since the ShortLex order is a well-ordering, all chains of reduction eventually terminate, that is, any string $u$ can be reduced to some *irreducible* string $v$, which cannot be further reduced. In this situation we say that $v$ is a *residue*, or $E$-residue, of $u$. The set of irreducible strings under $E$ is denoted by $\mathrm{Irr}_E \subset A^*$. The map $\pi : A^* \to \langle A/E \rangle$ restricts to a map $\pi : \mathrm{Irr}_E \to \langle A/E \rangle$, which is surjective.

In general, residues are not unique, that is, different chains of reduction can lead to different irreducible strings. A set of rules $E$ is *complete* if all strings have unique $E$-residues, and $k$*-complete* if all strings of length at most $k$ have unique $E$-residues. We allow the case $k = \infty$, so $\infty$-completeness is the same as completeness. A necessary and sufficient condition for completeness is that if $v \to v_1$ and $v \to v_2$, then there is a $u$ such that $v_1 \to^* u$ and $v_2 \to^* u$. This is called *confluence* of the rules. For necessity we can use for $u$ the unique residue of $v$. For sufficiency we prove the existence of unique residues by induction on the ordering.

**Lemma 6.2.1 (completeness and irreducibles).** *If $E$ is a complete set of rules, the set $\mathrm{Irr}_E$ is a semigroup, with multiplication defined by concatenation followed by taking residues. Furthermore, the map $\mathrm{Irr}_E \to \langle A/E \rangle$ is a semigroup isomorphism.*

*Proof of 6.2.1:* Completeness implies that multiplication in $\mathrm{Irr}_E$ is well-defined and associative. The map $A^* \to \mathrm{Irr}_E$ that takes each string to its residue is clearly a semigroup homomorphism, and therefore it factors through $\langle A/E \rangle$ (see Definition 6.1.1). The resulting homomorphism $\langle A/E \rangle \to \mathrm{Irr}_E$ is obviously inverse to the map $\mathrm{Irr}_E \to \langle A/E \rangle$ defined above. 6.2.1

This explains the importance of completeness: it allows one to find a unique "normal form" for elements of $\langle A/E \rangle$, starting from any string that represents the element. This assumes that $E$ is finite, or at least *recursive* (see the remark following the statement of Theorem 2.1.9). For the sequel, we remark that $E$ is recursive if it can listed in such a way that all rules with left-hand side of length at most $k$ appear among the first $n(k)$ rules, where $n(k)$ is a computable function of $k$.

In our context, we generally start from a semigroup $G$ with a finite set of generators $A$ and a set $E$ of relations, expressed as reduction rules. (Recall that a semigroup relation is a pair of words to be identified; to get the rule, we use the ShortLex-larger word as the left-hand side, and the smaller one as the right-hand side.) One generally assumes that $E$ is finite, but we will also find it convenient to be able to talk of infinite sets of rules. By construction, $G$ is identified with the quotient semigroup $\langle A/E \rangle$. In this situation, we call $E$ a set of *Knuth–Bendix rules* for $G$. By Lemma 6.2.1, if $E$ is a complete and finite (or complete and recursive) set of Knuth–Bendix rules for $G$, we can find ShortLex-minimal representatives for the elements of $G$, and, in particular, we can solve the word problem in $G$.

It should be pointed out that achieving this, although quite pleasant, is not equivalent to finding an automatic structure for the group $G$ (assuming we are dealing with a group). For although $\mathrm{ShortLex}(G, A)$ is a regular language if $E$ is finite, the group is not necessarily automatic with $\mathrm{ShortLex}(G, A)$ as the language of accepted words, as the next example shows. Nonetheless, a confluent set of rules gives us a great deal of information about a group, and, for many purposes, the information is easier to handle than an automatic structure, so we are normally content with this.

**Example 6.2.2 (confluent but not ShortLex-automatic).** We consider again the wreath product $G$ of $\mathbf{Z}$ with $\mathbf{Z}_2$, first introduced in Example 4.4.1 (order matters). As we saw in that example, $G$ is not ShortLex-automatic with respect to the set of generators $A = (x, y, Y, z, Z)$, in that order. However, the following set of Knuth–Bendix rules is complete, and reduces any word to a word in ShortLex$(G, A)$:

$$E = \{xx \Rightarrow \varepsilon,\ yY \Rightarrow \varepsilon,\ Yy \Rightarrow \varepsilon,\ zZ \Rightarrow \varepsilon,\ Zz \Rightarrow \varepsilon,\ zy \Rightarrow yz,\ Zy \Rightarrow yZ,$$
$$zY \Rightarrow Yz,\ ZY \Rightarrow YZ,\ yx \Rightarrow xz,\ Yx \Rightarrow xZ,\ zx \Rightarrow xy,\ Zx \Rightarrow xY\}.$$

**Open Question 6.2.3.** The interplay between confluence and the existence of an automatic structure needs further investigation. For example, is there a group with a finite confluent set of rules for the ShortLex order, which is not automatic (or ShortLex-automatic) for any choice of generators? It should not be hard to say what the situation is if we fix the set of generators, but allow changes in the order.

What can we do if we don't have a complete set of rules? The idea is to try to make the set complete in some algorithmic way. We first analyze how completeness can fail, and find that there are two basic reasons:

**Lemma 6.2.4 (completeness and rules).** *Let $k$ be a positive integer or infinity, and let $E$ be a set of rules over $A$. Then $E$ is $k$-complete if and only if, for all $u, y, v \in A^*$ with $|uyv| \leq k$, the following two conditions are satisfied:*

(1) *Suppose that $y \neq \varepsilon$ and that $uy \Rightarrow u'$ and $yv \Rightarrow v'$ are rules of $E$. Then there exists a string $t$ and reductions*

$$uyv \to u'v \to^* t \qquad \text{and} \qquad uyv \to uv' \to^* t.$$

(2) *Let $y \Rightarrow y'$ and $uyv \Rightarrow y''$ be rules of $E$. Then there exists a string $t$ and reductions*

$$uyv \to uy'v \to^* t \qquad \text{and} \qquad uyv \to y'' \to^* t.$$

In the first condition it is possible for $u$ or $v$ or both to be the nullstring. In the second condition, it is a consequence of the hypothesis that $y$ is not the nullstring, and we may assume that neither $u$ nor $v$ is the nullstring (otherwise we get a special case of the first condition).

*Proof of 6.2.4:* The two conditions are clearly implied by $k$-completeness. Conversely, if the two conditions are satisfied, we prove by induction on the ordering that each string has a unique residue.

We look at two different chains of reduction for a string of length at most $k$. We may assume that the first steps in the two chains are different. If the two left-hand sides of the rules used in the first steps are disjoint substrings,

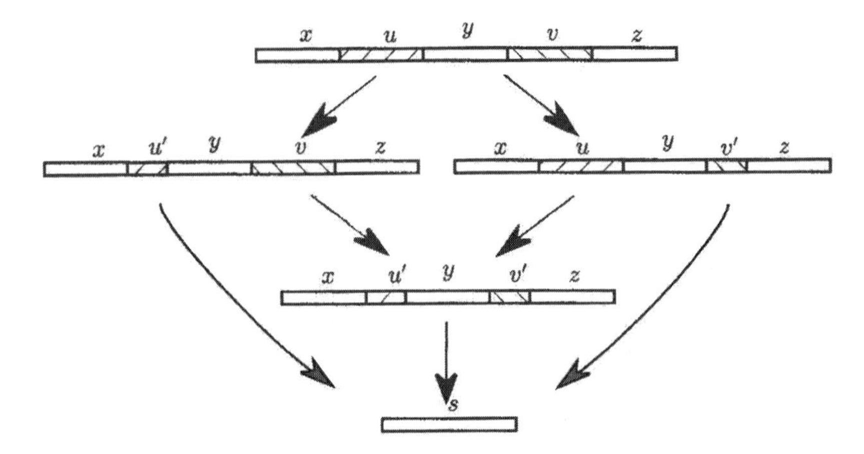


**Figure 6.1. Non-overlapping left-hand sides.** This illustrates one case in the proof of the "if" part of Lemma 6.2.4. Consider two reductions of the string $xuyvz$, starting with $u \Rightarrow u'$ and $v \Rightarrow v'$, respectively. The results $xu'yvz$ and $xuyv'z$ can both be reduced to $xu'yv'z$, and from there to some irreducible $s$. By induction, any chain of reductions of $xu'yvz$ to an irreducible must give $s$, and similarly for $xuyv'z$.

the results obviously have a common reduction, and therefore, by induction, lead to the same irreducible. This is illustrated in Figure 6.1.

If the two left-hand sides are not disjoint, they must overlap as in one of the two situations in the statement of the lemma, and once again we can apply induction. One of the two possible types of overlap is shown in Figure 6.2. 6.2.4

It follows from this lemma that a finite set of rules can be algorithmically checked for completeness and that a recursive set of rules can be algorithmically checked for $k$-completeness. All we have to do is consider in turn each pair of rules in the set $E$, examine in what ways the two left-hand sides can overlap (including overlaps of a left-hand side with itself), and then apply some sequence of reductions to the two relevant strings until we obtain irreducible strings, say $t_1$ and $t_2$. If $t_1$ and $t_2$ coincide for each possible overlap and for some sequence of reductions, $E$ is complete; otherwise it is not.

The lemma also tells us what to do when $E$ is not complete: If we find $t_1 \neq t_2$ at the end of one of the computations just described, we enlarge $E$ by adding to it one of the rules $t_1 \Rightarrow t_2$ or $t_2 \Rightarrow t_1$, depending on which word comes first in ShortLex order. We keep doing this as long as there are non-confluent rules. This is the essence of the Knuth–Bendix procedure.

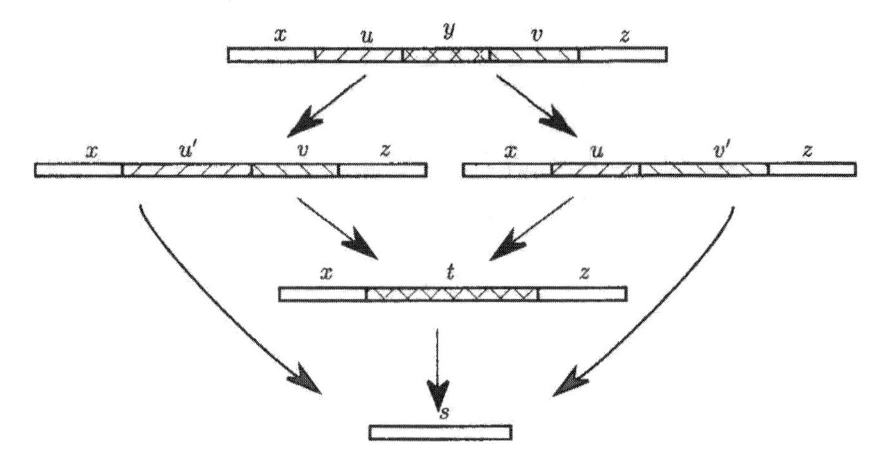


**Figure 6.2. Overlapping left-hand sides.** This illustrates another case in the proof of the "if" part of Lemma 6.2.4, where the initial rules in the two reductions being compared are $uy \Rightarrow u'$ and $yv \Rightarrow v'$, respectively. The results, $xu'vz$ and $xuv'z$, can be reduced to the same word $xtz$ by Condition (a) of the lemma. The rest of the argument is as in the caption of Figure 6.1. The argument for the case when the initial rules are of the form $y \Rightarrow y'$ and $uyv \Rightarrow y''$ (that is, one left-hand side is contained in another) is similar, and uses Condition (b) of the Lemma.

To flesh out this outline, we must eliminate an ambiguity in the procedure: how do we choose which rules to apply in obtaining $t_1$ and $t_2$? To this end, we start by defining the *canonical reduction* of a string $v$ with respect to $E$, as follows: Among all rules of $E$ that match some substring of $v$, we consider those where the first character of the match is as far to the left of $v$ as possible. Amongst these we select the rules having the shortest left-hand sides, and amongst these rules (which all have the same left-hand side) we select the one having the least right-hand side in ShortLex order. Then we apply this rule to the matching substring of $v$, and repeat the process until reaching an irreducible string. We call this irreducible string the *canonical residue* of $v$, and denote it by $\rho_E(v)$. Thus we get a map $\rho_E : A^* \to \mathrm{Irr}_E$.

**Algorithm 6.2.5 (Knuth–Bendix 1).** Let $G$ be a semigroup and $A$ a finite ordered set of generators. Given a finite set $E_0$ of Knuth–Bendix rules for $G$, we construct a sequence of finite sets of rules $E_i$, for each $i \geq 0$, by induction. To get $E_{i+1}$ from $E_i$, we add rules to make up for failures of the two confluence conditions of Lemma 6.2.4 (completeness and rules). More precisely, for each pair of rules in $E_i$, we check for the following situations:

(1) The rules are of the form $uv \Rightarrow x$ and $vw \Rightarrow y$, with $v \neq \varepsilon$. (There may be several ways to factor the left-hand sides, and they all have to be considered.) We reduce $uvw$ in two different ways, obtaining $t_1 = \rho_{E_i}(xw)$ and $t_2 = \rho_{E_i}(uy)$ in $\mathrm{Irr}_{E_i}$. If $t_1 \neq t_2$, we add either $t_1 \Rightarrow t_2$ or $t_2 \Rightarrow t_1$, depending on which string is ShortLex-smaller.

(2) The rules are of the form $v \Rightarrow x$ and $uvw \Rightarrow z$ (where again several overlaps may have to be considered). We reduce $uvw$ in two different ways, obtaining $t_1 = \rho_{E_i}(z)$ and $t_2 = \rho_{E_i}(uxw)$. Once again, if $t_1 \neq t_2$, we we add either $t_1 \Rightarrow t_2$ or $t_2 \Rightarrow t_1$.

In rare cases, the procedure stops with a finite complete set of rules. Most likely, it will continue forever—even if we are applying it to a group that is ShortLex automatic with respect to the chosen generators. (See Example 6.3.1.)

Clearly, $\langle A/E_i \rangle = \langle A/E_0 \rangle$ for all $i$, because we are only adding rules whose sides are already related. We also have $\mathrm{Irr}_{E_{i+1}} \subset \mathrm{Irr}_{E_i}$, since $E_i \subset E_{i+1}$.

**Lemma 6.2.6 ($k$-completeness).** *For each $k > 0$, there is a number $n(k)$ such that, for all $i \geq n(k)$, the set $E_i$ is $k$-complete, and canonical $E_i$-residues coincide with canonical $E_{n(k)}$-residues for strings of length at most $k$.*

*Proof of 6.2.6:* Let $\mathrm{Irr}_i(k)$ be the set of strings of length at most $k$ which are irreducible with respect to $E_i$. By Lemma 6.2.4 (completeness and rules), $E_i$ is complete if and only if no rules with left-hand side of length at most $k$ are added in going from $E_i$ to $E_{i+1}$, that is, if and only if $\mathrm{Irr}_i(k) = \mathrm{Irr}_{i+1}(k)$. Since the $\mathrm{Irr}_i(k)$ form a nested, non-increasing sequence of finite sets, there is a smallest positive number $n(k)$ such that $\mathrm{Irr}_i(k) = \mathrm{Irr}_{n(k)}(k)$ for $i \geq n(k)$. By Lemma 6.2.4 (completeness and rules) and the construction of $E_i$, $E_i$ is $k$-complete for $i \geq n(k)$. The following simple observation, which we state as a lemma for future reference, concludes the proof:

**Lemma 6.2.7 ($k$-residues coincide).** *If $E \subset E'$ are $k$-complete sets of rules and $\mathrm{Irr}_E(k) = \mathrm{Irr}_{E'}(k)$, canonical $E$- and $E'$-residues coincide for words of length at most $k$.* 6.2.7

6.2.6

It follows that the union $\bigcup_{i=0}^{\infty} E_i$ is a complete set of Knuth–Bendix rules for $G$. For each $w \in A^*$, the residue of $w$ with respect to this union equals the residue with respect to $E_i$, where $i \geq n(|w|)$. Note that $n(k)$ is generally not a computable function of $k$. If it is, the word problem is solvable in $G = \langle A/E_0 \rangle$.

Algorithm 6.2.5 (Knuth–Bendix 1) can only increase the number of rules and decrease the set of irreducibles. For the sake of efficiency, and for other theoretical reasons described later (see the remark following the proof of Theorem 6.2.12), we make two modifications to the algorithm's induction step: elimination of redundant rules, and introduction of shortcuts.

**Definition 6.2.8 (redundant rule).** A rule $u \Rightarrow v$ of $E$ is said to be *redundant* (in $E$) if $\rho_{E'}(u) = \rho_{E'}(v)$, where $E'$ is the set of rules obtained by omitting $u \Rightarrow v$ from $E$.

When $u \Rightarrow v$ is redundant, $u$ is necessarily reducible in $E'$, since $\rho_{E'}(u) = \rho_{E'}(v) \leq v < u$.

**Lemma 6.2.9 (omit redundant rule).** *Let $E$ and $E'$ be sets of rules, $E'$ being obtained from $E$ by omitting one redundant rule. Then $\langle A/E \rangle = \langle A/E' \rangle$ and $\mathrm{Irr}_E = \mathrm{Irr}_{E'}$. For any $k > 0$, $E$ is $k$-complete if and only if $E'$ is $k$-complete; in this case, $\rho_{E'}(w) = \rho_E(w)$ for any string $w$ of length at most $k$.*

*Proof of 6.2.9:* It is clear that $\langle A/E \rangle = \langle A/E' \rangle$, since the two sides of the removed rule are equivalent under $\leftrightarrow^*_{E'}$.

Let $u \Rightarrow v$ be the rule in $E$ that is missing from $E'$. Since it is redundant, $u$ must be $E'$-reducible. Therefore, if $x$ is $E'$-irreducible, it can not contain $u$ as a substring. But this means that $x$ is $E$-irreducible. This shows that $\mathrm{Irr}_{E'} \subset \mathrm{Irr}_E$; the opposite inclusion is trivial.

For any reduction $w \rightarrow^*_{E'} w'$ with $w' \in \mathrm{Irr}_{E'}$, we also have $w \rightarrow^*_E w'$ with $w' \in \mathrm{Irr}_E$, so the $k$-completeness of $E$ implies the $k$-completeness of $E'$.

Conversely, let $E'$ be $k$-complete; to prove the $k$-completeness of $E$, it is enough to show that, for any $E$-reduction $w \rightarrow^*_E w'$ with $|w| \leq k$ and $w' \in \mathrm{Irr}_E$, there is a $E'$-reduction $w \rightarrow^*_{E'} w'$. We can assume by induction that all strings $v < w$ in the ShortLex order have unique $E$-residues, equal to their $E'$-residues.

If the reduction $w \rightarrow^*_E w'$ is not an $E'$-reduction, we can assume that it starts with the rule that is missing from $E'$, say $u \Rightarrow v$. Thus $w$ is of the form $xuy$. Now $xvy < xuy = w$ in the ShortLex order, so, by the induction assumption, there is a chain of $E'$-reductions starting with $xvy \rightarrow^*_{E'} x\rho_{E'}(v)y$. But $\rho_{E'}(v) = \rho_{E'}(u)$ because $u \Rightarrow v$ is redundant; thus there is a chain of $E'$-reductions starting with $w = xuy \rightarrow^*_{E'} x\rho_{E'}(v)y$ and leading to $w'$. 6.2.9

**Lemma 6.2.10 (shortcut).** *Let $E'$ be obtained from $E$ by replacing a rule $u \Rightarrow v$, with $v \neq \rho_E(v)$, by $u \Rightarrow \rho_E(v)$. Then the conclusions of Lemma 6.2.9 hold.*

*Proof of 6.2.10:* Both the new and the old rule are "essentially redundant" in $E'' = E \cup E'$, in the sense that, in the complement of the rule, some residue of the rule's left-hand side equals some residue of its right-hand side. It is easy to see that the arguments in the proof of Lemma 6.2.9 still hold if the rule removed is almost redundant. 6.2.10

We now describe our adapted Knuth–Bendix procedure. We regard rules that have larger left-hand side in the ShortLex order as more significant than those with smaller left-hand sides. Among rules with the same left-hand side, we regard as more significant those that have smaller right-hand side. The idea is that more significant rules cause the greatest amount of reduction when applied.

**Algorithm 6.2.11 (Knuth–Bendix 2).** Let $G$ be a semigroup and $A$ an finite ordered set of generators. Given a finite set $E_0$ of Knuth–Bendix rules for $G$, we construct a sequence of finite set of rules $E_i$, for each $i \geq 0$, by induction, as follows:

(1) Given $E_i$, we replace each rule $u \Rightarrow v$ whose right-hand side is reducible (in the possibly already modified $E_i$) by the rule $u \Rightarrow v'$, where $v'$ is the canonical reduction of $v$. We call the set thus obtained $E_i'$.
(2) We remove redundant rules from $E_i'$, one at a time, always choosing the least significant, until there are no redundant rules left. We call the set thus obtained $E_i''$.
(3) We apply the induction step of Algorithm 6.2.5 (Knuth–Bendix 1) to $E_i''$, to obtain $E_{i+1}$.

We now come to the main result of this section:

**Theorem 6.2.12 (complete set of rules).** *Let $E_0$ be a set of Knuth–Bendix rules for $G$ over $A$, and let $E_i$ be constructed according to Algorithm 6.2.11. Then the set*

$$E = \bigcup_{r=0}^{\infty} \bigcap_{s=r}^{\infty} E_s$$

*of all rules which eventually stay in the $E_i$ is a complete set of Knuth–Bendix rules for $G$.*

*If $u$ and $v$ are strings, the rule $u \Rightarrow v$ is in $E$ if and only if $u$ and $v$ represent the same element of $G$, and $v$ and every proper substring of $u$ are elements of* ShortLex$(G, A)$. *In particular, there is at most one rule in $E$ with a given left-hand side $u$.*

*For each $k > 0$, there is a number $n(k)$ (not necessarily computable from $k$) such that $E_i$-residues and $E$-residues coincide for words of length at most $k$, for all $i \geq n(k)$.*

*Proof of 6.2.12:* By Lemmas 6.2.9 and 6.2.10, we have

$$\mathrm{Irr}_{E_i} = \mathrm{Irr}_{E'_i} = \mathrm{Irr}_{E''_i} \supset \mathrm{Irr}_{E_{i+1}}$$

for all $i$, so the proof and conclusions of Lemma 6.2.6 are valid for the modified algorithm as well. We define $n(k)$ as in that lemma. If $w \in A^*$ with length $k$, we have

$$\rho_{E_{n(k)}}(w) = \rho_{E'_{n(k)}}(w) = \rho_{E''_{n(k)}}(w) = \rho_{E_{n(k)+1}}(w) = \cdots,$$

by Lemmas 6.2.9 (omit redundant rule), 6.2.10 (shortcut) and 6.2.6 ($k$-completeness). We set $\rho(w)$ to this common residue. By construction, $\rho(w)$ is $E_i$-irreducible for all $i$.

Let $E_i(k)$ be the set of rules in $E_i$ with left-hand side having length at most $k$. We claim that, for $i > n(k)$, $E_i(k)$ is independent of $i$ and contains only rules of the form $u \Rightarrow \rho(u)$, with every proper substring of $u$ irreducible in $E$.

Let $u \Rightarrow v$ be a rule in $E_{n(k)}(k)$. By Lemma 6.2.6 ($k$-completeness), $\rho(u) = \rho(v)$. If $v \neq \rho(v)$, the rule $u \Rightarrow v$ gets replaced by $u \Rightarrow \rho(u)$ when we go from $E_{n(k)}(k)$ to $E'_{n(k)}(k)$. Therefore every rule of $E'_{n(k)}(k)$ has the form $u \Rightarrow \rho(u)$, and each left-hand side appears only once. Next, if there is a proper substring $x$ of $u$ such that $\rho(x) \neq x$, the rule $u \Rightarrow \rho(u)$ is redundant and gets dropped when we pass to $E''_{n(k)}(k)$. Therefore each rule of $E''_{n(k)}(k)$ has the form $u \Rightarrow \rho(u)$, where each proper substring $x$ of $u$ satisfies $\rho(x) = x$. Finally, $E_{n(k)+1}(k) = E''_{n(k)}(k)$. From there on it is now easy to see inductively that $E_i(k)$ is constant.

It follows that the set of rules in $E$ with left-hand sides of length at most $k$ equals $E_i(k)$ for $i > n(k)$. It also follows that $\rho = \rho_E$, so every rule of $E$ has the form $u \Rightarrow \rho_E(u)$, with all proper substrings of $u$ irreducible in $E$. In particular, there is at most one rule in $E$ with a given left-hand side $u$. By Lemma 6.2.6 ($k$-completeness), $E$ is complete.

We know that $\langle A/E_i \rangle = G$ for all $i$. To show that $\langle A/E \rangle = G$, notice that any rule in $E$ is a rule in some $E_i$, so its two sides represent the same element of $G$. Conversely, if two strings of length at most $k$ represent the same element of $G$, there are in $E_{n(k)}$ enough rules to relate them, and these rules remain in $E$.

Finally, we must show that if $u$ and $v$ represent the same element of $G$, and $v$ and every proper substring of $u$ are elements of $\mathsf{ShortLex}(G, A)$, the rule $u \Rightarrow v$ is in $E$. We have $\rho_E(u) = \rho_E(v)$ because $u$ and $v$ represent the same element, and $\rho_E(v) = v$ because $v$ is $E$-irreducible. Thus $u \to_E^* v$, and the first rule of this reduction is of the form $u \Rightarrow w$, since every proper substring of $u$ is $E$-irreducible. But then $w = \rho(u) = v$. $\boxed{6.2.12}$

We observe that if a complete set of Knuth–Bendix rules has the property that right-hand sides are irreducible and proper substrings of left-hand sides are irreducible, it must coincide with the set $E$ of Theorem 6.2.12. In addition, if a group is automatic, the set $E$, interpreted as a set of pairs of strings, constitutes a regular language. (Both of these statements can be easily checked.) This is another reason—in addition to efficiency—for preferring Algorithm 6.2.11 over a version of Knuth–Bendix that does not insist on eliminating redundant rules and taking shortcuts.

We have the following additional property if $G$ is a group and $A$ is a set of generators closed under inversion.

**Lemma 6.2.13 (final rules for groups).** *Let $G$ be a group, $A$ a finite ordered set of generators closed under inversion, and let $E$ be defined as in Theorem 6.2.12. Then $|v| \leq |u| \leq |v| + 2$ for every rule $u \Rightarrow v$ of $E$.*

*Proof of 6.2.13:* Suppose we have a rule in $E$ of the form $u \Rightarrow v$, where $|u| \geq |v| + 3$. Let $u = wx$ where $x$ is a generator and $|w| = |u| - 1 > |v| + 1$. Then $\overline{w} = \overline{vx^{-1}}$. It follows that $w \notin \mathsf{ShortLex}(G, A)$. But this contradicts Theorem 6.2.12 (complete set of rules), which says that each proper substring of $u$ is $E$-irreducible. $\boxed{6.2.13}$

## 6.3. Knuth–Bendix and Word Differences

Let $G$ be a group and let $A$ be an ordered set of generators closed under inversion. We suppose that $(G, A)$ is ShortLex automatic. The language of accepted strings is $\mathsf{ShortLex}(G, A)$, the set of ShortLex geodesics in the Cayley graph of $G$. Let $E$ be the (usually infinite) complete set of Knuth–Bendix rules as found in the preceding section.

Theorem 6.2.12 (complete set of rules) says that if $v \Rightarrow w$ is in $E$, then $w$ and all proper substrings of $v$ are ShortLex geodesics. Let $v = ux$, where $x \in A$ and $u$ is a ShortLex geodesic. It follows from the definition of an automatic group (Definition 2.3.1) that $(u, w) \in L(M_x)$, where $M_x$ is the $x$-multiplier. The difference in length between $u$ and $w$ is at most one. Lemma 2.3.2 shows that the *word difference* $\overline{u(t)}^{-1}\overline{w(t)}$ varies over a finite subset of $G$ as we range over the possibly infinite set of rules of $E$ and over all possible values of $t$. This indicates the way in which an infinite set of rules $E$ can be encoded by means of a finite amount of data.

**Example 6.3.1 ($\mathbf{Z}^2$ word differences).** Consider the free abelian group $G$ on two generators $x$ and $y$, with semigroup generators $A = \{x, y, X, Y\}$.

Let the order in $A$ be $x < X < y < Y$. Then we have the following finite complete set of rules:

$$\begin{array}{llll} Xx \Rightarrow \varepsilon, & xX \Rightarrow \varepsilon, & yY \Rightarrow \varepsilon, & Yy \Rightarrow \varepsilon, \\ yx \Rightarrow xy, & Yx \Rightarrow xY, & yX \Rightarrow Xy, & YX \Rightarrow XY. \end{array}$$

However, if we take the ordering $x < y < X < Y$, any complete set $E$ has to be infinite. To see this, notice that since $xy^n$ and $y^nX$ are ShortLex geodesic, they cannot contain a left-hand side of a rule of $E$. By taking $n$ longer than the left-hand side of any rule of the supposedly finite set $E$, we ensure that $xy^nX$ is irreducible. But we know that this string reduces to $y^n$, a contradiction.

In fact we may take $E$ to be the following infinite complete set of rules:

$$\textbf{6.3.2.} \quad \begin{cases} Xx \Rightarrow \varepsilon, & xX \Rightarrow \varepsilon, \quad yY \Rightarrow \varepsilon, \quad Yy \Rightarrow \varepsilon, \\ yx \Rightarrow xy, & Xy \Rightarrow yX, \quad YX \Rightarrow XY, \quad Yx \Rightarrow xY, \end{cases}$$

$$\textbf{6.3.3.} \quad \begin{cases} xy^nX \Rightarrow y^n & \text{for all } n > 0, \\ yX^nY \Rightarrow X^n & \text{for all } n > 0. \end{cases}$$

Only finitely many word differences arise: $\overline{\varepsilon}$, $\overline{x}$, $\overline{y}$, $\overline{X}$, $\overline{Y}$, $\overline{xy}$, $\overline{xY}$, $\overline{yX}$ and $\overline{XY}$.

We now build a partial deterministic finite state automaton $P$, called the *word difference machine*, to encode the word difference information. The machine is over $A \times B$, for $B = A \cup \{\$\}$, where \$ is an end-of-string symbol. The states of $P$ are the word differences resulting from the rules of $E$. The arrows are defined as follows: If $v \Rightarrow w$ is a rule of $E$, we can make $(v, w)$ into a padded string $u = (v_1, w_1)(v_2, w_2) \cdots (v_n, w_n)$, as explained in Definition 1.4.3. By Lemma 6.2.13 (final rules for groups), we may have $w_n = \$$ or $w_n = w_{n-1} = \$$, but all the other letters lie in $A$. We define inductively $s_0 = \varepsilon$ and $s_i = \rho_E(v_i^{-1} s_{i-1} w_i)$, where the inverse is formal and and the symbol \$ is interpreted as the nullstring. We define an arrow $(v_i, w_i) : s_{i-1} \to s_i$ for $1 \leq i \leq n$. Notice that $s_n = \varepsilon = s_0$. The start state for $P$ is $s_0$ and this is the only accept state.

Since $E$ is usually infinite, the above description is not constructive. We will return to this point later, because we are interested in practical programs. In the meantime, we will work in the mathematicians' usual non-constructive world, and this will at least tell us what to aim at when our point of view does become constructive.

The automaton we get for Example 6.3.1 with generators $x < y < X < Y$ is shown in Figure 6.3. Note that we obtain all the states and arrows of $P$ by considering only Rules 6.3.2 and 6.3.3 with $n \leq 2$. This situation is true in

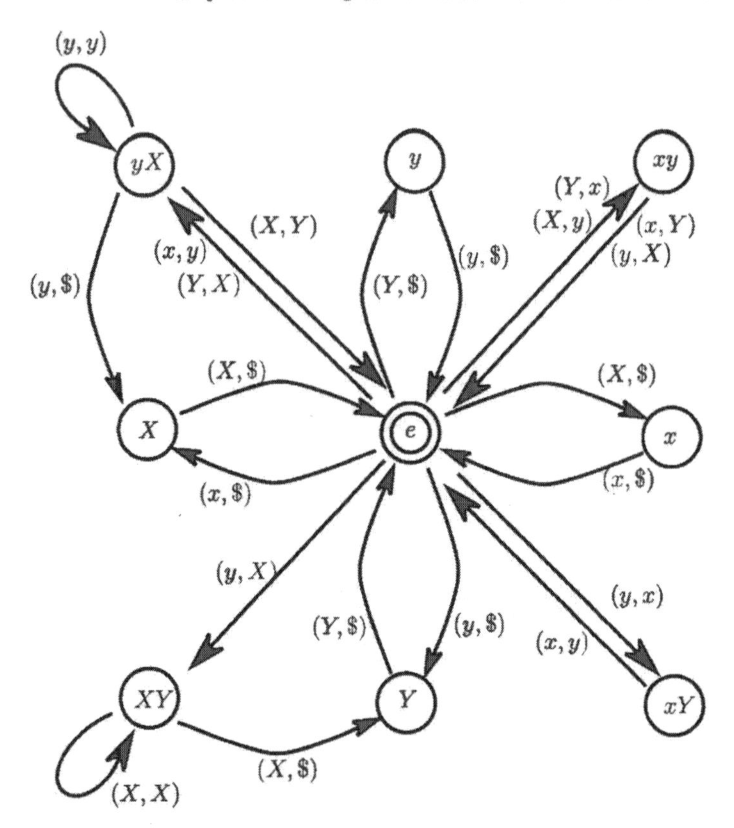


**Figure 6.3. The word difference machine.** Here is $P$ for the rank-two free abelian group, with generators $x < y < X < Y$ (see Example 6.3.1).

general: since $P$ is finite, all the information in it is, in principle, contained in a finite subset of $E$. (The question is knowing which subset.)

As described so far, $P$ accepts not only the padded strings coming from rules $v \Rightarrow w$ in $E$, but also strings that are not padded according to Definition 1.4.3 (such as $(x,\$)(X,\$)(X,y)(y,X)$ in Figure 6.3). We avoid this difficulty by combining $P$ with an automaton whose language is the set of all padded strings (see Lemma 1.4.1). We further combine $P$ with an automaton that accepts pairs of strings $(v, w)$ over $A$ if and only if $v$ is greater than $w$ in the ShortLex ordering. We denote by $Q$ the resulting automaton in two variables.

Notice that we can reduce strings using $Q$. For, if we ignore the second element of its labels, $Q$ becomes a non-deterministic automaton in one variable. A string $w$ is reducible if it has a non-trivial substring accepted by this non-deterministic automaton. We can check this for each substring by a long but constructive procedure. Once we know a path of arrows corresponding to such a substring of $w$, we obtain a reduction of the substring by following the same path of arrows, this time looking at the second element of the labels.This is essentially the same as the technique used in the proof of Theorem 2.3.10 (quadratic algorithm).

**Lemma 6.3.4 (fsa encodes rules).** *If $(v, w)$ is accepted by $Q$, then $v > w$ and $\overline{v} = \overline{w}$ in $G$, so $v$ is reducible. If $v \Rightarrow w$ is a rule of $E$, then $(v, w) \in L(Q)$.* 6.3.4

Let $L' \subset A^*$ be the set of $v$ such that $(v, w) \in L(Q)$ for some $w \in B^*$, and let $L = A^* \setminus A^* L' A^*$ be the language of strings that have no substrings in $L'$. Then $L$ is a regular language by Theorem 1.4.6.

**Lemma 6.3.5.** *$L$ is the language* ShortLex$(G, A)$.

*Proof of 6.3.5:* Let $v \in L$. By Lemma 6.3.4, $v$ is irreducible under $E$, so $v \in$ ShortLex$(G, A)$. Conversely, any $v \notin L$ has a substring in $L'$. It follows that $v$ is reducible and is therefore not in ShortLex$(G, A)$. 6.3.5

Having reached this seemingly satisfactory state of affairs, we now subject ourselves to a dose of reality. Since $E$ may be infinite, it is not actually possible in general to find the automaton $P$ directly. But we can construct automata $P_j$ from the approximations $E_j$ to $E$ obtained by Algorithm 6.2.11, as follows: Let $\rho_j$ be the canonical reduction (page 121) using $E_j$. Each state of $P_j$ is a word difference arising from $E_j$, regarded as a string over $A$ (and thus not to be confused with the element of $G$ which it represents). We set $s_0 = \varepsilon$. Using the same notation for a rule of $E_j$ as we did above for a rule of $E$, we set $s_i = \rho_i(v_i^{-1} s_{i-1} w_i)$. (Note that, while $s_n$ must represent the identity element in $G$, for $n = |v|$, we do not necessarily have $s_n = \varepsilon$. A slightly different treatment would at this point introduce a new Knuth–Bendix rule equating $s_n$ with $\varepsilon$, but we will not bother to do this.) Arrows in $P_j$ are defined in the same way as for $P$, and the unique start and accept state of $P_j$ is $\varepsilon$.

$P_j$ is a finite state automaton over $A \times B$, where $B = A \cup \{\$\}$. We form $Q_j$ from $P_j$ by insisting that an accepted string $(v, w)$ must be padded (see Definition 1.4.3) and satisfy $v > w$ in the ShortLex ordering. Let $L'_j$ and $L_j$ be defined from $Q_j$ in the same way that $L'$ and $L$ are defined from $Q$.

**Theorem 6.3.6 (fsa j encodes rules).** *Let $(v, w)$ be a string over $A \times B$ accepted by $P_j$. Then $\overline{v} = \overline{w}$, where the end-of-string symbol is sent to the identity element in $G$. If $(G, A)$ is ShortLex automatic, $L_j = L$ for $j$ sufficiently large.*

*Proof of 6.3.6:* From the construction, it follows by induction that if $(v, w)$ leads from the start state $\varepsilon$ of $P_j$ to $s$, then $\overline{v^{-1}w} = \overline{s}$. It follows that if $(v, w)$ is accepted by $Q_j$, then $\overline{v} = \overline{w}$ and $v > w$ in the ShortLex ordering.

Each state or arrow of the machine $P$ results from consideration of some rule $v_i \Rightarrow w_i$ of $E$, where $i$ ranges over some finite index set $I$. We choose $k$ so that $k > 2|v_i|$ for each $i \in I$. Each word difference that arises from the rule $v_i \Rightarrow w_i$ is represented by a string of length less than $k$. We adopt the notation of Theorem 6.2.12 (complete set of rules), and suppose that $j \geq n(k)$. Then each of the rules $v_i \Rightarrow w_i$ lies in $E_j$ and reduction of the corresponding word differences using the rules of $E_j$ gives the same answer as reduction using the rules of $E$. Recall from Proposition 6.2.12 that $E$-reduction allows us to calculate as though the strings are actually elements of $G$. It follows that $P$ is a subautomaton of $P_j$.

By the definition of $L$ (see after Lemma 6.3.4), any string $u \notin L$ has a substring $v$ such that $(v, w)$ is accepted by $Q$ for some $w$. Then $(v, w)$ is also accepted by $Q_j$, and $u \notin L_j$.

Conversely, suppose that $u \notin L_j$. Then $u$ has a substring $v$ such that $(v, w)$ is accepted by $Q_j$ for some $w$. We know that $\overline{v} = \overline{w}$, so $v$ must be $E$-reducible. Then $v$ contains the left-hand side of some rule in $E$, and hence so does $u$. This shows that $u \notin L$, and therefore that $L = L_j$. $\boxed{6.3.6}$

Thus, if we start with a ShortLex automatic group and apply Algorithm 6.2.11 (Knuth–Bendix 2), we will eventually obtain a set of rules $E_j$ large enough to give us the word acceptor $W$ for the group. We now consider the construction of the multiplier automata. Recall from Definition 2.3.1 that, for $a \in A$, the multiplier automaton $M_a$ is a finite state automaton over $(A, A)$ that accepts the pairs $(v, w)$ such that $\overline{va} = \overline{w}$ and $v, w \in L(W)$ (that is, $v$ and $w$ are irreducible).

We first get a couple of easy special cases out of the way. The equality recognizer $M_0 = M_\varepsilon$ is trivial to find, because $L = L(W)$ has a unique representative for each group element; we just make $M_\varepsilon$ accept all pairs $(w, w)$ with $w \in L$. Also, if a generator $a$ reduces to $b < a$, with $b \in A$ or $b = \varepsilon$, we set $M_a = M_b$. Since no word containing a reducible generator is accepted by $W$, we may as well restrict the alphabet to irreducible generators, which we do from on.

We will be able to find all the $M_a$'s at once if we can find an automaton $C$ that accepts all pairs $(va, w)$ satisfying the following conditions: $v$ and $w$ have no common prefixes; $\overline{va} = \overline{w}$; and $v, w \in L(W)$. For we have

$$L(M_a) = \{(xva, xw) \mid xv \in L(W) \wedge xw \in L(W) \wedge (va, w) \in L(C)\},$$

so $L(M_a)$ is regular if $L(C)$ is.

Let $E$ be the set of Knuth-Bendix rules derived in Theorem 6.2.12. According to that result, if a generator is reducible the only guise in which it can appear in a rule is as the entire left-hand side. It is therefore no loss to exclude such rules, in conformity with our earlier decision to exclude reducible generators from consideration. Let $\rho$ be reduction according to the rules of $E$. Let $X_0 = E$, and define $X_j$ recursively as follows: At each step, we look for overlaps of left-hand sides of rules in $E$ with right-hand sides of rules in $X_{j-1}$. Explicitly, we look for rules of the form $ta \Rightarrow yz$ in $X_{j-1}$ and $xy \Rightarrow u$ in $E$, with $t$ and $y$ non-trivial and $a \in A$ (see Figure 6.4). We take

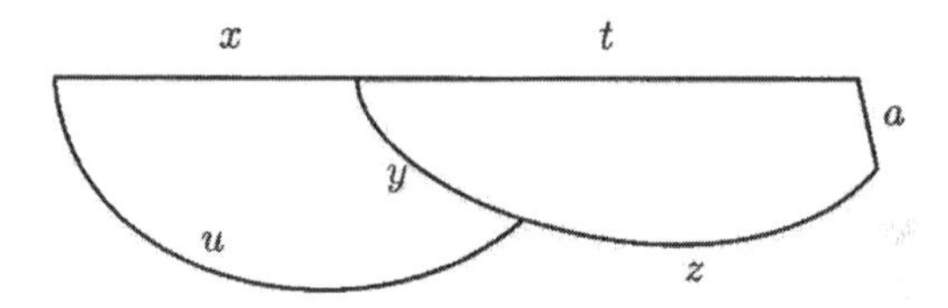


**Figure 6.4. Creating new rules.** We have rules $xy \Rightarrow u$ in $E$ and $ta \Rightarrow yz$ in $X_{j-1}$. This may lead to a new rule in $X_j$, obtained from reducing $xt$ and $uz$ and eliminating common prefixes.

the residues $\rho(xt)$ and $\rho(uz)$, strip common prefixes from them to get strings $\alpha$ and $\beta$, then form the rule $\alpha a \Rightarrow \beta$. We repeat this for all possible matches between rules of $X_{j-1}$ and rules of $E$, and define $X_j$ as the union of $X_{j-1}$ with the set of rules thus created. Finally, we let $X$ be the union of $\bigcup_j X_j$ with rules of the form $a \Rightarrow a$, for $a \in A$.

**Lemma 6.3.7 (enlarging $E$).** *A rule $va \Rightarrow W$ is in $X$ if and only if $v$ and $w$ have no common prefixes, $\overline{va} = \overline{w}$, and $v, w \in L(W)$. In other words, $X$ is the language $L(C)$.*

*Proof of 6.3.7:* That every rule in $X$ gives a pair in $L(C)$ is obvious from the construction. Now take $(va, w) \in L(C)$; if $v = \varepsilon$ we have $w = a$, and the rule $a \Rightarrow a$ is in $X$. If $v$ is non-trivial, $\rho(va) = w$ and $va \neq w$, since $v$ and $w$ have no common prefix. Since $v$ is irreducible, there is some non-trivial suffix $t$ of $v$ and some irreducible $r$ such that $ta \Rightarrow r$ is a rule in $E = X_0$ (Figure 6.5). We define $s$ so that $v = st$. If $s \neq \varepsilon$, then, since $s$ is irreducible, has no common prefix with $w$ and $\rho(sr) = w$, there is again a rule $xy \Rightarrow u$ in $E$, where $x$ is a non-trivial suffix of $s$, $y$ is non-trivial, and $r = yz$ for some $z$. If there is a choice, we choose this rule with $x$ of minimal length. By construction, the presence of $ta \Rightarrow yz$ in $X_0$ and of $xy \Rightarrow u$ in $E$ implies

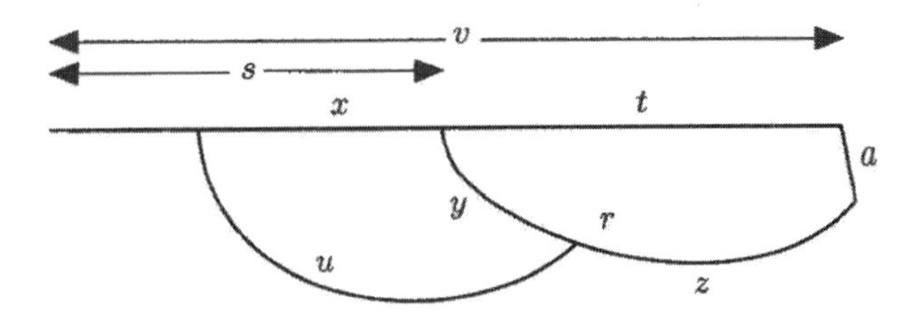


**Figure 6.5. Using the new rules**

the presence of $xta \Rightarrow \rho(uz)$ in $X_1$. Since $x$ is minimal and $xy$ is reducible, $x$ and $\rho(uz)$ have no common prefix. We repeat the process with $xt$ in place of $t$, until, in at most $j \leq |v|$ steps, we must have $t = v$, and the rule $va \Rightarrow w$ is in $X_j$. 6.3.7

If our group is ShortLex automatic with respect to the given ordered set of generators $A$, the rules in $X$ give rise to only a finite set of word differences. Thus we can define the desired automaton $C$ as the word-difference automaton based on $X$, just as we created $P$ from $E$ (see page 127). As before, since $X$ is generally infinite, it is not actually possible to find $C$ constructively; worse yet, we cannot even start building $X$, because we do not know $E$. But also as before, we can work from approximations.

More precisely, we compute the rules $E_k$, until we guess we have reached a high enough value of $k$ that the word acceptor $W_k$ obtained from $E_k$ is identical to the real word acceptor $W$ (a reasonable time is when no new word differences have appeared for a long time). If we guessed right, we now know the language accepted by $W$, and also how to do $\rho$-reduction (see page 129). Next we enlarge $E_k$ into sets $X_{k,j}$, starting with $X_{k,0} = E_k$ and following the same process described just before Lemma 6.3.7. Again, we stop when it seems appropriate, and compute the word-difference machine $C_{k,j}$ based on $X_{k,j}$, and candidate multipliers $M_{a,j,k}$. Then we check the conditions of Theorem 6.1.2 (specialized axioms). If they are true, we have found the ShortLex automatic structure; if not, we go back to computing more of the rules $E_j$ and $X_{k,j}$.

We are guaranteed to succeed if the group is ShortLex automatic and we are patient enough, by the following result:

**Theorem 6.3.8.** *Suppose $(G, A)$ is ShortLex automatic and $a \in A$ is irreducible. Then, for $k$ and $j$ sufficiently large, $M_{a,k,j}$ and $M_a$ accept the same set of pairs.*

*Proof of 6.3.8:* It is clear that if $(v, w)$ is accepted by $M_{a,k,j}$, then $v$ and $w$ are accepted by $W$ and $\overline{va} = \overline{w}$. Therefore $(v, w)$ is accepted by $M_a$.

The idea of the converse is the same as in the proof of Theorem 6.3.6. Each state and arrow of $C$ arises from some rule $v \Rightarrow w$ in $X$; if $j$ is greater than the length of all the left-hand sides, each of the rules will be in $X_j$. Furthermore, each such rule is obtained from a finite number of rules of $E$; by taking $k$ large enough, we can assume that all these finitely many component rules lie in $E_k$. It follows that, for these values of $j$ and $k$, $C_{k,j}$ is a subautomaton of $C$, and every pair accepted by $C$ will also be accepted by $C_{k,j}$. 6.3.8

Many optimizations not discussed here can be used to make the process above more efficient. Some of them have been explained in [EHR91]. Some of us intend to write more fully on this subject in the future.

# Chapter 7

# Asynchronous Automatic Groups

In this chapter we will describe a more general class of groups than the automatic groups we have looked at so far. As before, we have a word acceptor automaton and multiplier automata, and the multipliers read two strings, which we can imagine are written on two tapes. However, in the case of (synchronous) automatic groups, the two tapes are read at the same speed, while for asynchronous groups, they may be read at very different speeds.

There are many similarities, but also many differences, between asynchronous and synchronous automatic groups. The class of asynchronous automatic groups is sufficiently large to enable some interesting examples to be built easily. This makes it possible to show that the isomorphism problem is undecidable for automatic groups, and that there are asynchronous automatic groups for which the conjugacy problem is undecidable [BGSS] (compare Open Questions 2.3.11 and 2.5.8, and Theorem 2.5.7). The word problem, on the other hand, is still solvable; this is a consequence of Theorems Theorem 2.2.5 and Theorem 7.3.4.

**Open Question 7.0.1.** Can the word problem be solved in polynomial time for asynchronous automatic groups? The best algorithm we know for reducing a word to normal form is exponential in the length of the input, instead of quadratic as in the synchronous case (Theorem 2.3.10).

**Open Question 7.0.2.** In Section 5.1 (Axiom Checking) we produced an algorithm that takes as input a collection of automata and finds out whether this collection gives a synchronous automatic structure. We do not know how to do this in the asynchronous case. It is conceivable that the question is undecidable. We do not know how to make any sort of systematic attempt to find an asynchronous automatic structure, starting with generators and relators for a group. Known asynchronous structures are all built according to special recipes, which depend on exactly how the group is given.

Consideration of asynchronous automatic groups was first proposed by Paterson. The first example of a group that is asynchronously but not synchronously automatic was worked out by Thurston, and is discussed in Section 7.4. Most of the results in this chapter are due jointly to Epstein and Thurston. Theorem 7.2.8 (characterizing asynchronous) was proved by Epstein in a weaker version; its current stronger form is due to Levy.

## 7.1. Asynchronous Automata

Informally, an *asynchronous automaton* is a machine that reads from two tapes, each containing a string of characters. The machine remembers which tape should be read next, that is, the machine's state contains information to that effect. After an end-of-string symbol \$ is read on one tape, only the other tape can be read from. When a second end-of-string symbol is read, the machine goes to an accept state.

To translate this in terms of ordinary finite state automata, which read from a single string, we introduce the idea of a *shuffle*, which is indeed analogous to taking two decks of playing cards and shuffling them once into a single deck. If $A$ is an alphabet and $w_{\mathtt{L}}, w_{\mathtt{R}} \in A^*$ are words over $A$, a *shuffle* of $(w_{\mathtt{L}}, w_{\mathtt{R}})$ is a string $w \in A^*$ and a map $[1, |w|] \to \{\mathtt{L}, \mathtt{R}\}$ such that, if we substitute the nullstring $\varepsilon$ in $w$ for each element that maps to $\mathtt{R}$, we get $w_{\mathtt{L}}$, and if we substitute $\varepsilon$ in $w$ for each element that maps to $\mathtt{L}$, we get $w_{\mathtt{R}}$. For example, "carpenter" is a shuffle of (rent, caper); the map can be deduced from the order of the characters. By contrast, "aa," that longstanding favourite of Scrabble players, is a shuffle of (a, a) in two different ways.

**Definition 7.1.1 (asynchronous automaton).** An *asynchronous deterministic two-tape automaton*, or simply *asynchronous automaton*, over $A$ is a partial deterministic automaton over $A \cup \{\$\}$, in which the set of states is partitioned into five subsets, denoted by $S_{\mathtt{L}}$, $S_{\mathtt{L}}^{\$}$, $S_{\mathtt{R}}$, $S_{\mathtt{R}}^{\$}$ and $S^{\$}$. The only state in $S^{\$}$ is the unique accept state, called $s_{\$}$; no arrows originate from it. An arrow labelled by an element of $A$ and having its source in $S_{\mathtt{L}} \cup S_{\mathtt{R}}$ has its target in $S_{\mathtt{L}} \cup S_{\mathtt{R}}$; if its source is in $S_{\mathtt{L}}^{\$}$ or $S_{\mathtt{R}}^{\$}$, its target is in the same set. A \$ arrow with source in $S_{\mathtt{L}}$ has target in $S_{\mathtt{R}}^{\$}$; one with source in $S_{\mathtt{R}}$ has target in $S_{\mathtt{L}}^{\$}$; and one with source in $S_{\mathtt{L}}^{\$} \cup S_{\mathtt{R}}^{\$}$ has target $s^{\$}$. The start state is in $S_{\mathtt{L}} \cup S_{\mathtt{R}}$.

The interpretation of all this is the following: The subscript indicates what tape will be read from next. Thus, if the automaton is in a state in $S_{\mathtt{L}} \cup S_{\mathtt{L}}^{\$}$, the next character will be read from the left tape. A superscript \$ indicates whether an end-of-string symbol has been seen: initially the state is in $S_{\mathtt{L}} \cup S_{\mathtt{R}}$, but when a \$ is read from (say) the right tape, that is, when we follow a \$ arrow with source in $S_{\mathtt{R}}$, we go to $S_{\mathtt{L}}^{\$}$. After that only the left tape is read, that is, the state remains in $S_{\mathtt{L}}^{\$}$. Another \$ causes the automaton to go to the accept state. An analogous situation obtains with L and R interchanged. Thus any accepted string contains exactly two \$'s, one from each tape.

If there is danger of confusion about what automaton the $S$'s refer to, we write $S_{\mathtt{L}}^{\$}(M)$ instead of $S_{\mathtt{L}}^{\$}$ for an automaton $M$, and so on.

We say that an asynchronous automaton *accepts* a pair of strings $(w_{\mathtt{L}}, w_{\mathtt{R}}) \in A^* \times A^*$ if there is a shuffle $w$ of $(w_{\mathtt{L}}\$, w_{\mathtt{R}}\$)$ which is accepted by the automaton. Note that, although there are many shuffles of a fixed pair of strings, at most one can be accepted by the automaton—this follows easily from the decomposition of the set of states. (Remember that a shuffle includes information about which string each character comes from.)

The next lemma tells us when an asynchronous automaton may be regarded as an ordinary finite state automaton over a pair of alphabets. The proof amounts to just changing the technique of bookkeeping. The details are not particularly enlightening and can easily be worked out.

**Lemma 7.1.2.** *Let $L$ be a two-variable language over $(A, A)$ (see Definition 1.4.2). A necessary and sufficient condition that it should be regular, in the sense of Definition 1.4.5, is that it should be accepted by an asynchronous automaton such that every arrow in $A$ with source in $S_{\mathtt{L}}$ has target in $S_{\mathtt{R}}$, and vice versa. In other words, until one of the two strings is entirely consumed, the two tapes are read alternately.* [7.1.2]

Languages accepted by asynchronous automata do not enjoy all the pleasant properties of regular languages.

**Example 7.1.3 (predicates not closed).** The set of strings of the form $(x^n, x^{2n})$, as $n$ varies over the positive integers, is recognized by an asynchronous automaton, and similarly for the set of strings of the form $(x^{2n}, x^n)$. But the union of these sets is not recognizable by any asynchronous automaton, because an automaton could not know which of the two strings to scan faster.

We do, however, have the following lemma:

**Lemma 7.1.4.** *Complementation in $A^* \times A^*$ preserves the class of languages accepted by asynchronous automata.*

*Proof of 7.1.4:* Let $W$ be an asynchronous automaton accepting a language $L \subset A^* \times A^*$. We will construct an automaton $W'$ such that every failure of $W$ is a success of $W'$ and vice versa. We use the following recipe.

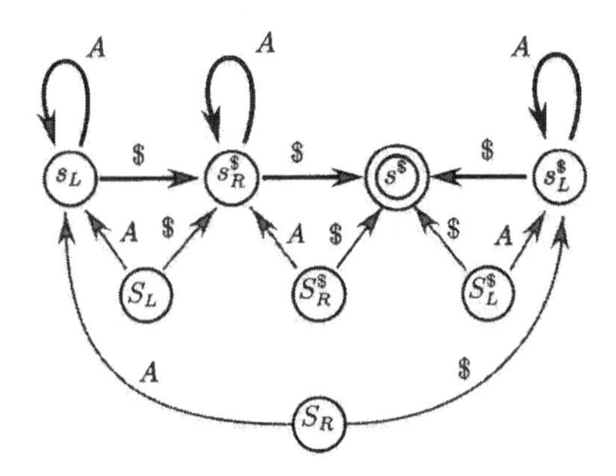


**Figure 7.1. Additional states.** We show the additional states $s_L \in S_L$, $s^\$_R \in S^\$_R$ and $s^\$_L \in S^\$_L$ needed to construct an asynchronous automaton which accepts the complement of a given asynchronous language. We have shown all arrows with these states as source. Possible arrows with these states as targets are symbolically represented by grey arrows. Such arrows may or may not exist.

We start by removing all arrows in $W$ with target $s^\$$. We then add three new states $s_L \in S_L$, $s^\$_R \in S^\$_R$ and $s^\$_L \in S^\$_L$ as in Figure 7.1. All arrows with labels in $A$ are defined on these three states, and loop back to their source. There are \$ arrows going from $s_L$ to $s^\$_R$, from $s^\$_R$ to $s^\$$ and from $s^\$_L$ to $s^\$$.

Whenever an arrow is not defined on a non-accept state of $W$, a corresponding arrow is defined from the same source to one of the states $s_L$, $s^\$_L$, $s^\$_R$ and $s^\$$. The target of the arrow is uniquely determined by the rules in Definition 7.1.1, and is also indicated by the dotted arrows in Figure 7.1. For example, if $s$ is in $S^\$_L(W)$ and there is no arrow labelled \$ with source $s$, we install in $W'$ an arrow \$ from $s$ to $s^\$$.

Finally, in order to get back to the original definition of an asynchronous automaton, we remove all inaccessible and all dead states from $W'$ and all arrows with such states as source or target. It is easy to see that a pair of strings $(w_L, w_R)$ is accepted by $W$ if and only if it is not accepted by $W'$.

7.1.4

**Lemma 7.1.5.** *The $\exists$ operator and the $\forall$ operator convert a language accepted by an asynchronous automaton to a regular one-variable language.*

*Proof of 7.1.5:* According to the previous lemma, we need only prove the result for $\exists$. To handle this, we replace each arrow with source in $S_L \cup S^\$_L$, and each \$ arrow, by an $\varepsilon$ arrow. This gives a non-deterministic automaton over $A$. The result follows. 7.1.5

**Lemma 7.1.6 (fixed string).** *Let $W$ be an asychronous automaton. If $w_{\mathrm{L}}$ is a fixed string over $A$, the language $\{w_{\mathrm{R}} \mid (w_{\mathrm{L}}, w_{\mathrm{R}}) \in L(W)\}$ over $A$ is regular. The same is true with* L *and* R *interchanged.*

*Proof of 7.1.6:* Let $W$ be the given asynchronous automaton over $(A, A)$, and let $w_{\mathrm{R}}$ be fixed. We define a deterministic partial automaton $W'$ over $A \cup \{\$\}$ which accepts the language $\{w_{\mathrm{L}} \mid (w_{\mathrm{L}}, w_{\mathrm{R}}) \in L(W)\}$.

Let $Y$ be the set of suffixes of $w_{\mathrm{R}}\$$. The idea of the proof is that the states of $W'$ will remember the element of $Y$ that remains to be read (by the original automaton). We define the state set of $W'$ to be $Z \times Y$, where $Z = S_{\mathrm{L}}(W) \cup S_{\mathrm{L}}^{\$}(W) \cup S^{\$}(W)$.

We define the arrows so that each state of $W'$ corresponds to a situation where $W$ would have read all but the suffix $y$ of $w_{\mathrm{R}}\$$ and would be waiting to read a symbol from the left string. Let $(s, y)$ be a state of $W'$. If $x : s \to t$ is an arrow of $W$, we define $(s, y)x = (sq, p)$, where $q$ is the shortest non-trivial prefix $q$ of $xy = qp$, such that $sq \in Z$. If there is no such $q$, then $(s, y)x$ is not defined. The unique accept state of the new partial automaton is $(s^{\$}, \varepsilon)$.

Let $s_0$ be the initial state of $W$, and let $q$ be the shortest (possibly trivial) prefix of $w_{\mathrm{R}}\$ = qp$, such that $s_0 q \in Z$. The initial state of $W'$ is defined to be is $(s_0 q, p)$. If there is no such $q$, our partial automaton is empty and the corresponding regular language is the empty set. We complete the construction of $W'$ by removing all inaccessible states and all dead states.

7.1.6

## 7.2. Asynchronous Automatic Groups: Definition

**Definition 7.2.1 (asynchronous automatic group).** Let $G$ be a group. An *asynchronous automatic structure* on $G$ consists of a set $A$ of semigroup generators of $G$, a finite state automaton $W$ over $A$, and asynchronous automata $M_x$ over $A$, for $x \in A \cup \{\varepsilon\}$, satisfying the following conditions:

(1) The map $\pi : L(W) \to G$ is surjective.

(2) For $x \in A \cup \{\varepsilon\}$, we have $(w_L, w_R) \in L(M_x)$ if and only if $\overline{w_L x} = \overline{w_R}$ and both $w_L$ and $w_R$ are elements of $L(W)$.

The expressions *word acceptor*, *equality recognizer* and *multiplier automaton* have the same meaning as in the synchronous case.

Just as we did in Theorem 2.3.4 (standard automata theorem) for synchronous automatic structures, we will show that the existence of an asynchronous automatic structure with language of accepted strings $L$ allows us to define certain standard multiplier automata, based only on the structure of $L$ and a neighbourhood of the identity in $G$. The proof is rather longer than in the case of a synchronous automatic structure.

Suppose $M$ is an asynchronous automaton over $A$, and that the pair $(w_L, w_R)$, for $w_L, w_R \in A^*$, is accepted by $M$. We imagine that $M$ starts reading this pair of strings at time zero, and reads one letter in each unit of time. We define functions $t_L$ and $t_R$ of $t$, by setting $w_L(t_L(t))$ and $w_R(t_R(t))$ to be the prefixes of $w_L$ and $w_R$ read after time $t$. In other words, after time $t$, we have advanced a distance $t_L(t)$ along the left string and a distance $t_R(t)$ along the right string.

**Lemma 7.2.2 (word difference function).** *Let $G$ be a group and let $A$ be a set of semigroup generators for $G$. Let $g$ be a fixed element of $G$. Let $M_g$ be an asynchronous automaton, such that each accepted pair of strings $(w_L, w_R)$ satisfies $\overline{w_L}g = \overline{w_R}$ in $G$. Then there is a map $f$ from the set of states of $M_g$ to $G$ such that $\overline{w_L(t_L)}^{-1}\overline{w_R(t_R)} = f(s)$, where $(w_L, w_R)$ is any pair accepted by $M_g$ and $s$ is the state of $M_g$ on reading $(w_L(t_L), w_R(t_R))$.*

In other words, the state $s$ of $M_g$ records, amongst other things, the *word difference* between the prefixes of $w_L$ and $w_R$, regarded as group elements, as they are read.

*Proof of 7.2.2:* By the definition of an asynchronous automaton, each state of $M_g$ is live. Therefore, from each state $s$ of $M_g$ there is a shortest path of arrows to the unique accept state, whose length is bounded by the number of states. Let this path of arrows be given by a shuffle of $(u_s, v_s)$. We define $f(s) = \overline{u_s} g \overline{v_s}^{-1}$.

Now suppose that at time $t$, the machine has read the prefixes $w_L(t_L)$ of $w_L$ and $w_R(t_R)$ of $w_R$, and has arrived at the state $s$. Then the machine accepts the input $(w_L(t_L)u_s, w_R(t_R)v_s)$. This means that $\overline{w_L(t_L)u_s}g = \overline{w_R(t_R)v_s}$ in $G$. It follows that $\overline{w_L(t_L)}^{-1}\overline{w_R(t_R)} = f(s)$. 7.2.2

It is possible for the multiplier automata in an asynchronous structure to show extreme favouritism between the two strings being read. An example is any finite group $G$, with all elements as generators and all strings as accepted strings. We can have multiplier automata that operate by first reading $w_L$ completely, and then $w_R$, while always keeping track of the difference $\overline{w_L}^{-1}\overline{w_R}$. Our first step in trying to find standard multiplier automata for an asynchronous automatic structure is to modify the automata to ensure that they are not so biased:

**Definition 7.2.3.** We say that an asynchronous automaton is *boundedly asynchronous* if there is an integer $k$ such that the automaton never reads more that $k$ letters in a row from either string. The number $k$ is called the *asynchronous factor* of the automaton. We also talk of a *boundedly asynchronous automatic structure* when all the multiplier automata are boundedly asynchronous.

A boundedly asynchronous automaton reads one string at most $k$ times faster than the other. Furthermore, after reading an end-of-string symbol on one string, such an automaton can read at most $k$ symbols from the other.

**Theorem 7.2.4 (boundedly asynchronous).** *Let $G$ be a group with an asynchronous automatic structure given by a set of generators $A$, a word acceptor $W$ and multiplier automata $M_x$, for $x \in A$. Then $G$ has a boundedly asynchronous structure over $A$, with a language that is a subset of the language accepted by $W$.*

*Proof of 7.2.4:* Let $c$ be the supremum of the number of states in any of the structure's automata. We start by changing $W$ to an automaton that remembers the previous $c^2$ steps in its history. In general, given a deterministic finite state automaton $M$ and a positive integer $k$, we can form an *associated $k$-history automaton* $M'$ over the same alphabet as follows: The states of $M'$ are paths of arrows of the form

**7.2.5.** $$(s_1, a_1, s_2, \ldots, a_n, s_{n+1}),$$

where $n \leq k$ and each $a_i$, for $1 \leq i \leq n$, is an arrow of $M$ with source $s_i$ and target $s_{i+1}$. We require that the length $n$ of the path equal $k$ unless $s_1$ is the initial state; the case $n < k$ accounts for shallow histories (one can't remember something that hasn't happened). The accept states of $M'$ are those for which $s_{n+1}$ is an accept state in $M$. The initial state in $M'$ is a path of arrows $(s_1)$ of length zero, where $s_1$ is the initial state of $M$. For each arrow $a$ leading out of $s_n$ in $M$, with target $s$, there is an arrow $a$ in $M'$ going out of state 7.2.5 to the state

$$\begin{aligned} &(s_1, \ldots, a_{n-1}, s_{n+1}, a, s) \quad \text{if } n < k, \\ &(s_2, \ldots, a_{n-1}, s_{n+1}, a, s) \quad \text{if } n = k. \end{aligned}$$

Clearly $M'$ accepts the same language as $M$.

So we form the $c^2$-history automaton associated with $W$. We next make into a failure state any state of this automaton that includes a loop

$$(\ldots, s_i, a_i, \ldots, a_{j-1}, s_j = s_i, \ldots)$$

such that $a_i a_{i+1} \ldots a_{j-1}$ represents the trivial element of $G$. (To check if a loop satisfies these conditions, we use the equality recognizer $M_\varepsilon$ to compare a word that includes the loop and is accepted by $W$ with the same word minus the loop.) After removing the new failure states and any inaccessible states created as a consequence, as well as the arrows leading to such states, we obtain an automaton that we call $W'$. We also build an automaton $W''$ that accepts the language $L(W')\$$, and which we assume normalized, so it has a unique accept state $s^{\$}$. In Definition 7.2.6 below, we will refer to $W'$ and $W''$ as the *auxiliary automata* obtained from $W$.

$W'$ does not necessarily accept all the words accepted by $W$; the point of this construction is that words containing trivial loops of length at most $c^2$ are no longer accepted. However, the language accepted by $W$ still maps onto $G$. For, given $g \in G$, let $w$ be a shortest string accepted by $W$ and representing $g$. If $W'$, on reading $w$, were to arrive at one of the new failure states, $w$ would have a substring equivalent to the identity and leading from some state $s \in W$ to itself, which would contradict the minimality of $w$.

We now build new multiplier automata $M_x$, for $x \in A \cup \{\varepsilon\}$, that accept a pair $(w_L, w_R)$ if and only if $(w_L, w_R)$ is accepted by $M_x$ and both $w_L$ and $w_R$ are accepted by $W'$. We do this by multiplying the state set of $M_x$ by the state set of $W''$ twice, so $M'_x$ can keep track of the states in $W''$ corresponding to the two strings $w_L\$$ and $w_R\$$ being read. This is the same construction used for taking the intersection of two synchronous automata: see the remark after Lemma 1.4.1.

Now the new automata $W'$ and $M'_x$, for $x \in A \cup \{\varepsilon\}$, satisfy the conditions of Definition 7.2.1 (asynchronous automatic group), so that we still have an automatic structure. We claim that this structure is boundedly asynchronous, with factor $c^2$. For assume that $M'_x$ reads $c^2$ letters consecutively from, say, $w_L$. Then there exist $t$ and $t+p > t$, within this interval of time, such that the state of $W$ as it reads $w_L$ is the same at times $t$ and $t+p$, and likewise for $M_x$; this is because $W$ and $M_x$ have no more than $c$ states each. Furthermore, $t_R(t+p) = t_R(t)$ and $t_L(t+p) = t_L(t) + p$. Applying Lemma 7.2.2 (word difference function) to $M'_x$, we get

$$\begin{aligned}\overline{w_L(t_L(t))}^{-1}\overline{w_R(t_R(t))} &= \overline{w_L(t_L(t+p))}^{-1}\overline{w_R(t_R(t+p))}\\ &= \overline{w_L(t_L(t+p))}^{-1}\overline{w_R(t_R(t))}.\end{aligned}$$

It follows that $\overline{w_L(t_L(t))} = \overline{w_L(t_L(t+p))} = \overline{w_L(t_L(t)+p)}$, which shows that $w_L$ has a trivial loop of length at most $c^2$, in contradiction with the construction of $W'$. $\boxed{7.2.4}$

**Definition 7.2.6 (standard asynchronous automata).** Let $G$ be a group with a set of semigroup generators $A$, and let $W$ be a finite state automaton over $A$. Given $y \in G$ and three positive numbers $R_1 < R_2 < R_3$, we define the *standard multiplier automaton* $M_y$ *based on* $(W, R_1, R_2, R_3)$, an asynchronous automaton over $A$. The idea is that $M_y$ should be able to recognize right multiplication by $y$ in $G$ of elements represented by strings in $L(W)$, but this will only happen when $R_1$, $R_2$ and $R_3$ are chosen appropriately (see Lemma 7.2.9).

Let $W'$ and $W''$ be the auxiliary automata derived from $W$ (see page 142), and let $S$ be the state set of $W''$. Let $N$ be the open ball centred at the identity with radius $R_3$ in the Cayley graph of $G$. The states of $M_y$ are quintuples

$$(h, s_L, s_R, g, T) \in \{\mathtt{L}, \mathtt{R}\} \times S \times S \times N \times \{\mathtt{A}, \mathtt{B}\}.$$

As $M_y$ reads a pair of words $(w_L, w_R)$, the component $h$ says which tape should be read from next. The components $s_L$ and $s_R$ say what state $W''$ would be in after reading $w_L(t_L)$ and $w_R(t_R)$, respectively (see page 140 for the meaning of $t_L$ and $t_R$). They also tell whether end-of-string symbols have been encountered. The component $g$ records the word difference $\overline{w_L(t_L)}^{-1}\overline{w_R(t_R)}$. The last component, $T$, records whether the tape currently being read is "ahead" ($\mathtt{A}$) or "behind" ($\mathtt{B}$) the other tape, in a sense to be made precise later.

We partition the states into the sets $S_L$, $S_L^\$$, $S_R$, $S_R^\$$ and $S^\$$ of Definition 7.1.1 according to the values of $h$, $s_L$ and $s_R$. We discard all quintuples of the form $(h, s^\$, s^\$, g, T)$, where $s^\$$ is the accept state of $W''$—except that $(\mathtt{L}, s^\$, s^\$, y, \mathtt{A})$ and $(\mathtt{R}, s^\$, s^\$, y, \mathtt{A})$ are preserved, and amalgamated to become the unique accept state of $M_y$. The initial state of $M_y$ is $(L, s_0, s_0, e, \mathtt{A})$, where $e$ is the identity and $s_0$ is the initial state of $W''$.

The action of the arrows on $s_L$ and $s_R$ is the obvious one. For instance, an arrow labelled $x \in A$ and originating in a state with $h = \mathtt{L}$ leaves $s_R$ unchanged and acts on $s_L$ as the $x$ arrow would in $W''$; if $x$ is undefined at $s_L$ in $W''$, it is also undefined in $M_y$.

The action on $g$ is also easy to define: an arrow labelled $x \in A$ and originating in a state with $h = \mathtt{L}$ takes $g$ to $g' = x^{-1}g$, and one originating in a state with $h = \mathtt{R}$ takes $g$ to $g' = gx$. A \$ arrow has no effect on $g$. We let $r$ be the distance from $g'$ to the identity in the Cayley graph of $G$; if $r \geq R_3$, that is, if an arrow would cause $g'$ to fall outside $N$, that arrow is undefined and attempting the transition leads to failure.

The components $h$ and $T$ remain unchanged under the action of any arrows, except in the following cases (see also Figure 7.2 and Figure 7.3):

(1) If $r$ changes from $R_1 - 1$ to $R_1$, we leave $h$ unchanged and make $T$ equal to $\mathtt{A}$, no matter what its previous value. (This is the optimistic point

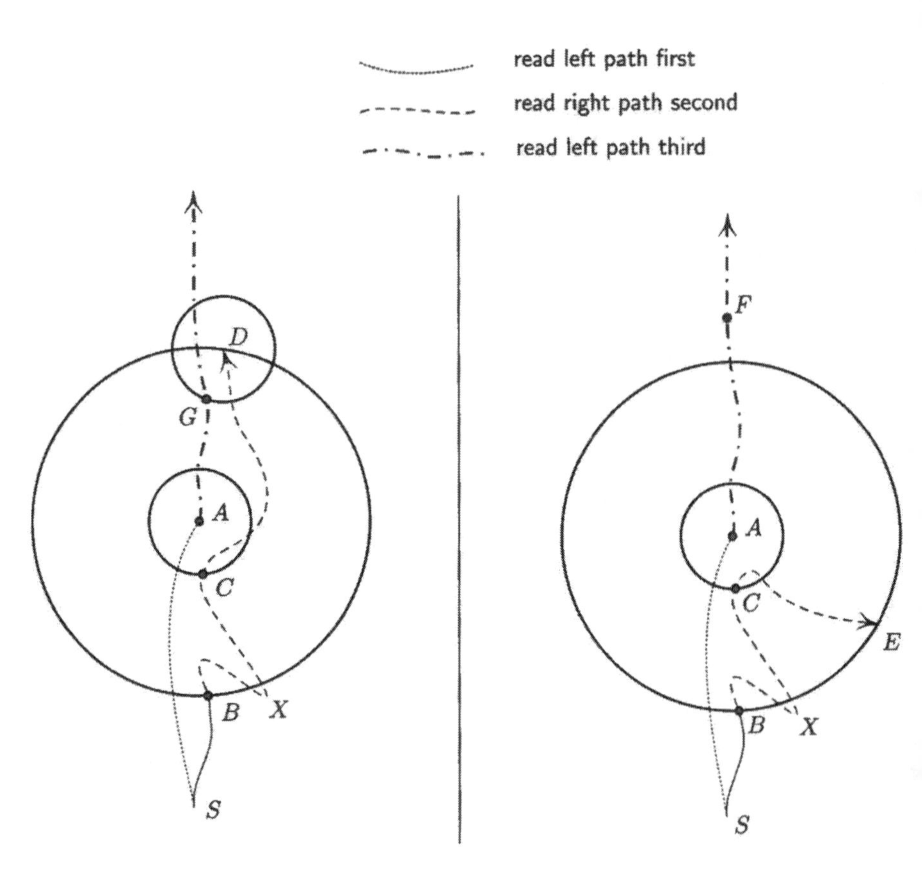


**Figure 7.2. The standard asynchronous multipliers.** The diagram is drawn in the Cayley graph. The point $S$ is basepoint for the Cayley graph, namely the identity element of the group. The small circles have radius $R_1$ and the large circles radius $R_2$. We start by reading along the left string, with the right string stopped at $B$ and the left string "ahead." When we reach $A$, we start to read the right string. There is no change at $X$, since the right string is "behind", even though the distance to $A$ has increased to $R_2$ again. At $C$ the state of the right string changes to "ahead" and we continue to read it. At $D$ or $E$, we stop reading the right string and start to read the left string, which is still at $A$. The left string is "behind". In the left picture at $G$ the left string changes from "behind" to "ahead", and the left string continues to be read. In the right picture at $F$, the distance to $E$ reaches $R_3$, and the left string is "behind". We are unable to continue and the process fails.

of view adopted by politicians, who say they are ahead whenever the opponent is in sight.)

(2) If $r$ changes from $R_2 - 1$ to $R_2$ and $T = \mathtt{A}$, we set $T$ to $\mathtt{B}$; also, we change $h$, unless $s_L = s^\$$ or $s_R = s^\$$.

(3) If $s_L = s^\$$ we set $h = \mathtt{R}$, and if $s_R = s^\$$ we set $h = \mathtt{L}$.

It follows from this description that if $(w_L, w_R)$ is accepted by $M_y$, then $w_L$ and $w_R$ are accepted by $W'$ and $\overline{w_L}y = \overline{w_R}$. The converse is not necessarily true, but the only way in which it may fail is if the word difference $g$ were to fall outside the neighbourhood $N$, that is, if its distance $r$ to the origin were to exceed $R_3$. We will see in Lemma 7.2.9 conditions for this not to happen.

Notice that we have defined standard asynchronous multiplier automata for all $y$, not just $y \in A$. This is because, unlike the synchronous case, we cannot obtain a multiplier $M_{ab}$ from the multipliers $M_a$ and $M_b$ by "composing" the respective languages (in the sense of composing the relations $L(M_a), L(M_b) \subset A^* \times A^*$). Composition does not work well with languages accepted by asynchronous automata. For example, let $A$ be an alphabet with three letters, namely $a$, 2 and 3. Consider the language $L_1$ which is the union of the set of all pairs of strings of the form $(a^n, 2a^{2n})$ and the set of all pairs of strings of the form $(a^n, 3a^{3n})$. Let $L_2$ be the set of all pairs of strings of the form $(xa^j, a^j)$, where $x = 2$ or $x = 3$. Clearly $L_1$ and $L_2$ are languages accepted by asynchronous automata. But the composition of $L_1$ and $L_2$ is the union of the set of all pairs of strings of the form $(a^n, a^{2n})$ with the set of all pairs of strings of the form $(a^n, a^{3n})$, which is not the language of any asynchronous automaton.

The following concept will help us state the main results of this section, Theorem 7.2.8 (characterizing asynchronous) and Lemma 7.2.9 (standard asynchronous multiplier theorem), in a manageable fashion:

**Definition 7.2.7 (departure function).** Let $G$ be a group, $A$ a set of semigroup generators for $G$, $\Gamma = \Gamma(G, A)$ the Cayley graph, and $L$ a regular language over $A$ that maps onto $G$. A *departure function* for $(G, L)$ is any function $D : \mathbf{R} \to \mathbf{R}$ such that, if $w \in L$, $r, s \geq 0$, $t \geq D(r)$ and $s + t \leq |w|$, then $d_\Gamma(\widehat{w}(s), \widehat{w}(s+t)) > r$.

If there is a departure function, an accepted string is going to eventually depart from every finite neighbourhood. We could have defined a departure function on the positive integers only; given a function defined this way, one can easily increase it a little to get a departure function according to Definition 7.2.7.

**Theorem 7.2.8 (characterizing asynchronous).** *Let $G$ be a group, $A$ a set of semigroup generators for $G$, and $L$ a regular language over $A$ that* *maps onto $G$. Then $(A, L)$ is part of a bounded asynchronous structure on $G$ if and only if the following conditions are satisfied:*

(1) *there exist a departure function $D$ for $G$; and*

(2) *there exists a constant $K > 0$ such that, for every pair of strings $w, w' \in L$ whose images under the map $\pi : A^* \to G$ are a distance at most one apart in the Cayley graph of $G$, the paths $\widehat{w}$ and $\widehat{w'}$ are at most a hausdorff distance $K$ from each other.*

Saying that the images of $\widehat{w}$ and $\widehat{w'}$ are at most a hausdorff distance $K$ from each other means that, for every $s$, there exists $s'$ such that the distance from $\overline{w(s)}$ to $\overline{w'(s')}$ is at most $K$, and likewise with $w$ and $w'$ interchanged. The constant $K$ is called an *asynchronous lipschitz constant*, by analogy with Lemma 2.3.2.

As the first paragraph of the proof will make clear, we can choose this relation between $s$ and $s'$ in such a way that, as $s$ increases, $s'$ does not decrease, and likewise with $s$ and $s'$ interchanged. Thus, there is a relation $\mathcal{R}$ between the domains of $\widehat{w}$ and $\widehat{w'}$ that involves all of both paths, relates points that are at most a distance $K$ apart, and is *weakly monotonic*, that is, if $(s, s')$ and $(t, t')$ are two pairs in $\mathcal{R}$, $s > t$ implies $s' \geq t'$ and $s' > t'$ implies $s \geq t$. We will call such a relation an *asynchronous correspondence* (with constant $K$) between $w$ and $w'$. It is true, but not entirely trivial, that if there is an asynchronous correspondence relating $w$ and $w'$ and one relating $w'$ and $w''$, there is also one relating $w$ and $w''$ (with a constant twice as big).

*Proof of 7.2.8:* Let $G$ have a boundedly asynchronous automatic structure, with asynchronous factor $k$. We first show the existence of the lipschitz constant $K$. Let $w_L, w_R \in L$ be strings satisfying $\overline{w_L x} = \overline{w_R}$, where $x \in A$. Then $(w_L, w_R)$ is accepted by the multiplier automaton $M_x$. By Lemma 7.2.2 (word difference function), the set of word differences $\overline{w_L(t_L(t))}^{-1} \overline{w_R(t_R(t))}$ is finite, where $t_L$ and $t_R$ are defined just before the statement of that lemma. If we choose $K$ greater than any of the distances from these word differences to the origin, it is clear that condition (2) is satisfied.

We next show that there is a departure function $D$ for $G$. Let $W$ be the word acceptor for the automatic structure, and let $r$ be a positive integer. For each element $g \in G$ whose distance to the identity in the Cayley graph is at most $r$, and for each pair of states $s_1$ and $s_2$ of $W$, we consider a shortest path of arrows in $W$ (if there is any such path) going from $s_1$ to $s_2$ and representing $g$ in $G$. Let $f$ be the supremum of the lengths of the paths thus obtained.

If $v$ is a substring of an accepted string, $v$ gives a path of arrows from a state $s_1$ of $W$ to a state $s_2$. Let $u_1$ be a shortest string leading from the initial state to $s_1$, and $u_2$ a shortest string from $u_2$ to an accept state. The lengths of $u_1$ and $u_2$ are bounded by the number $c$ of states in $W$. Let $v'$ be a shortest string from $s_1$ to $s_2$ representing the same element of $G$ as $v$; if $\overline{v}$ is at distance at most $r$ from the identity, the length of $v'$ is at most $f$. Now $(u_1vu_2, u_1v'u_2)$ is accepted by $M_\varepsilon$, and therefore

$$|u_1| + |v| + |u_2| \leq k(|u_1| + |v'| + |u_2|),$$

from which it follows that $|v| \leq k(f + 2c)$.

The "if" part of the theorem is implied by the following result:

**Lemma 7.2.9 (standard asynchronous multiplier theorem).** *Assume that the conditions of Theorem 7.2.8 are satisfied. Let $y \in G$ lie at a distance $p$ from the identity in the Cayley graph of $G$, and choose integers*

$$\begin{aligned} R_0 &> Kp + K + 1 \\ R_1 &> \tfrac{1}{2}D(2R_0) + R_0, \\ R_2 &> \tfrac{1}{2}D(2R_1) + R_1, \\ R_3 &> \tfrac{1}{2}D(2R_2) + R_2. \end{aligned}$$

*Then the standard multiplier automaton $M_y$ based on $(W, R_1, R_2, R_3)$, where $W$ is a finite state automaton accepting $L$, is a true multiplier: it will accept a pair of strings $(w_L, w_R)$ if and only if $w_L, w_R \in L$ and $\overline{w_L}y = \overline{w_R}$. Furthermore, $M_y$ is boundedly asynchronous with asynchronous factor $D(R_3)$.*

*Proof of 7.2.9:* Suppose that $w_L$ and $w_R$ are accepted by $W$ and that $\overline{w_L}y = \overline{w_R}$ in $G$. As remarked after Definition 7.2.6, we will have proved the lemma (except for the last statement) if we show that the distance between $\overline{w_L(t_L(t))}$ and $\overline{w_R(t_R(t))}$ in the Cayley graph never exceeds $R_3$.

Let $I_L = [0, |w_L|]$ and $I_R = [0, |w_R|]$ be intervals in $\mathbf{R}$, and let $d : I_L \times I_R \to \mathbf{R}$ be the function that assigns to a pair $(s,t)$ the distance from $\overline{w_L(s)}$ to $\overline{w_R(t)}$ in the Cayley graph. Set $S_i = d^{-1}([0, R_i])$, for $i = 0, 1, 2, 3$.

We first observe that if there are two points with the same value of $d$, say $d_0$, on the same horizontal or vertical line, the value of $d$ at all points in between is at most $\frac{1}{2}D(2d_0) + d_0$. For otherwise the image of one path would be moving from within a $d_0$-neighbourhood of a fixed point to outside a $(\frac{1}{2}D(2d_0) + d_0)$-neighbourhood of the same point and back to the $d_0$-neighbourhood. It would need more than $D(2d_0)$ steps to do that, contradicting the definition of the departure function $D$. We will apply these considerations to the cases where $d_0$ is equal to $R_0$, $R_1$ and $R_2$, and this explains our choice of $R_1$, $R_2$ and $R_3$.

We claim that the corners $(0,0)$ and $(|w_L|, |w_R|)$ of the rectangle $I_L \times I_R$ belong to the same connected component of $S_1$. To show this, we choose a

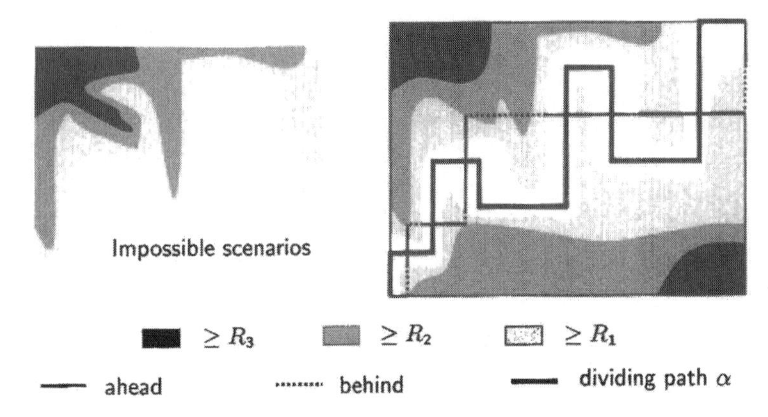


**Figure 7.3. Why word differences remain bounded.** This picture illustrates the proof of Lemma 7.2.9. The picture on the right shows the dividing path $\alpha$, drawn thick. The thinner line indicates the history of $(t_L, t_R)$ as $M_y$ reads its input: it is solid when $T = \mathtt{A}$, and dotted when $T = \mathtt{B}$. We are moving up when $h = \mathtt{R}$ and right when $h = \mathtt{L}$. The white area (or, rather, its closure) denotes $S_1$, the light-shaded area denotes $S_2 \setminus S_1$, and the medium-shaded area $S_3 \setminus S_2$. To avoid complicating the picture, $S_0$ is not shown: each horizontal segment of $\alpha$ is contained in some component of $S_0$. The situation on the left, where a segment crosses a region two shades darker than the segment's endpoints, cannot occur.

point $(s, t(s)) \in S_0$ for every $s \in I_L$; this is possible because the hausdorff distance in $\Gamma$ between the images of $\widehat{w_L}$ and $\widehat{w_R}$ is at most $Kp+K$ (if $p = 0$ so that $y$ is the identity, we need hausdorff distance $K$). Since $R_0 > Kp+K+1$, and we have $(s+r, t(s)) \in S_0$ for $\max(-s, -1) \leq r \leq 1$. The reader should be cautioned that the assignment $s \mapsto t(s)$ will in general be neither continuous nor monotonic.

We may take $t(0) = 0$ and $t(|w_L|) = |w_R|$. The entire segment $[s, s+1] \times \{t(s+1)\}$ lies in $S_0$ and so do the endpoints of the segment $\{s\} \times [t(s), t(s+1)]$. By the definition of $R_1$ each of these segments therefore lies in $S_1$. Thus we get a path $\alpha$ in $S_1$ joining the two corners, which we call the *dividing path*, denoted by a thick line in Figure 7.3.

We now look at what $M_y$ does as it scans $w_L$ and $w_R$—it is helpful to look at Figure 7.3. We start at the corner $(0, 0)$, moving right, with $T = \mathtt{A}$. We continue doing so until we first get out of $S_2$ or arrive at the right wall; then we turn up and set $T = \mathtt{B}$. Because of our choice of $R_2$, at the moment of the turn there cannot be any portions of $S_1$ to our right on the current row. This means that we must encounter the dividing path $\alpha$ again on our way up the current column. At that point $T$ is set to A, and our position becomes entirely analogous to the initial one, with L and R interchanged. During all this time we've remained inside $S_3$, because of our choice of $R_3$: we were first moving horizontally from a point with $d \leq R_1$ to a point with $d = R_2$, then vertically to another point with $d \leq R_1$.

By induction, we never get out of $S_3$. Since we are always moving right or up, we eventually get to the accept state at $(|w_L|, |w_R|)$.

We still have to prove the last statement of the lemma. If $M_y$ were to read more than $D(2R_3)$ symbols from one tape without reading anything from the other, it would change the word difference to a length greater than $R_3$, by the definition of $D$. But this contradicts the fact that the word difference is circumscribed to a ball of radius $R_3$. 7.2.9

7.2.8

## 7.3. Properties of Asynchronous Automatic Groups

In this section we prove the counterparts for asynchronous automatic structures of several results from the previous chapters. Many of the extensions are straightforward, but one of the most interesting—the possibility of imposing uniqueness—is more subtle. The proof of the synchronous version, Theorem 2.5.1, used the fact that a synchronous two-variable automaton can decide which of its two inputs comes first in the ShortLex order. This does not help us here, because predicate calculus is not available. It turns out, however, that dictionary order can be checked synchronously on a pair of strings accepted by a *bounded* asynchronous automaton.

**Lemma 7.3.1 (common prefix).** *Let $W$ be a boundedly asynchronous automaton. There is some integer $K$ such that, if $(w_L, w_R)$ is an accepted pair of strings whose prefixes $w_L(T)$ and $w_R(T)$ are identical for some $T \geq 0$, we have $|t_L - t_R| < K$ if $t_L \leq T$ or $t_R \leq T$. In other words, the two strings cannot get too far out of step, so long as $W$ is reading a prefix common to both.*

*Proof of 7.3.1:* Suppose that, at some point, $t_L$ characters have been read from $w_L$ and $t_R$ from $w_R$, and that $w_L(t_L)$ is a prefix of $w_R(t_R) = w$ (the same reasoning applies if $w_R(t_R)$ is a prefix of $w_L(t_L)$). Let $u$ be a shortest string such that $wu$ is accepted. Now $W$ accepts $(wu, wu)$ and, while doing so, it must behave exactly as if it were scanning $(w_L, w_R)$ for the first $t_L + t_R$ steps. After that, $W$ reads the final $|u|$ characters from the right-hand string, while it reads $t_R - t_L + |u|$ characters from the left-hand string. If $W$ has asynchronous factor $k$, this implies that $t_R - t_L + |u| \leq k|u|$. But $|u|$ is bounded by the number of states of $W$, so $t_R - t_L$ is also bounded by some constant $K$. 7.3.1

**Theorem 7.3.2 (asynchronous uniqueness).** *Let $G$ have an asynchronous automatic structure with respect to some set $A$ of generators, and let $L$ be the language of accepted strings. Then $G$ has an asynchronous automatic structure with respect to $A$, and such that the language $L'$ of accepted strings is a subset of $L$ and contains a unique representative for each element of $G$.*

*Proof of 7.3.2:* By Theorem 7.2.4 (boundedly asynchronous), we may assume that the automatic structure is boundedly asynchronous.

We let $L'$ be the language that contains, for each element $g \in G$, the representative of $g$ in $L$ that comes earliest in dictionary order. This is well-defined because there are only finitely many representatives of $g$ in $L$: if $w$ is a shortest representative, any other representative must have length at most $k|w|$, where $k$ is the asynchronous factor for the equality recognizer. (This would not be the case if the structure were unboundedly asynchronous, or even synchronous: take the example of a finite group $G$ with $A = G$ and $L = A^*$, the identity being considered the smallest generator.)

We just have to show that $L'$ is a regular language; the equality recognizer and multiplier automata can then be adapted easily to reject strings not in $L'$. We will build a non-deterministic automaton that accepts the language $L \setminus L'$. The idea is to start with the equality recognizer $M_\varepsilon$, and make it capable of "inventing" one of its strings, say the left-hand one, by replacing arrows from states in $S_L \cup S_L^\$$ by $\varepsilon$-arrows. Naturally, if we just do this, the language accepted by the resulting automaton is simply $L$; instead, we must make the automaton $Y$ keep track of the difference between the phantom string and the real string until it can decide which is less in dictionary order. By Lemma 7.3.1 (common prefix), only a bounded amount of memory is necessary for that. After the strings diverge, if the phantom string is less, $Y$ forges ahead, trying to reach an accept state as it reads the remainder of the real input string; if it can do that, the input string is not the least representative of its class, and it should be accepted. If the phantom string is greater than the real string in dictionary order, nothing is learned from the exercise, and $Y$ can stop right at the point of divergence.

To formalize this, we first describe $Y$ deterministically, as an asynchronous automaton. Let $S$ be the set of states of $M_\varepsilon$, and $A_K$ the set of strings over $A \cup \{\$\}$ of length at most $K$, where $K$ is the bound given by Lemma 7.3.1. Let $P$ be the subset of $S \times A_K \times A_K$ consisting of triples $(s, \lambda, \rho)$ where $\lambda$ and $\rho$ have no common prefix. Each state of $Y$ is a collection of states of elements of $P$, as follows: For each $s \in S$, all triples $(s, \lambda, \rho) \in P$

with $\varepsilon < \lambda < \rho$ in dictionary order get lumped into a state denoted $[s]$. All triples $(s, \lambda, \rho) \in P$, with $s$ arbitrary and $\lambda > \rho > \varepsilon$ in dictionary order get lumped into a failure state, together with $(s^\$, \varepsilon, \varepsilon)$, where $s^\$$ is the accept state of $M_\varepsilon$. In these comparisons, the symbol \$ precedes all others. The other elements of $P$ are not lumped together.

The initial state of $Y$ is $(s_0, \varepsilon, \varepsilon)$, where $s_0$ is the initial state of $M_\varepsilon$; the accept state of $Y$ is the lumped state $[s^\$]$. For each arrow $x : s \to t$ of $M_\varepsilon$ there is an arrow $x : [s] \to [t]$ in $Y$, and these are all the arrows originating at lumped states. It remains to define the arrows from states of the form $(s, \lambda, \rho)$. An input $x \in A \cup \{\$\}$ takes $Y$ from $(s, \lambda, \rho)$ to $(sx, \lambda', \rho')$, where $\lambda'$ and $\rho'$ are obtained by removing common prefixes from $\lambda x$ and $\rho$, if $s \in S_{\mathrm{L}} \cup S_{\mathrm{L}}^\$$, or from $\lambda$ and $\rho x$, if $s \in S_{\mathrm{R}} \cup S_{\mathrm{R}}^\$$. Of course, $(sx, \lambda', \rho')$ should be interpreted according to the identifications above. If $sx$ is undefined or if $\lambda'$ or $\rho'$ has length greater than $K$, the transition results in failure.

As a consequence of these rules and of Lemma 7.3.1, $Y$ can reach an unlumped state $(s, \lambda, \rho)$ if and only if the following conditions are satisfied: at least one of $\lambda$ and $\rho$ is null; $Y$ has read the partial input $(w\lambda, w\rho)$ for some string $w$; after reading the same input, $M_\varepsilon$ would be in state $s$; and $s \neq s^\$$. The reason $s \neq s^\$$ is that $s^\$$ is only reached when a \$ has been read in both left and right strings. This would force the state to be lumped. Likewise, $Y$ can reach a state $[s]$ if and only if it has read the partial input $(w_{\mathrm{L}}, w_{\mathrm{R}})$, where $w_{\mathrm{L}} < w_{\mathrm{R}}$ in dictionary order and $w_L$ is not a prefix of $w_R$, and $M_\varepsilon$ would be in state $s$ after reading the same input.

To conclude our construction, we eliminate all dead and inaccessible states of $Y$, and all arrows to and from them. We then replace by $\varepsilon$-arrows all \$-arrows and all arrows whose source is either some $[s]$ or some $(s, \lambda, \rho)$, where $s \in S_{\mathrm{L}} \cup S_{\mathrm{L}}^\$$. This corresponds to making $Y$ guess all of its left string, as well as the end-of-string symbol for the right string (which is necessary because the input string is not terminated by a \$). It follows from the preceding paragraph that the non-deterministic automaton so defined accepts exactly the language $L \setminus L'$, as desired. 7.3.2

**Theorem 7.3.3 (asynchronous change of generators).** *Let $G$ have an asynchronous automatic structure with respect to one set of semigroup generators. Then it has an asynchronous structure with respect to any other set of generators.*

*Proof of 7.3.3:* Suppose we have an asynchronous automatic structure for the set of generators $A$, with language of accepted strings $L$. We want to find a structure for the alternative set of generators $B$. By Theorem 7.2.4, we may assume that the given structure is boundedly asynchronous, and thus that $L$ satisfies the conditions of Theorem 7.2.8 (characterizing asynchronous).

For each $x \in A$, choose a shortest string $w_x$ over $B$ representing the same element of $G$ as $x$, and, likewise, choose a shortest representative $v_y$ over $A$ for $y \in B$. Let $l$ be an upper bound for the lengths of the strings $w_x$ and $v_y$. Let $L'$ be the language over $B$ obtained by substituting $w_x$ for $x$ in each string of $L$; we must show that $L'$ also satisfies the conditions of Theorem 7.2.8. We will need the observation that the metrics $d_A$ and $d_B$ on the Cayley graphs of $G$ with respect to $A$ and $B$ are related by

$$l^{-1}d_B(g,g') \leq d_A(g,g') \leq l d_B(g,g').$$

Let $K$ be a lipschitz constant for $L$, and consider two strings over $B$ representing elements of $G$ that differ by right multiplication by $y \in B$. Then these strings come from strings in $A$ that differ by $v_y$, and therefore the hausdorff distance between the corresponding graphs in the Cayley graph with respect to $A$ is at most $Kl$. In the Cayley graph with respect to $B$ the hausdorff distance is at most $Kl^2$, so $Kl^2$ is a lipschitz constant for $L'$.

Similarly, if $D$ is a departure function for $L$, we show that there is a departure function for $L'$. Let $w$ be a substring of a string of $L'$ such that $|w| > 2l + D(rl + 2l^2)l$. By removing a substring of some $w_x$ from each end, we get a string that comes from a substring of a string of $L$ of length at least $D(rl + 2l^2)$. Such a string represents an element $g \in G$ such that $d_A(g,e) \geq rl + 2l^2$, where $e \in G$ is the identity. Then $d_B(g,e) \geq r + 2l$, and $d_B(\overline{w},e) \geq r$, proving the claim. 7.3.3

A number of results about automatic groups proved in the previous chapters also hold for asynchronous automatic groups, often with very similar proofs (mostly replacing "uniform distance" by "hausdorff distance," "lipschitz constant" by "asynchronous lipschitz constant," and so on). We limit ourselves to two of the most important:

**Theorem 7.3.4 (asynchronous isoperimetric inequality).** *Every asynchronous automatic group is finitely presented and satisfies an exponential isoperimetric inequality.*

*Proof of 7.3.4:* We can essentially repeat the proof of Theorem 2.3.12 (quadratic isoperimetric inequality); the only difference is that, instead of invoking Lemma 2.3.9 (bounded length difference), we use Theorem 7.2.4 (boundedly asynchronous) to choose a bounded asynchronous automatic structure. Then, when we choose accepted representatives $u_t$ for the prefixes of the word $w$ under consideration (that is, when we divide a loop into wedges, as in Figure 2.16), the lengths of $u_t$ and $u_{t+1}$ are within a bounded factor of each other. Thus the maximum length of the $u_t$ grows at most exponentially with the length of $w$. 7.3.4

The following facts are proved almost exactly like their synchronous counterparts, Theorems 4.1.1 (direct product) and 4.1.4 (finite index). The changes consist mostly in replacing "uniform distance" by "hausdorff distance," "lipschitz constant" by "asynchronous lipschitz constant," and so on.

**Theorem 7.3.5.** (1) *The direct product of two asynchronous automatic groups is asynchronous automatic.*

(2) *Let $G$ be a group and let $H$ be a subgroup of finite index. $G$ is asynchronous automatic if and only if $H$ is.* 7.3.5

Combable groups (Section 3.6), which, as we have seen, are more general than automatic groups, can also be generalized along the lines of this chapter.

Let $X$ be a metric space with a basepoint. In the notation of Section 3.6, consider the endpoint map $P^*_{k,\varepsilon} \to X$, for some $k \geq 1$ and some $\varepsilon \geq 0$. Suppose that $\alpha : X \to P^*_{k,\varepsilon}$ is a (not necessarily continuous) right inverse to this endpoint map. Thus $\alpha(x)$ is a broken path for each $x$, and we denote its domain by $[0, b_x]$. We extend $\alpha_x$ to $[0, \infty)$, by setting it equal to a constant on $[b_x, \infty)$.

We say that $\alpha$ is an *asynchronous combing* of $X$ if, for any two points $x, y \in X$, the broken path $\alpha_x$ can be reparametrized with a bounded amount of distortion in the domain so as to lie within a bounded distance of $\alpha_y$, with both bounds depending solely on the distance $d(x, y)$. More precisely, for each $n > 0$, there should exist $N > 0$ and $K \geq 1$ such that, for any two points $x, y \in X$ with $d(x, y) < N$, we can find a monotonic homeomorphism $\rho_{x,y} : [0, \infty) \to [0, \infty)$ such that $d_P(\alpha_y, \alpha_x \rho_{x,y}) \leq K$ (see Definition 3.6.1 for the definition of $d_P$), and

$$(t_1 - t_2)/K \leq \rho_{x,y}(t_1) - \rho_{x,y}(t_2) \leq (t_1 - t_2)K$$

for all $0 \leq t_2 \leq t_1$. If $X$ is a path-metric space, we need only find $N = N_0$ and $K = K_0$ for $n = 1$. Knowing these values, we can find values for an arbitrary value of $n$, by rounding $n$ up to an integer and setting $N = nN_0$ and $K = K_0^n$. An *asynchronous combable group*, of course, is one whose Cayley graph has an asynchronous combing.

**Theorem 7.3.6.** *Any asynchronous automatic group is asynchronous combable.*

*Proof of 7.3.6:* Using theorems 7.2.4 and 7.3.2, we choose a boundedly asynchronous structure with uniqueness. The paths given by this asynchronous structure, when parametrized by pathlength, give an asynchronous combing: the bilipschitz reparametrizations $\rho_{x,y}$ which appear in the definition are constructed using the boundedness of the automatic structure. Details are left to the reader. 7.3.6

Combining the arguments used to prove Theorems 3.6.6 (combable isoperimetric inequality) and 7.3.4 (asynchronous isoperimetric inequality), we get:

**Theorem 7.3.7 (asynchronous combable isoperimetric inequality).** *Every asynchronous combable group satisfies an exponential isoperimetric inequality. Every combable group has a solvable word problem.* 7.3.7

This result will be further generalized by Theorem 10.2.1 (mass times diameter estimate).

## 7.4. Asynchronous but not synchronous

In this section we prove that the so-called Baumslag–Solitar groups are asynchronously, but not synchronously, automatic. This result is due to Thurston. The proof given here uses Theorem 2.3.12 (quadratic isoperimetric inequality), but Thurston had an earlier direct proof. Gersten [Ger91] independently proved that these groups have an exponential isoperimetric inequality, a fact which also follows from Theorem 7.3.4.

**Example 7.4.1 (asynchronous but not synchronous).** For integers $p, q > 0$, let $G_{p,q}$ be the Baumslag–Solitar group

$$G_{p,q} = \langle \{x, y\} / \{yx^p y^{-1} x^{-q}\} \rangle.$$

We will show that $G_{p,q}$ is asynchronously automatic, but not automatic, for $p \neq q$. For $p = q$, the group is automatic.

We can assume without loss of generality that $p \leq q$. We start by drawing the unique relator $yx^p y^{-1} x^{-q}$ as a rectangular loop with vertical sides labelled $y$, the lower horizontal side divided into $q$ equal lengths, each labelled $x$, and the upper horizontal side divided into $p$ equal lengths, each labelled $x$. Figure 7.4 shows the situation for $p = 1$ and $q = 2$. We will glue together copies of this rectangular loop to build the Cayley graph $\Gamma = \Gamma(G_{p,q}, \{x, y\})$, and we will fill in the rectangles to get a two-dimensional complex $\tilde{K}$ with the property that any trivial word gives a null-homotopic loop in $\tilde{K}$. We will call $\tilde{K}$ the *filled Cayley graph* of the group.

We start by gluing together copies of the model rectangle along the vertical sides, obtaining an infinite horizontal strip. This takes care of incoming and outgoing $x$ edges for each vertex. To obtain the entire Cayley graph, we must also have one incoming and one outgoing $y$ edge for each vertex. Thus along the top edge of our horizontal strip, we have to glue a fan of $q$ new strips, joined along their bottom edge and staggered (Figure 7.5). Similarly, there must be $p$ strips fanning down from this common edge. If $p = q = 1$, we obtain the familiar tiling of the euclidean plane by rectangles.

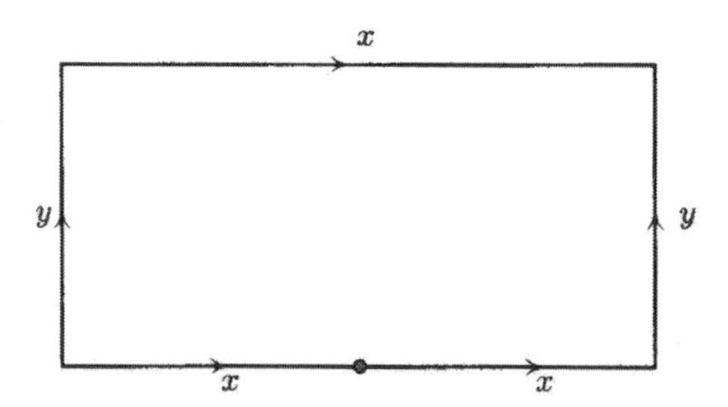


**Figure 7.4. The defining relator for $G_{1,2}$**

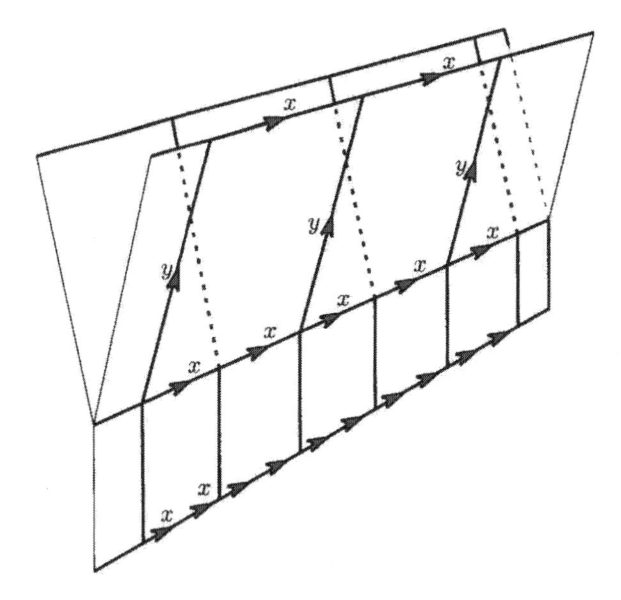


**Figure 7.5. A portion of the Cayley graph of $G_{1,2}$.** Every vertex is the endpoint of four edges, one for each generator and inverse generator. Each rectangle is a copy of the one in Figure 7.4. The bottom horizontal strip contains five full such rectangles, and two partial ones. Each of the two upper strips contains two full rectangles and two partial ones. Each horizontal edge is labelled with an $x$ pointing from left to right and each (roughly) vertical edge is labelled with a $y$ pointing upwards. The two upper strips are displaced relative to each other in the $x$ direction.

If we look at the Cayley graph from the side, we see an infinite tree $T$ (Figure 7.6), where each vertex has $q$ edges pointing up and $p$ pointing down. The filled Cayley graph $\tilde{K}$ has the topology of $T \times \mathbf{R}$, and is therefore contractible.

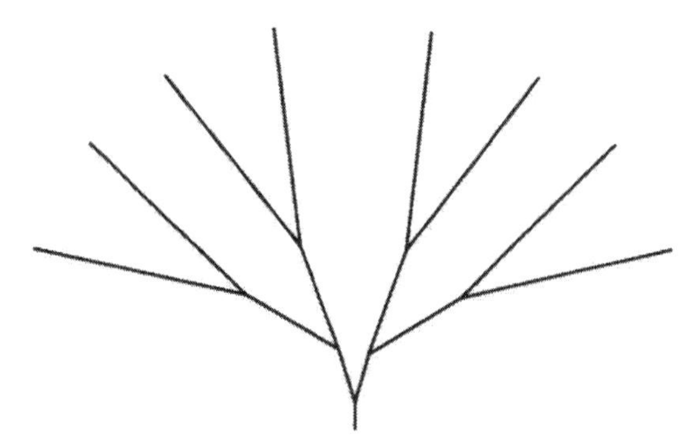

**Figure 7.6. The side view of the Cayley graph.** Seen from the side, the Cayley graph of $G_{1,2}$ is a binary tree, extending infinitely both up and down. In general, we have a more complicated acyclic graph, where each vertex is the endpoint of $(p+q)$ edges.

We now show that $G_{p,q}$ is asynchronously automatic, and also automatic when $p = q$. We do this by finding a normal form for $G_{p,q}$. We use the following complete set of Knuth–Bendix rules (see page 117):

$$E = \{xX \Rightarrow \varepsilon,\ Xx \Rightarrow \varepsilon,\ yY \Rightarrow \varepsilon,\ Yy \Rightarrow \varepsilon,\\ x^q y \Rightarrow yx^p,\ Xy \Rightarrow x^{q-1}yX^p,\ x^pY \Rightarrow Yx^q,\ XY \Rightarrow x^{p-1}YX^q\}.$$

One can easily see that a string is in normal form (that is, fully reduced) according to these rules if and only if it matches the regular expression

$$(((\varepsilon \vee x \vee \cdots \vee x^{q-1})y) \vee ((\varepsilon \vee x \vee \cdots \vee x^{p-1})Y))^*(x^* \vee X^*),$$

and it contains no adjacent inverse generators. Note that choosing one of $y$, $xy$, ..., $x^{q-1}y$ corresponds to choosing a particular upward horizontal strip, relative to one's location in the Cayley graph. We define the *suffix* of a reduced string as the substring that matches $(x^* \vee X^*)$ in the expression above, and the *prefix* as the rest.

The rules for the normal form clearly give a regular language. To prove that this language gives an asynchronous automatic structure, we must check that the hypotheses of Theorem 7.2.8 (characterizing asynchronous) are satisfied. A departure function exists because the proposed structure has unique representatives. To show the existence of a lipschitz constant, we take a word in normal form, multiply it on the right by a generator, and compute the new normal form. Projecting the corresponding paths into the tree $T$, we see that, if the generator is $x$ or $X$, the prefix does not change at all and the suffix changes only at its last entry. Now suppose we multiply the word by $y$. If the suffix is $x^{i+nq}$, for $i, n \geq 0$, we have the reduction $x^{i+nq}y \to x^i y x^{np}$, so the prefix gets longer by one group and the suffix gets shortened by (roughly) the factor $p/q$. (If $i = 0$ and the prefix ends with $Y$, we have a cancellation, so the prefix actually gets shortened by one group; but this doesn't change the basic idea.) Geometrically, the old and new suffixes run parallel to and immediately opposite each other, on the two sides of a horizontal strip. The case of multiplication by $Y$, or of a suffix of the form $X^i$, is analyzed similarly. In all cases the result is that the hausdorff distance between the two paths is small.

If $p = q$, the uniform distance between the paths is bounded as well, since the old and new suffixes are approximately the same length. This shows that the language of normal forms of elements of $G_{p,p}$ is part of a synchronous automatic structure.

Finally, we want to show that $G_{p,q}$ does not have a synchronous automatic structure if $p \neq q$. The criterion that we use is the existence of a quadratic isoperimetric inequality (Theorem 2.3.12). We prove that no such inequality exists by constructing a sequence of loops whose length increases linearly but whose combinatorial area increases exponentially.

To construct such loops, we return to the description of the filled Cayley graph $\tilde{K}$. Recall that $\tilde{K}$ projects onto a tree $T$ that has $q$ edges pointing up and $p$ edges pointing down at each node. A *sheet* in $\tilde{K}$ is the inverse image (under the projection) of a path in $T$ that is infinite in each direction and travels steadily upwards along the tree. A sheet is homeomorphic to $\mathbf{R}^2$, and can be mapped onto the complex upper half-plane $U = \{z \in \mathbf{C} \mid \operatorname{Im} z > 0\}$ in such a way that the homothety $z \mapsto qz/p$ takes each horizontal strip to the one immediately above it (see Figure 7.7).

(Readers familiar with hyperbolic geometry will recognize that Figure 7.7 really shows the hyperbolic plane, in the upper half-plane model. $G_{p,q}$ acts on the hyperbolic plane by isometries, the action of the generators being (in this model) $\sigma_x : z \mapsto z + 1$ and $\sigma_y : z \mapsto qz/p$, for example. Figure 7.7 then shows the image of a sheet under the equivariant map $\Gamma \to U$ defined by this action and by the additional condition that edges are mapped to euclidean straight-line segments.)

An *upper sheet* is the portion of a sheet above some horizontal line (or, in terms of the tree, the inverse image of a certain semi-infinite path). To prove that $G_{p,q}$ cannot satisfy a polynomial isoperimetric inequality, the idea is to

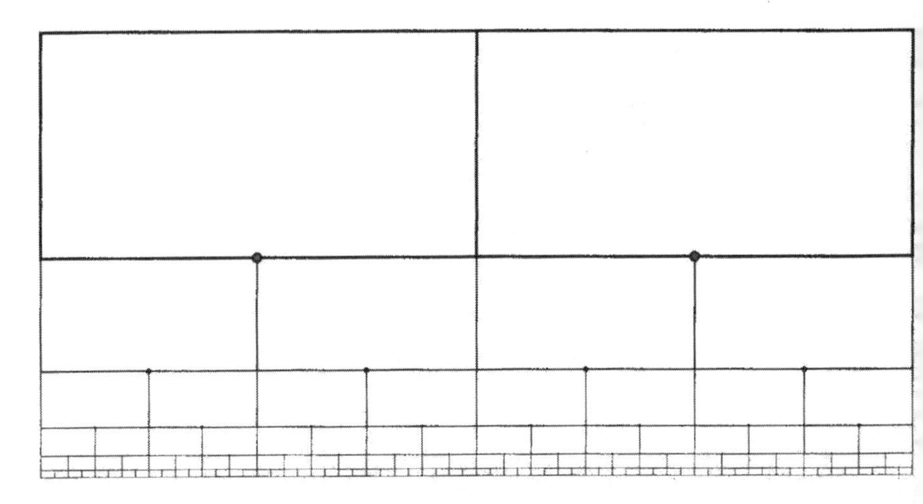

**Figure 7.7. The upper half-plane.** Choosing an infinite sequence of adjacent horizontal strips at different heights in the filled Cayley graph gives the upper half of the complex plane. The rectangles converge to the bounding real axis, becoming smaller and smaller as they do so. If we give the upper half-plane a hyperbolic metric, the rectangles become congruent.

take two upper sheets $S_1$ and $S_2$ whose intersection is a horizontal line $L$ at the common lower boundary, and to open them flat like a book (Figure 7.8). Horizontally, the two half-sheets are displaced relative to one another by some fraction of the width of the smallest rectangles (Figure 7.9). If we choose two vertices $P$ and $Q$ on $L$ that are far enough apart, they can be joined efficiently within either sheet by going far up and then down. But the area enclosed between these two paths is very large in terms of the number of rectangles (this corresponds to the fact that the hyperbolic area of a circle grows exponentially with the radius).

To make this idea precise, let $u_n$ be the shortest word matching the regular expression

$$y^n x^p Y((\varepsilon \vee x \vee \cdots \vee x^{p-1})Y)^*$$

and such that $\overline{u_n} = \overline{x^{Np}}$ for some integer $N$. That is, the path $\widehat{u_n}$ goes up straight for $n$ steps, defining a sheet as it goes along; it then moves horizontally along the top of a single rectangle, and finally cascades down along the same sheet in as straight a fashion as possible, but never going backwards along the $x$ direction. In particular, if $p$ divides $q$, we have simply $u_n = y^n x^p Y^n$, as in Figures 7.8 and 7.9. One easily checks that $N \geq (q/p)^n$ and $|u_n| \leq n(p+2)$.

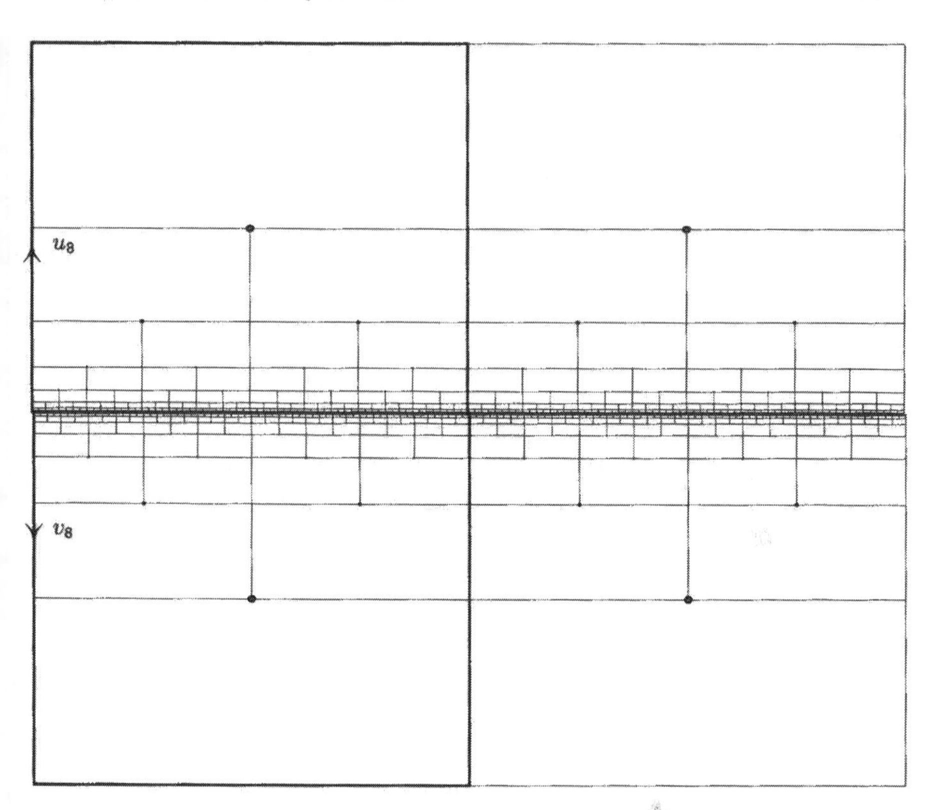


**Figure 7.8. A large area with a small perimeter.** We show the loop $w_n = u_n v_n^{-1}$, with $n = 8$. The length of the curve is 36 and the number of rectangles bounded is 1022. In general, for $p = 1$ and $q = 2$, $|w_n| = 4n + 4$ and the number of rectangles bounded is $2(2^{n+1} - 1)$. See also Figure 7.9.

We set $v_n = x u_n x^{-1}$ and $w_n = u_n v_n^{-1}$, so that $\widehat{u_n}$ and $\widehat{v_n}$ lie on different upper sheets $S_1$ and $S_2$ whose intersection is the horizontal line through the basepoint, and $\widehat{w_n}$ is a loop contained in $S_1 \cup S_2$. We will show that every disk spanning $w_n$ must contain at least one copy of each rectangle inside the loop $\widehat{w_n}$ in the plane $S_1 \cup S_2$. Since the number of such rectangles is certainly greater than $N$, it grows exponentially with $n$, while $|w_n|$ grows linearly with $n$, as already remarked. This will prove the impossibility of a polynomial isoperimetric inequality.

Let $V$ be the projection of $S_1 \cup S_2$ in the tree $T$. Then $V$ is a chain of edges, joined end to end. Let $p : T \to V$ be the map taking every point of $T$ to the point of $V$ nearest to it (in the tree metric of $T$). This map

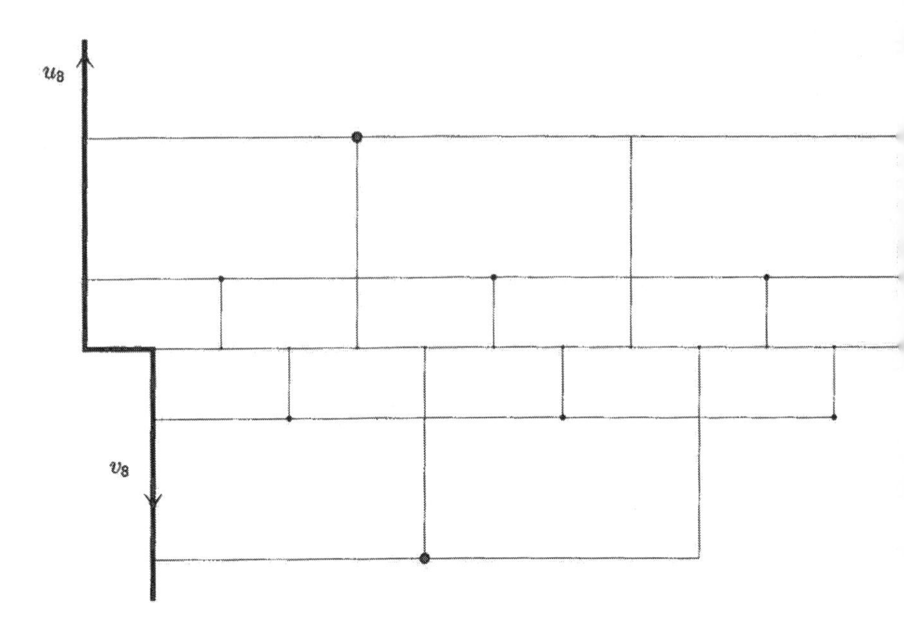


**Figure 7.9. Blow-up of part of the large area with small perimeter.** A blow-up of Figure 7.8 in the region where $u_n$ and $v_n$ start. We see here that the two paths $u_n$ and $v_n$ are offset with respect to one another by one $x$ unit.

is continuous and equals the identity on $V$ (that is, $p$ is a retraction). We extend $p$ to a retraction from the filled Cayley graph $\tilde{K}$ to $S_1 \cup S_2$, which we also call $p$. The image of any rectangle in $\tilde{K}$ is itself, if the rectangle lies in $S_1 \cup S_2$; otherwise it is a straight-line segment (not necessarily an edge in the Cayley graph, however). If $D$ is any disk spanning $w_n$, the image $p(w_n)$ must contain the disk bounded by $w_n$ in $S_1 \cup S_2$, by the Jordan curve theorem, and therefore all the rectangles in this disk must also be present in $D$, as we wished to show.

# Chapter 8

# Nilpotent Groups

So far we have concentrated on proving that various groups are automatic. Here we prove an important result, showing that nilpotent groups are not in general automatic. This shows how strong a restriction it is to insist that a group be automatic. In particular, a nilpotent group has a decidable word problem, so the theory of automatic groups is very far from being able to tackle all groups with solvable word problems.

It would be interesting to generalize the concept of an automatic structure, so that it would apply to nilpotent groups. An important feature of any successful definition would be the ability to generalize the results of Section 5.1 (Axiom Checking) in some appropriate way.

In Section 8.1, by way of motivation, we give a simple proof that the Heisenberg group is not automatic. In Section 8.2 we show that an automatic nilpotent group is virtually abelian. This was proved by Holt for the synchronous case; Epstein massaged Holt's proof to make it work also for asynchronous groups.

Section 8.3 gives results of Gersten and Short [GS00a] concerning subgroups of automatic, and especially biautomatic, groups. In particular, we prove that every finitely generated nilpotent subgroup of a biautomatic group is virtually abelian. We do not know if a nilpotent subgroup of a merely automatic group has to be virtually abelian; conceivably, this might not be the case, in which case we would have an automatic group that is not biautomatic (Open Question 2.5.6).

## 8.1. The Heisenberg Group

We denote the centre of a group $G$ by $Z(G)$, and the commutator $x^{-1}y^{-1}xy$ of $x$ and $y$ by $[x, y]$. We recall that the *upper central series*

$$Z_0 \subset Z_1 \subset \cdots \subset Z_i \subset \cdots$$

of $G$ is defined inductively by $Z_0 = \{e\}$ (where $e$ is the identity) and $Z_i/Z_{i-1} = Z(G/Z_{i-1})$. We say that $G$ is *nilpotent* (of class $c$) if $Z_c = G$. Thus a group is abelian if and only if it is nilpotent of class 1.

**Example 8.1.1 (the Heisenberg group).** The next more complicated example of an infinite nilpotent group is probably the *Heisenberg group* $H$, which has presentation

**8.1.2.** $$\langle\{\alpha, \beta, \gamma\}/\{[\alpha, \beta]\gamma^{-1}, [\alpha, \gamma], [\beta, \gamma]\}\rangle.$$

We will show that $H$ does not have a quadratic isoperimetric inequality, and therefore, by Theorem 2.3.12, is not automatic. In the next section we will see that nilpotent groups are not even asynchronously automatic (Theorem 8.2.8). However, the technique of this section is particularly simple, and can be applied to certain non-nilpotent groups as well (see Theorem 8.1.3). These results are due to Thurston; Gersten proved independently the properties of the isoperimetric functions of the groups discussed in this section [Ger91].

To understand $H$ better, we need a geometric description for it. We will show that the group $G$ generated by the affine transformations

$$\alpha : (x, y, z) \mapsto (x + 1, y, z + y),$$
$$\beta : (x, y, z) \mapsto (x, y + 1, z),$$
$$\gamma : (x, y, z) \mapsto (x, y, z + 1)$$

on $\mathbf{R}^3$ is, in fact, the same as $H$. One can see directly that $G$ acts properly discontinuously on $\mathbf{R}^3$; we will study the topology of the quotient of the action, and show that its fundamental group has the presentation 8.1.2. Since $\mathbf{R}^3$ is simply connected, this will imply that $H$ and $G$ are identical.

The subgroup generated by $\beta$ and $\gamma$ is free abelian of rank two and normal in $G$. It acts on $\mathbf{R}^3$ with quotient $\mathbf{R} \times T^2$, where $T^2$ is the torus. The action of $\alpha$ on this quotient maps the torus at $x = 0$ to the one at $x = 1$ by a shearing transformation $(y, z) \mapsto (y, z + y)$. Thus the quotient $\mathbf{R}^3/G$ can be seen as the product $[0, 1] \times T^2$ with the two toruses at the ends identified by means of this shear, as shown in Figure 8.1. In other words, $\mathbf{R}^3/G$ is a fibre bundle, with base space the circle and fibre the torus. (This roughly means that there is a map from $\mathbf{R}^3/G$ to the circle such that the inverse image of each point is a torus, and the map looks locally like the projection onto one of the factors of a product. The local projection maps come from the projection $\mathbf{R} \times T^2 \to \mathbf{R}$.)

In this situation, the fundamental group of the bundle is the fundamental group of the fibre (namely, the free abelian group on $\beta$ and $\gamma$), extended by

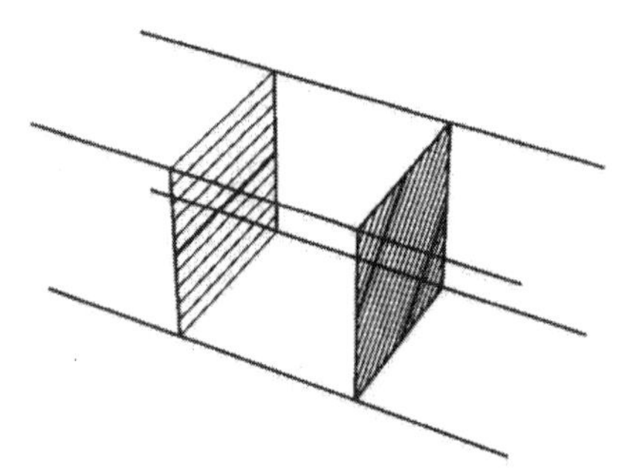

**Figure 8.1. The quotient of $\mathbf{R}^3$ by the Heisenberg group.** The quotient of $\mathbf{R}^3$ by $H$ is also the quotient of $\mathbf{R} \times T^2$ by the cyclic group generated by the map $(x, y, z) \mapsto (x+1, y, z+y)$. The figure shows $\mathbf{R} \times T^2$ (opposite walls of the cylinder should be thought of as being identified); the shaded squares represent the toruses at $x = 0$ and $x = 1$. We get $\mathbf{R}^3/H$ by taking the region between the two toruses and identifying the toruses so the shadings match.

a generator corresponding to a loop around the base (namely $\alpha$). Relators must be added for the effect that going around the base has on the fibre (this is called the monodromy of the bundle): they are $\alpha\beta\alpha^{-1} = \beta\gamma$ and $\alpha\gamma\alpha^{-1} = \gamma$. The resulting presentation is clearly equivalent to 8.1.2.

(We remark that $H$ is also the group of Example 2.1.6. To see this, notice that the map

$$(x, y, z) \mapsto \begin{pmatrix} 1 & 0 & 0 \\ x & 1 & 0 \\ z & y & 1 \end{pmatrix}$$

gives a one-to-one correspondence between the set of images of the origin under the group $H$ in the description above and the set of lower triangular matrices with ones in the diagonal; and that, under this correspondence, multiplication on the left by $(1, 0, 0)$, $(0, 1, 0)$ and $(0, 0, 1)$ is the same as the action of $\alpha$, $\beta$ and $\gamma$.)

We now embed the Cayley graph of $H$ in $\mathbf{R}^3$ in an equivariant way under the transformations $\alpha$, $\beta$ and $\gamma$. There is a unique way to do that, if we fix the position of the basepoint (at the origin, say) and insist that the edges be line segments. The $\alpha$ edges will all be parallel to the $x$-axis, and the $\gamma$ edges parallel to the $z$-axis, while the $\beta$ edges will lie on planes parallel to the $yz$-plane, and have slope $x$. Figure 8.2 shows three loops in the graph, corresponding to the relators from 8.1.2. Caution: A label $\alpha$ on an edge does not mean that the transformation $\alpha \in H$ takes one endpoint of the edge to

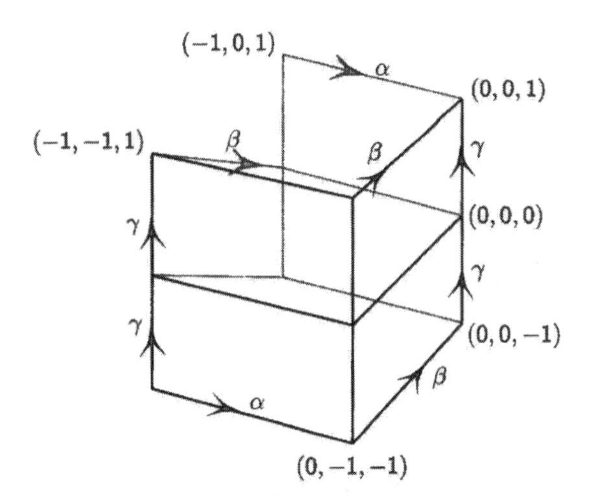


**Figure 8.2. Relators in the Cayley graph of $H$.** Spanning surfaces for the three basic relators in $H$ can be chosen as parallelograms (for the commuting generators) or the union of a parallelogram with a triangle (for $\alpha\beta\alpha^{-1}\beta^{-1}\gamma^{-1}$). Each contributes a fixed amount to the area in the $yz$-plane.

the other; rather it means that this edge is the image of other $\alpha$ edges under transformations in $H$. In other words, the action of the group on the graph is on the left, whereas a label on an edge refers to multiplication on the right.

Notice that the path corresponding to the word $u_n = \alpha^n\beta^n\alpha^{-n}\beta^{-n}$ fails to close by an amount proportional to $n^2$: if the path starts at the origin, it ends at $(0, 0, n^2)$. A similar statement is true about $u_{-n}$. We will see that $w_n = u_n u_{-n}^{-1}$ is a loop that encloses an area proportional to $n^3$. This will be the basis of our proof that $H$ does not have a quadratic isoperimetric inequality.

We form three surfaces $S_1$, $S_2$ and $S_3$ spanning the loops based at the origin given by $[\alpha, \beta]\gamma^{-1}$, $[\alpha, \gamma]$ and $[\beta, \gamma]$. While $S_2$ and $S_3$ are squares, $S_1$ is the union of a sloping rectangle and a triangle. We take all the images of $S_1$, $S_2$ and $S_3$ under $H$, and glue them to the Cayley graph forming the filled Cayley graph. Any loop in the Cayley graph can be spanned by a union of these images. We now consider the projection of such a loop and of the spanning surface in the $yz$-plane. Images of $S_2$ are squares parallel to the $xz$-plane, so their projection has no area. Images of $S_3$ are parallelograms, and their projection has unit area. Images of $S_1$ are composed of a sloping rectangle, whose projection has no area, and of a triangle, whose projection has area $\frac{1}{2}$. The loop $w_n$ of the previous paragraph is formed by line segments visiting the points

$$(0,0,0),\ (n,0,0),\ (n,n,n^2),\ (0,n,n^2),\ (0,-n,n^2),\ (-n,-n,n^2),\ (-n,0,0),$$

and back to $(0,0,0)$, so its projection onto the $yz$-plane encloses an area $n^3$. It follows that at least $n^3$ images of the elementary surfaces $S_1$, $S_2$ and $S_3$ are needed to span this loop. Since the length of $w_n$ grows linearly with $n$, we have proved that the isoperimetric inequality in $H$ must be at least cubic.

A more sophisticated way to rephrase the argument above is to say that the maps $\alpha$, $\beta$ and $\gamma$ preserve the form $dy \wedge dz$, which therefore defines a closed two-form $\omega$ on $\mathbf{R}^3/H$. We can integrate $\omega$ on each $S_i$ directly, and the results are $\omega(S_1) = \pm\frac{1}{2}$, $\omega(S_2) = \pm 1$ and $\omega(S_3) = 0$. Now, by Stokes' Theorem, the integral of $\omega$ on any surface $S$ spanning the loop $w_n$ equals the integral of the one-form $y\,dz$ along $w_n$, because $\omega = d(y\,dz)$. But $\int_{w_n} y\,dz = n^3$, by direct calculation.

We have shown that the isoperimetric function of $H$ is at least cubic. We now show that it is, in fact, cubic. (The argument here is a version of our original argument, now simplified by Benson Farb.) Given a word $w$ of length $n$ over $\{\alpha, \beta, \gamma\}$, we will rewrite it in the form $\alpha^i\beta^j\gamma^k r(n^3)$, where $r(m)$ is the product of no more than $m$ conjugates of the defining relators and their inverses. (By "rewriting," we mean finding an equivalent word in the free group generated by $\alpha$, $\beta$ and $\gamma$, so the only cancellations allowed are those of generators adjacent to their inverses.) In particular, if $w$ represents the identity in $H$, we have $i = j = k = 0$, and $w$ can be expressed as the product of at most $n^3$ conjugates of relators and their inverses.

Notice first that a word of the form $urv$, where $r$ is the conjugate of a relator, can be rewritten as $uv(v^{-1}rv)$, so we can move conjugates of relators freely without increasing their count.

Now starting with $w$, we move occurrences of $\gamma$ and $\gamma^{-1}$ all the way to the right, creating one relator for each move. (That is, we replace $\gamma\alpha$ by $\alpha\gamma[\gamma, \alpha]$, and so on.) The relator created is immediately moved to the right. After fewer than $n^2$ steps, we obtain a word $u\gamma^k r(n^2)$, where $u$ is free of $\gamma$ and $\gamma^{-1}$ and has length at most $n$.

Next we move each occurrence of $\beta$ and $\beta^{-1}$ to the end of $u$. Each time we interchange $\alpha$ and $\beta$, we introduce a $\gamma$ and a relator. We immediately move the relator to the right, and then move the $\gamma$ to join the other $\gamma$'s on the right, which introduces at most $n$ new relators. This is done fewer than $n^2$ times, so the total number of relators introduced is fewer than $n^3$.

The method of proof used above is relevant for the fundamental group of any torus bundle over the circle. We may always assume that the torus is glued to itself by the quotient of a linear map $\mathbf{R}^2 \to \mathbf{R}^2$. Such a map has determinant $\pm 1$, since it and its inverse preserve the integer lattice. In either case, there is an invariant two-form on the three-manifold that can be used to measure the area of a relation. An important application is the following result.

**Theorem 8.1.3 (soluble implies not automatic).** *Let $G$ be an extension of a free abelian group on two generators $\beta$ and $\gamma$ by an infinite cyclic group with generator $\alpha$. Let the action of $\alpha$ on $\langle\beta,\gamma\rangle$ by conjugation be given by a matrix $A$ with two real eigenvalues not equal to $\pm 1$. Then any isoperimetric inequality in $G$ is at least exponential, so $G$ is not automatic.*

*Proof of 8.1.3:* By taking a subgroup of index two, we may assume that the $A$ has determinant one. This will not affect the conclusion that $G$ is not automatic, by Theorem 4.1.4 (finite index). Let the eigenvalues of $A$ be $\lambda$ and $\lambda^{-1}$, with $\lambda > 1$.

We consider the following action of $G$ on $\mathbf{R}^3$:

$$\begin{aligned} \alpha &: (x,y,z) \mapsto (x+1, y', z'), \\ \beta &: (x,y,z) \mapsto (x, y+1, z), \\ \gamma &: (x,y,z) \mapsto (x, y, z+1), \end{aligned}$$

where $(y', z')$ is the image of $(y, z)$ under $A$. As in Example 8.1.1, we embed the Cayley graph of $G$ into $\mathbf{R}^3$ equivariantly, with the basepoint at the origin and all edges straight. Note that every edge labelled $\alpha$ is parallel to the $x$-axis in this embedding, because all three generators preserve the $x$ direction.

Consider the word

$$w_n = (\alpha^n\beta\alpha^{-n})(\alpha^{-n}\gamma\alpha^n)(\alpha^n\beta^{-1}\alpha^{-n})(\alpha^{-n}\gamma^{-1}\alpha^n),$$

which represents the identity. Regarded as a path in the Cayley graph in $\mathbf{R}^3$, $w_n$ visits successively the following points (where the second coordinate is a vector in $\mathbf{R}^2$): $(0,0)$, $(n,0)$, $(n, A^n e_1)$, $(0, A^n e_1)$, $(-n, A^n e_1)$, $(-n, A^n e_1 + A^{-n} e_2)$, $(0, A^n e_1 + A^{-n} e_2)$, $(n, A^n e_1 + A^{-n} e_2)$, $(n, A^{-n} e_2)$, $(0, A^{-n} e_2)$, $(-n, A^{-n} e_2)$, $(-n, 0)$, $(0,0)$.

We claim that any disk with $w_n$ as boundary has area at least a positive constant times $\lambda^{2n}$. To see this, we once again consider the two-form $dy \wedge dz$ on $\mathbf{R}^3$. This form is preserved by the action of $G$. Also, it is the pullback of the area form under the projection onto the $yz$-plane. The path $w_n$ is sent by this projection to the parallelogram with vertices $0$, $A^n e_1$, $A^n e_1 + A^{-n} e_2$ and $A^{-n} e_2$ in the $yz$-plane, and the area of this parallelogram is approximately a constant times $\lambda^{2n}$ for $n$ large. Thus the integral of $dy \wedge dz$ over a disk spanning $w_n$ is approximately grows exponentially with $n$.

We use a finite set of standard relators for $G$, each relator represented by some piecewise smooth surface. The contribution of a relator to the integral of $dy \wedge dz$ is a constant. It follows that the number of relators needed to give a disk whose boundary is $w_n$ is at least a positive constant times $\lambda^{2n}$ for $n$ large. 8.1.3

## 8.2. Nilpotent Groups Are Not Automatic

This section is devoted to proving that any nilpotent automatic group is virtually abelian. This was proved by Holt for synchronous groups; Holt's proof was adapted by Epstein to the asynchronous case.

**Example 8.2.1.** Thurston has proved that the nilpotent groups

$$\langle\{\alpha_1, \beta_1, \ldots, \alpha_n, \beta_n, \gamma\}/\{[\alpha_i, \beta_i]\gamma^{-1}, [\alpha_i, \gamma], [\beta_i, \gamma], [u_i, u_j]\}\rangle,$$

where $u_i$ stands for either $\alpha_i$ or $\beta_i$ and $i \neq j$, have quadratic isoperimetric inequalities if $n \geq 2$ (compare Example 8.1.1). It follows that a group with a quadratic isoperimetric inequality is not necessarily automatic.

Before proving the main theorem, we collect some necessary facts about nilpotent groups. Every subgroup of a nilpotent group is nilpotent (easy). Every subgroup of a finitely generated nilpotent group is finitely generated [Mac68, Theorem 9.16]. Every nilpotent group generated by finitely many elements of finite order is finite [Mac68, Theorem 9.17].

**Lemma 8.2.2 (torsion subgroup).** *The elements of finite order in a nilpotent group $G$ form a normal subgroup $T$, called the torsion subgroup of $G$. If $G$ is finitely generated, $T$ is finite. The quotient $G/T$ is torsionfree, that is, its only element of finite order is the identity.*

*Proof of 8.2.2:* By the paragraph preceding the lemma, if $x, y \in G$ have finite order, the subgroup generated by $x$ and $y$ is finite, so $xy$ has finite order. Therefore $T$ is a group. The order of an element is invariant under conjugation, so $T$ is normal. If $G$ is finitely generated, so is $T$, which implies that $T$ is finite, since its generators have finite order. Finally, if $(gT)^n$ is the identity in $G/T$ for some coset $gT$, we have $g^n \in T$ and $g$ is itself in $T$, so $gT = T$. 8.2.2

**Lemma 8.2.3 (torsionfree).** *If $G$ is torsionfree and nilpotent, $G/Z_i$ is torsionfree for each $i \geq 0$. In other words, $g^m \in Z_i$ implies $g \in Z_i$, for any $i \geq 0$ and any $m > 0$.*

*Proof of 8.2.3:* We show first that $Z_{i+1}/Z_i$ is torsionfree for $i \geq 0$. For $i = 0$, there is nothing to show. For other values of $i$, we work modulo $Z_{i-1}$, that is, with equivalence classes in $G/Z_{i-1}$. Take $g \in Z_{i+1}/Z_{i-1}$; then every commutator involving $g$ is in $Z_i/Z_{i-1}$, and therefore central. If $g^m \in Z_i/Z_{i-1}$, we have $g^m h = hg^m$ for every $h \in G/Z_{i-1}$. We also have $g^m h = h(g[g,h])^m = hg^m[g,h]^m$, so $[g,h]$ is the identity because $Z_i/Z_{i-1}$ is torsionfree (by the induction assumption). Therefore $g \in Z_i/Z_{i-1}$, completing the induction step.

It follows that $G/Z_i$ is also torsionfree, because the torsion of an element of $G/Z_i$ would manifest itself in at least one of the quotients $Z_{j+1}/Z_j$, for $j \geq i$. 8.2.3

**Lemma 8.2.4 (abelian quotient implies abelian subgroup).** *If $G$ is nilpotent and finitely generated, and the quotient $Q = G/T$ of $G$ by its torsion group is abelian, $G$ is virtually abelian.*

*Proof of 8.2.4:* Replacing $G$, if necessary, by the centralizer of $T$, we can assume that $T$ is central. By Lemma 8.2.2, $Q$ is a finitely generated free abelian group. We choose a set of generators for $Q$, and coset representatives $x_1, \dots, x_n$ for them in $G$. Now the commutators $[x_i, x_j]$, for $1 \leq i, j \leq n$, are in $T$, and therefore central and of finite order. It follows that appropriate multiples of $x_i, \dots, x_n$ commute. The group generated by these multiples is the desired finite-index subgroup of $G$. 8.2.4

**Lemma 8.2.5 (powers and conjugates).** *Let $G$ be nilpotent and torsion-free. Take $g \in Z_i$ and $h \in G$. Then $h$ commutes with $g$ modulo $Z_{i-2}$ if and only if $h'$ commutes with $g'$ modulo $Z_{i-2}$, where $h'$ and $g'$ are non-zero powers, roots or conjugates of $h$ and $g$. In particular, $g \in Z_i \setminus Z_{i-1}$ if and only if $g' \in Z_i \setminus Z_{i-1}$ is.*

*Proof of 8.2.5:* We work modulo $Z_{i-2}$, and interpret $g$, $g'$ and $h$ as elements of $G/Z_{i-2}$. We have $g, g' \in Z_i/Z_{i-2}$ (in the case of a root, because $G/Z_i$ is torsionfree), so any commutator involving $g$ or $g'$ is central. We have the equations

$$\begin{aligned}
[g, h^m] &= g^{-1}h^{-m}gh^m = g^{-1}g(h^{-1}[h^{-1}, g])^m h^m = [h^{-1}, g]^m, \\
[g^m, h] &= g^{-m}h^{-1}g^m h = [h^{-1}, g]^m, \\
[v^{-1}gv, h] &= [g[g, v], h] = (g[g, v])^{-1}h(g[g, v])h = [g, h], \\
[g, vhv^{-1}] &= v[v^{-1}gv, h]v^{-1}.
\end{aligned}$$

Since $G/Z_{i-2}$ is torsionfree, this proves the equivalence of $[g,h] = 0$, $[g, h^m] = 0$, $[g^m, h] = 0$ and $[v^{-1}gv, h] = 0$. The last statement of the theorem follows by choosing $h$ such that $[g,h] \notin Z_{i-2}$. 8.2.5

**Lemma 8.2.6 (nilpotent implies polynomial growth).** *A finitely generated nilpotent group $G$ has polynomial growth, that is, the number of elements in a neighbourhood of radius $n$ in the Cayley graph of $n$ with respect to any finite set of generators is bounded by a polynomial in $n$.*

*Proof of 8.2.6:* Let $A = A^{(0)}$ be a set of generators for $G$, closed under inversion. We define $A^{(i)}$ by induction to be the set of commutators $[x, y]$, for $x \in A^{(i-1)}$ and $y \in A^{(0)} \cup \cdots \cup A^{(i-1)}$. (The set $A$ being outside $G$, these commutators should be regarded just as new symbols, which map into $G$ in the obvious way.) By induction on $i$, commutators in $A^{(i)}$ represent elements of $Z_{c-i}$ if $G$ is nilpotent of class $c$. We set $\bar{A} = A^{(0)} \cup \cdots \cup A^{(c-1)}$, and we give $\bar{A}$ any total order such that elements of $A^{(i)}$ are less than elements of $A^{(j)}$, for $i < j$.

Given a string $w$ over $\bar{A}$, there is a procedure to find a string representing the same group element and having the form

**8.2.7.**
$$\prod_{a \in \bar{A}} a^{n(a)},$$

with the $a$'s appearing in increasing order. This is very similar to the procedure used to put an element of the Heisenberg group in normal form (see page 165): starting from the right, we transpose any pair $xy$ that is out of order, creating in the process a new $[x, y]$ that must be recursively moved right until it finds its place. Commutators in $A^{(c)}$ can be discarded, since we are interested in equality in $G$, rather than in the free group.

One can show that the length of the string thus obtained is bounded by a polynomial of degree $c$ in the length of $w$. Details can be worked out by the reader, or looked up in [Wol68]. It follows that $G$ has polynomial growth with respect to $A$, since the number of distinct expressions of the form $\prod a^{n(a)}$ and having less than a given length also grows polynomially with the length. 8.2.6

**Theorem 8.2.8 (nilpotent implies not automatic).** *A group $G$ that is asynchronously automatic and virtually nilpotent is virtually abelian. A group $G$ that is asynchronously automatic, torsionfree and nilpotent is abelian.*

*Proof of 8.2.8:* We first notice that the second statement implies the first, by Theorem 7.3.5(2) (asynchronous finite index) and Lemmas 8.2.2 and 8.2.4.

To prove the second statement, we assume that $G$ is nilpotent, torsionfree, asynchronously automatic, and not abelian, and seek a contradiction.

We choose the automatic structure $(A, L)$ of $G$ so that $L$ is boundedly asynchronous, and has the uniqueness property (Theorem 7.3.2). By Lemma 8.2.6 (nilpotent implies polynomial growth) and Proposition 1.3.8

(polynomial growth condition), there is a regular expression for $L$ of the form

**8.2.9.** $R = R_1 \vee \ldots \vee R_I$ with $R_i = v_{i,0}u_{i,1}^{*}v_{i,1}\ldots v_{i,P_i-1}u_{i,P_i}^{*}v_{i,P_i}$,

where the $u_{i,p}$ and $v_{i,p}$ are strings over $A$, and the $u_{i,p}$ are non-null. We call $P_i$ the *star length* of $R_i$, and assume, without loss of generality, that $P_1 \geq \cdots \geq P_I$. We associate with $R$ the sequence $(P_1, \ldots, P_I)$. Among all possible regular expressions $R$ of the form 8.2.9 and such that $L = L(R)$, we choose one such that $(P_1, \ldots, P_I)$ is minimal in dictionary order.

We call an $R_i$ or a string that matches it *star central* if $\overline{u_{i,p}}$ is central for $1 \leq p \leq P_i$. It cannot be the case that all the $R_i$ are star central; if they were, the strings $v_{i,0}v_{i,1}\ldots v_{i,P_i}$, for $1 \leq i \leq I$, would represent all the cosets of $Z$, and $G/Z$ would be finite. By Lemma 8.2.3 (torsionfree), $G/Z$ would be trivial, contradicting our assumption that $G$ is not abelian.

We now consider two accepted strings $w$ and $w'$, matching the regular expressions $R_i$ and $R_{i'}$ in Equation 8.2.9, for some $1 \leq i, i' \leq I$. (The case $i = i'$ is allowed.) To simplify the notation, we set $u_p = u_{i,p}$, $u'_{p'} = u_{i',p'}$, and so on. We also set $P = \max_i P_i$.

We let $n_p$ be the exponent of $u_p$ in $w$, and say that $w$ *repeats* $N$ *times* if $n_p \geq N$ for all $p$. We define the *p-th prefix* of $w$ to be $w_p = v_0u_1^{n_1}\ldots u_{p-1}^{n_{p-1}}v_{p-1}$. Thus $w_pu_p^{n_p}v_p = w_{p+1}$. We make similar definitions for $w'$.

Assume the distance in the Cayley graph between the elements of $G$ represented by $w$ and $w'$ is at most $I$. (This $I$ is the same as in Equation 8.2.9, for reasons that will become clear later.) By the discussion preceding the proof of Theorem 7.2.8 (characterizing asynchronous), there is an asynchronous correspondence $\mathcal{R}$ between $w$ and $w'$ with constant $IK$, where $K$ is an asynchronous lipschitz constant for the structure $(A, L)$. If $w = abc$ and $w' = a'b'c'$, we say that $b$ and $b'$ *match* if $a$ is $\mathcal{R}$-related to $a'$ and $ab$ is $\mathcal{R}$-related to $a'b'$. By increasing the constant from $IK$ to $IK + l$, where $l$ is more than the length of any $u_{i,p}$ or $v_{i,p}$, we can assume that any prefix of the form $w_pu_p^n$ matches a prefix of the form $w_{p'}u_{p'}^{n'}$. Let $N$ be the number of elements of $G$ in a neighbourhood of the identity of radius $IK + l$; then among $N+1$ matching pairs of prefixes, there is at least one repeating word difference. We observe also that a substring of length $n$ can match a substring of length at most $k(n+1)$, where $k$ is the boundedness constant of the automatic structure (or slightly more, since we have tampered with $\mathcal{R}$).

In the next three lemmas, $w$ and $w'$ are accepted strings matching $R_i$ and $R_{i'}$ and representing elements of the group whose distance is at most $I$, and $\mathcal{R}$ and $N$ are as in the previous paragraph.

**Lemma 8.2.10 (conjugates).** *If the substring $u_p^N$ of $w$ matches the substring ${u'_{p'}}^M$ of $w'$, for $M \geq 0$, some $u_p$ is a root of a conjugate of a power of $u'_{p'}$.*

*Proof of 8.2.10:* For $N+1$ contiguous values of $n$, and appropriate values of $n'$, we have $w_p u_p^n$ matched to $w'_{p'} {u'_{p'}}^{n'}$. The word difference between these two strings is bound to repeat, as we observed above. The corresponding values of $n'$ must be different, by the uniqueness property of the word acceptor. We deduce that $\overline{{u_p}^{n-m}}$ and $\overline{{u'_{p'}}^{n'-m'}}$ are conjugate, with $m \neq n$ and $m' \neq n'$. The situation is illustrated in Figure 8.3. 8.2.10

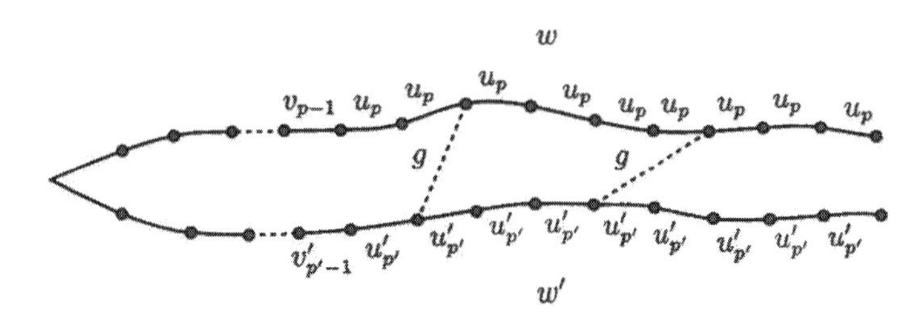


**Figure 8.3. Conjugating a power of $\overline{u_p}$ to a power of $\overline{u'_{p'}}$.** If a high power of $u_p$ matches some power of $u'_{p'}$, there are conjugate powers of $u_p$ and $u'_{p'}$, because there must be repeated word differences.

A slight strengthening of this argument gives the following result (see Figure 8.4):

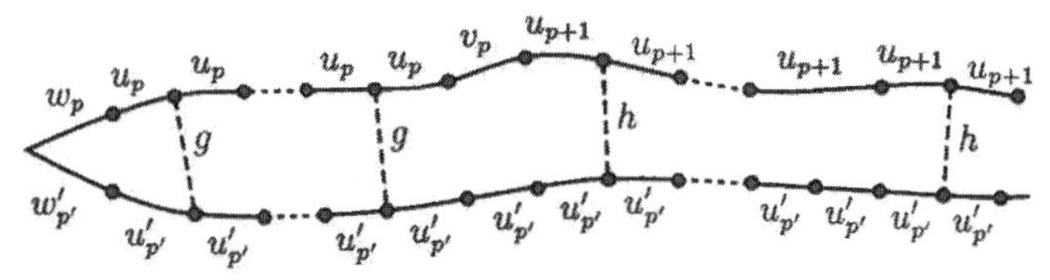


**Figure 8.4. Two accepted words.** This figure, which takes place within the Cayley graph, illustrates the proof of Lemma 8.2.11. If some power of $u'_{p'}$ matches $u_p^N v_p u_{p+1}^N$, for $N$ large, the word differences $g$ and $h$ must repeat. We can therefore express $g$ in two different ways in terms of the components of $w$ and $w'$ and similarly for $h$. The resulting equations lead to a contradiction.

**Lemma 8.2.11 (two against one).** *A substring of $w$ of the form $u_p^N v_p u_{p+1}^N$ cannot match a substring of $w'$ of the form ${u'_{p'}}^{N'}$, for $N' \geq 0$.*

*Proof of 8.2.11:* Because of the repeating word differences (see Figure 8.4), we have

$$\overline{{u'_{p'}}^{m'-n'}} = \overline{(w'_{p'})^{-1} w_p u_p^{n-m} w_p^{-1} w'_{p'}},$$
$$\overline{{u'_{p'}}^{\mu'-\nu'}} = \overline{(w'_{p'})^{-1} w_{p+1} u_{p+1}^{\nu-\mu} w_{p+1})^{-1} w'_{p'}},$$

with all the exponents non-zero. Raising these expressions to suitable powers, so as to get equal powers of $u'_{p'}$, and using the fact that $w_{p+1} = w_p u_p^{n_p} v_p$, we get an equation of the form

$$\overline{u_p^r v_p} = \overline{v_p u_{p+1}^s},$$

for some $r, s > 0$. From the uniqueness property of the word acceptor, we obtain an equality of strings $u_p^r v_p = v_p u_{p+1}^s$. But this means that we could replace $R_i$ by a finite number of similar regular expressions, each with smaller star length. This contradicts our choice of $R$. $\boxed{8.2.11}$

**Lemma 8.2.12 (related chunks).** *Assume that the regular expression $R_i$ matched by $w$ is not star central, and that it has maximal star length subject to this condition. Given $M' > 0$, there exists $M = f(M')$ such that, if $w$ repeats $M$ times, we have the following situation: $R_i$ and $R_{i'}$ have the same star length; for each $p$, there exist $n, n' \geq M'$ such that some substring $u_p^n$ of $w$ matches a substring ${u'_p}^{n'}$ of $w'$ (in particular, $w'$ repeats $M'$ times); and some power of $u_p$ is conjugate to some power of $u'_p$ (in particular, $u_p$ is central if and only if $u'_p$ is).*

*Proof of 8.2.12:* Consider a substring of $w$ of the form $u_p^M$. Deleting from it any segments that might match the $v'_{p'}$, we are left with up to $P$ segments that match powers of $u'_{p'}$. Among themselves these segments add up to at least $M - Pkl$ copies of $u_p$, where we recall that $l$ is an upper bound for the $|v_{i,j}|$, and $k$ is the asynchronous bound. We now define $M = Pkl + Pn$. If we take $n > N$, there is at least one segment $u_p^N$ that must match a power of $u'_{p'}$, for some $p'$. Applying Lemma 8.2.11 (two against one), we conclude that different values of $p$ cannot correspond to the same value of $p'$; thus the correspondence $p \mapsto p'$ is injective. If $\overline{u_p}$ is not central, Lemmas 8.2.10 (conjugates) and 8.2.5 (powers and conjugates) show that $\overline{u'_{p'}}$ is not central. Thus $R_{i'}$ is not star central, so it has the same star length as $R_i$, and the correspondence $p \mapsto p'$ is, in fact, bijective, so $p' = p$.

In addition, if $u_p^n$ matches $u_p^{n'}$, we must have $n' \geq n/(kl)$, by the boundedness of the asynchronous structure. Thus, by increasing $n$ (and $M = f(M')$) accordingly, we can ensure that $n, n' \geq M'$. $\boxed{8.2.12}$

From now on we fix $i$ so that $R_i$ is not star central, and has maximal star length subject to this condition, and we let $p$ be the maximal index such that

$u_p = u_{i,p}$ is not central. (For the remainder of this proof we will drop the distinction between strings and group elements, as no confusion can arise.) We define $q > 1$ so that $u_p \in Z_q \setminus Z_{q-1}$, and fix a generator $x \in A$ such that $[u_p, x] \notin Z_{q-2}$. Our strategy to conclude the proof will be to show that some power of $x$ commutes with some conjugate of $u_p$, in contradiction with Lemma 8.2.5 (powers and conjugates).

Let $w$ match $R_i$ and repeat $f(f(N))$ times, where $f$ is the function defined in Lemma 8.2.12. We look at the unique representatives $w_j$ of $\overline{w}x^j$, for $0 \leq j \leq I$. At least two of them, say $w'$ and $w''$, must match the same regular expression $R_{i'}$, and each must repeat $f(N)$ times. By our choice of $p$ and Lemma 8.2.12, $u'_p = u_{i',p}$ is not central, so another application of Lemma 8.2.12 shows that there are matching substrings ${u'_p}^n$ of $w'$ and ${u'_p}^{n'}$ of $w''$, with $n, n' \geq N$. We know that $u'_p$ is a root of a conjugate of a power of $u_p$, so $u'_p \in Z_q \setminus Z_{q-1}$ by Lemma 8.2.5 (powers and conjugates). At the same time, the repeating word differences argument gives

$$ {u'_p}^{-r} g {u'_p}^s = g $$

for some $r, s > 0$ and $g \in G$; since $u'_p \in Z_q$, we get ${u'_p}^{r-s} \in Z_{q-1}$, which, together with $u'_p \notin Z_{q-1}$, implies that $r = s$ by Lemma 8.2.3 (torsionfree). Thus $g$ commutes with ${u'_p}^r$, and, by Lemma 8.2.5, also with $u'_p$.

We also know that some power of $x$ equals

$$ {w'}^{-1} w'' = ({u'_p}^a v'_p \cdots v'_{P'})^{-1} g\, {u'_p}^b v'_p \cdots v'_{P'} z = v^{-1} {u'_p}^{-a} g\, {u'_p}^b v z = v^{-1} {u'_p}^{b-a} g z v, $$

where $z$ is central, $a, b \geq 0$, and $v = v'_p \ldots v'_r$. The right-hand side of this chain of equalities commutes with $v^{-1} {u'_p}^{b-a} v$. Therefore a power of $x$ commutes with a conjugate of a power of $u'_p$. But by Lemma 8.2.5 (powers and conjugates), this is impossible, because $u'_p$ and $u_p$ have conjugate powers, and we assumed that $u_p$ does not commute with $x$ modulo $Z_{q-2}$. 8.2.8

## 8.3. Regular Subgroups and Nilpotency

In this section we prove that every finitely generated nilpotent subgroup of a biautomatic group is virtually abelian, a result due to Steve Gersten and Hamish Short [GS00a]. Their proof involves the important concepts of regular subgroups (which they call *rational subgroups*) and translation numbers. We will discuss these concepts and related results only briefly, to the extent that they are necessary to the proof, but they are important in their own right, and have been used to prove several other results.

Let $G$ be a group, $A$ a set of semigroup generators of $G$ and $L$ a regular language over $A$ such that the natural map $\pi : L \to G$ is onto. In this situation we say that $(A, L)$ is a *regular structure* for $G$. We say that a subgroup $H$ of $G$ is *regular* (or *L-regular*) if $L \cap \pi^{-1}(H)$ is a regular language. For example, $G$ is regularly generated (Theorem 2.1.9) if and only if the trivial subgroup is regular.

**Theorem 8.3.1 (characterizing regular subgroups).** *Let $(A, L)$ be a regular structure for a group $G$. A subgroup $H$ of $G$ is $L$-regular if and only if there is a constant $r \geq 0$ such that, for each $w \in L \cap \pi^{-1}(H)$, the path $\widehat{w}$ in the Cayley graph $\Gamma(G, A)$ lies within a $r$-neighbourhood of $H$.*

The condition that there exists such a constant is sometimes called *quasiconvexity*, following Gromov [Gro87].

**Example 8.3.2 (finite index is regular).** If $H$ has finite index in $G$, $H$ is regular. This is proved by fixing strings representing each coset. If $r$ is larger than the length of any of these strings, $G$ is contained in an $r$-neighbourhood of $H$.

*Proof of 8.3.1:* Let $(A, L)$ be the regular structure of $G$, and suppose that $H$ is regular. Let $M$ be a finite state automaton accepting $L \cap \pi^{-1}(H)$. For each live state $s$ of $M$, let $\alpha_s$ be a shortest path in $M$ from $s$ to an accept state. The length of $\alpha_s$ is bounded by the number of states of $M$, so we take $r$ equal to this number. If $w \in L$ represents an element of $H$ and $v$ is a vertex of $\Gamma(G, A)$ on $\widehat{w}$, some path $\alpha_s$ leads from $v$ to an element of $H$, so $d(v, H) < r$. This proves the "only if" part.

Suppose, conversely, that there is a constant $r$ as in the statement of the theorem. We construct a non-deterministic finite state automaton $M$ over $A$ that accepts $L \cap \pi^{-1}(H)$. Let $W$ be a finite state automaton accepting $L$, and having state set $S$. The state set of $M$ is $B \times S$, where $B$ is the neighbourhood of radius $r$ around the identity in $\Gamma(G, A)$. We give $M$ an arrow labelled $x \in A$ from $(g, s)$ to $(g', s')$ if and only if $g^{-1}xg' \in H$ and $s' = sx$ in $W$. The initial state of $M$ is $(e, s_0)$, where $e$ is the identity in $G$ and $s_0$ is the initial state of $M$. The accept states of $M$ are those of the form $(e, s)$, where $s$ is an accept state of $M$. It is straightforward to check that $M$ accepts $L \cap \pi^{-1}(H)$, as claimed (see also Figure 8.5). [8.3.1]

This proof also shows that $H$ is generated by the finite set $A_H$ of triples $(g, x, g') \in B \times A \times B$ such that $g^{-1}xg' \in H$. (Here, of course, the map $p$ of Definition 2.1.1 takes $(g, x, g')$ to $g^{-1}xg' \in H$.) We can consider the finite state automaton $M$ of the proof as a deterministic automaton over $A_H$, if we give an arrow from $(g, s)$ to $(g', sx)$ the label $(g, x, g')$ instead of $x$. This automaton accepts a language $L_H$ over $A_H$ that maps to $L \cap \pi^{-1}(H)$ under the projection $B \times A \times B \to A$. In particular, given a word over $A$

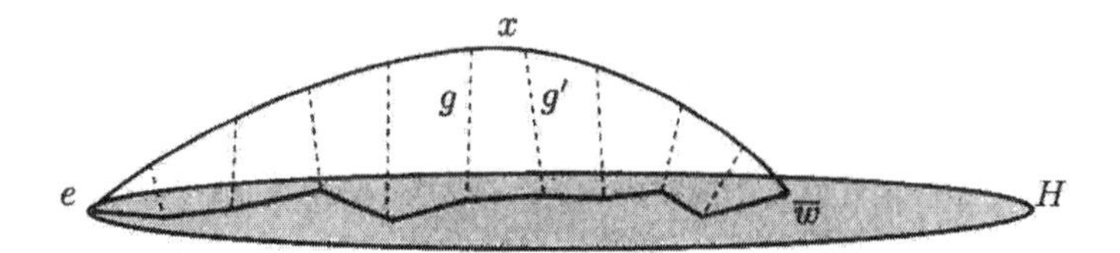


**Figure 8.5. The regular structure inherited by a subgroup.** The upper path represents an accepted string of $L$, drawn in the Cayley graph $\Gamma(G, A)$, and representing an element of $H$. An edge $x$ corresponds to a transition in $M$ from $(g, s)$ to $(g', sx)$, where $g^{-1}xg' \in H$. The lower path represents a path in $\Gamma(H, A_H)$ having the same length and going to same point; the two paths correspond under the map $(g, x, g') \mapsto x$, and are within a distance $r$ of each other.

representing an element of $H$, there is a word over $A_H$ of the same length and representing the same element: just take any inverse image under the projection (Figure 8.5). The corresponding paths are uniformly within a distance $r$ of each other.

We call $(A_H, L_H)$ the regular structure *inherited* by $H$. This definition depends on the choice of $r$ in the statement of Theorem 8.3.1 (characterizing regular subgroups), but this does not affect any of the arguments below.

**Lemma 8.3.3 (subgroup pseudoisometry).** *The identity map is a pseudoisometry between $H \subset \Gamma(H, A_H)$ and $H \subset \Gamma(G, A)$.*

*Proof of 8.3.3:* Each element of $A_H$ can be expressed as string over $A$ of length at most $2r+1$, so one direction is obvious. To prove the other direction, let $w$ be a word over $A$ representing an element of $H$. By Lemma 2.3.9 (bounded length difference), there is an accepted word in $L$ representing $\overline{w}$ and having length at most (roughly) proportional to $|w|$. As just observed, there is an equivalent accepted word of same length in $L_H$. 8.3.3

**Theorem 8.3.4 (regular subgroups inherit automation).** *Let $(A, L)$ be a regular structure on $G$ and let $H$ be a regular subgroup of $G$. If $(A, L)$ is an automatic, biautomatic or boundedly asynchronous automatic structure (for $G$), the regular structure inherited by $H$ is likewise (for $H$).*

*Proof of 8.3.4:* Suppose $(A, L)$ is automatic, and let $w_1, w_2 \in L_H$ satisfy $\overline{w_1 y} = \overline{w_2}$, for some $y \in A_H$. We need to show that the uniform distance between the paths $\widehat{w_1}$ and $\widehat{w_2}$ in the Cayley graph $\Gamma(H, A_H)$ is bounded by a constant independent of $w_1$ and $w_2$. Let $u_1, u_2 \in L$ be the images of $w_1$ and $w_2$ under the map $(g, x, g') \mapsto x$. Then $\overline{u_1} = \overline{w_1}$ and $\overline{u_2} = \overline{w_2}$, so the two endpoints in the Cayley graph of $G$ are within a distance $2r+1$ of each other.

If $k$ is a lipschitz constant for $(A, L)$, the uniform distance in $\Gamma(G, A)$ between $\widehat{u_1}$ and $\widehat{u_2}$ is bounded by $k(2r+1)$, and the uniform distance between $\widehat{w_1}$ and $\widehat{w_2}$ is bounded by $2r+k(2r+1)$. By Lemma 8.3.3 (subgroup pseudoisometry), this implies that the uniform distance between $\widehat{w_1}$ and $\widehat{w_2}$ is bounded.

The same method is used in the biautomatic and bounded asynchronous cases, using Lemma 2.5.5 (characterizing biautomatic) and Theorem 7.2.8 (characterizing asynchronous) instead of Theorem 2.3.5 (characterizing synchronous). 8.3.4

**Corollary 8.3.5 (centralizer).** *Let $(A, L)$ be a biautomatic structure on a group $G$, and let $X \subset G$ be finite. The centralizer $C(X)$ and the centre $Z(G)$ of $G$ are regular subgroups of $G$; in particular, their inherited regular structure is biautomatic.*

*Proof of 8.3.5:* Given any $g \in G$, we see by induction on the length of a word representing $g$ that the language

$$L_g = \{w \in L \mid (\exists v \in L)(\overline{w}g = v = g\overline{w})\}$$

is regular, the case $|g| = 1$ following immediately from the definition of a biautomatic group and Theorem 1.4.6 (predicate calculus). Therefore $C(g)$ is regular, since $L_g = L \cap \pi^{-1}(C(g))$. It follows that $C(X) = \bigcap_{g \in X} C(g)$ is regular, and therefore biautomatic by Theorem 8.3.4 (regular subgroups inherit automation). The case of $Z(G)$ follows by by taking $X = A$. 8.3.5

If $G$ is a group and $A$ a set of semigroup generators for $G$, we define the *translation number* of $g \in G$ as

$$\tau(g) = \tau_A(g) = \lim_{n \to \infty} |g^n|/n,$$

where $|g^n|$, as usual, is the word length of $g^n$ (page 37). A standard argument, using the fact that $|g^{m+n}| \leq |g^m| + |g^n|$, shows that the limit is well-defined; clearly $0 \leq \tau(g) \leq |g|$. It follows from Lemma 3.3.3 (change of generators is pseudoisometry) that, if we replace $A$ by another set $B$ of semigroup generators, the translation numbers with respect to the two sets satisfy

$$\lambda^{-1} \tau_B(g) \leq \tau_A(g) \leq \lambda \tau_B(g)$$

for some $\lambda \geq 1$ and all $g \in G$. Similarly, the next lemma is an immediate consequence of Lemma 8.3.3 (subgroup pseudoisometry):

**Lemma 8.3.6 (translation numbers and subgroups).** *If $(A, L)$ is an automatic structure on a group $G$ and $H$ is a regular subgroup of $G$, there exists $\lambda \geq 1$ such that*

$$\lambda^{-1} \tau_{A_H}(h) \leq \tau_A(h) \leq \lambda \tau_{A_H}(h).$$

8.3.6

**Lemma 8.3.7 (infinite order).** *In a biautomatic group, every element of infinite order has positive translation number.*

*Proof of 8.3.7:* Let $g$ be of infinite order in the biautomatic group $G$. Let $C(g)$ be the centralizer of $g$ in $G$, and let $Z(C(g))$ be the centre of $C(g)$. By Corollary 8.3.5 (centralizer), $Z(g)$ and also $Z(C(g))$ are biautomatic; in particular, $Z(C(g))$ is finitely generated. It is also abelian, so we can find a free abelian subgroup $H$ of finite index in $Z(C(g))$ containing $g$.

Now the translation number of a non-zero element in $H$ cannot be zero. To see this, embed $H$ in a real vector space $V$, as in the proof of Theorem 4.2.1 (automatic structure with invariance), and let $K$ be the convex hull of $A$ in $V$. Then $\tau(h) = \|h\|_K$, where the norm is defined by making $K$ the unit ball. In particular, $\tau(h) = 0$ if and only if $h$ is the identity, so $\tau_{A_H}(g) > 0$.

By Example 8.3.2 (finite index is regular) and Lemma 8.3.6 (translation numbers and subgroups), the translation number of $x$ in $Z(Z(x))$ is non-zero. The same is true about $\tau_A(x)$, by Corollary 8.3.5 (centralizer) and Lemma 8.3.6. 8.3.7

We are now ready for the main result of this section:

**Theorem 8.3.8 (nilpotent subgroups).** *A finitely generated nilpotent subgroup of a biautomatic group is virtually abelian.*

This should be compared with Holt's result in the previous section, Theorem 8.2.8 (nilpotent implies not automatic). We remark that Theorem 8.3.8 also follows from a more general result of Gersten and Short in [GS00a], saying that every polycyclic subgroup of a biautomatic group is virtually abelian.

*Proof of 8.3.8:* Let $H$ be the subgroup in question. By Lemma 8.2.2 (torsion subgroup), we can assume that $H$ is torsionfree, by passing to a finite index subgroup. We will deduce a contradiction from the assumption that $H$ is not abelian.

Let $Z$ be the centre of $H$, and $Z_2$ the second term in the upper central series (so that $Z_2/Z$ is the centre of $H/Z$). Take $x \in Z_2 \setminus Z$, and a $y \in H$ that does not commute with $x$. Then $[x, y]$ is central in $H$, and a simple calculation shows that

$$[x,y][x^n,y] = [x^n,y][x,y] = [x^{n+1},y],$$
$$[x,y][x,y^n] = [x,y^{n+1}].$$

Since $x^m \in Z_2$ for each $m$, we can replace $x$ by $x^m$ in the second equality. It follows easily that $[x^m, y^n] = [x,y]^{mn}$. Then

$$\tau([x,y]) = \lim_{n\to\infty} \frac{|[x,y]^{n^2}|}{n^2} = \lim_{n\to\infty} \frac{|[x^n,y^n]|}{n^2} \le \lim_{n\to\infty} \frac{4n(|x|+|y|)}{n^2} = 0.$$

By Lemma 8.3.7 (infinite order), $[x, y]$ is the identity, contrary to our choice of $x$ and $y$. $\boxed{8.3.8}$

# Part II

# Topics in the Theory of Automatic Groups

# Chapter 9

# Braid Groups

Braid groups describe the physically intuitive concept of isotopy classes of braids, which are collections of intertwining strands whose endpoints are kept fixed. They are fairly well understood groups, with simple presentations. Algorithms have been known for a while for the word problem and conjugacy problem [Art47, Gar69, Bir75].

In addition to their intrinsic interest, braid groups are important because they are closely related to the mapping class groups for the $n$-punctured sphere, and algorithms that work in the mapping class groups can be specialized for the braid group. Algorithms for the word problem and the conjugacy problem in the mapping class groups are also known [Hem79, Thu88, Mos87, Mos].

Despite this information, the braid groups still have a certain mystery. They are not small cancellation groups, so the most straightforward algorithms from combinatorial group theory do not apply. Also, the algorithms for the word problem in the references above either are exponential in the length of the input, or their complexity has not been analyzed (and appears to be exponential).

Our approach is close in spirit to Garside's [Gar69], but goes further. In particular, we show in Section 9.3 that the braid groups are automatic. One consequence is that there is a canonical form for elements of a braid group that can be found in quadratic time as a function of the length of the initial word, as described in Theorem 2.3.10 (quadratic algorithm).

We also show (Corollary 9.3.7) that the braid group has the structure of a lattice, invariant under multiplication on the left, whose partial ordering $a \prec b$ is the condition that $a^{-1}b$ can be represented as a positive braid. Like multiplication in the braid group, the lattice operations $\wedge$ and $\vee$ can be computed in quadratic time, and the inequality $a \prec b$ can be tested in quadratic time. These results refer to a fixed number of strings and a variable number of crossings.

The results in this chapter are due to Thurston, except where otherwise stated in the text, and have been circulated in preprint form. Minor errors have been corrected by Epstein in the course of writing this book.

## 9.1. The Braid Group and the Symmetric Group

A *braid* is obtained by laying down a number of parallel pieces of string and intertwining them, without losing track of the fact that they run essentially in the same direction. In our pictures this direction will be horizontal—we imagine that the braid starts on the right, and that crossings get added as we move left. (Why start on the right, rather than the left? For the same reason that we put the argument of a function on the right of the function—two arbitrary decisions, but it's nice to have them arbitrary in the same way. The connection will be obvious in a minute.) We number strands at each horizontal position from the top down.

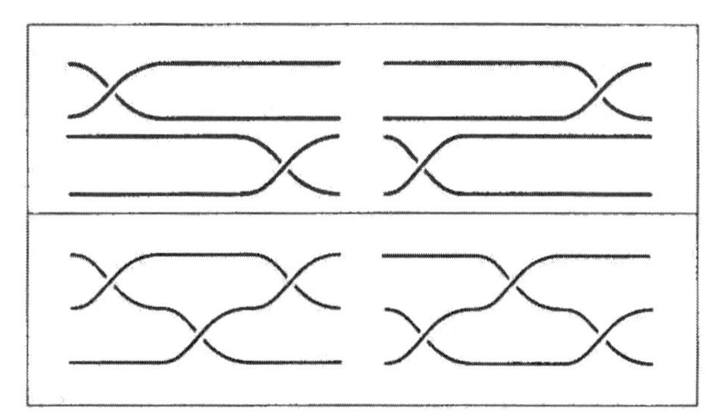

**Figure 9.1. Braid relations.** The two types of elementary moves for braids: interchanging the horizontal position of two adjacent crossings that do not share a strand (top), and moving one strand past a crossing of two others (bottom).

If we lay down two braids $u$ and $v$ in a row, so the end of $v$ matches the beginning of $u$ strand by strand, we get another braid $uv$; this operation defines a product in the set of all $n$-strand braids, for $n$ fixed. We consider two braids equivalent if there is an isotopy between them (an isotopy corresponds physically to a motion of the strands in space, with the endpoints being kept fixed). We insist that the isotopy moves the braid through braids—for example a full twist of a pair of adjacent strands is not equivalent to the trivial braid, but it can be isotoped to the trivial braid if we move it out of the set of braids. The set $B = B_n$ of isotopy classes of $n$-strand braids has a group structure, because if we concatenate a braid with its mirror image in a vertical plane, the result is isotopic to the trivial braid (the one with no crossings). We call $B_n$ the *n-strand braid group*.

We will use as generators for $B_n$ the set of *positive crossings*, that is, crossings between two (necessarily adjacent) strands, with the front strand having a positive slope. We denote these generators by $t_1, \ldots, t_{n-1}$, and their inverses by $T_1, \ldots, T_{n-1}$. Thus $t_1$ is a crossing of the strand currently at the top with the one immediately below it, regardless of these strands' positions at the beginning of the braid. To write down a presentation for $B_n$, we observe that any isotopy can be broken down into "elementary moves" of two types, shown in Figure 9.1.

The top part of the figure shows that two crossings commute if they do not involve the same strand:

**9.1.1.** $$t_i t_j = t_j t_i \quad \text{if } |i - j| > 1.$$

In the elementary move at the bottom, one strand goes past a crossing of two others, which gives the relation

**9.1.2.** $$t_i t_{i+1} t_i = t_{i+1} t_i t_{i+1}$$

(the analogous move involving the same strands, but with the other possible strand in front gives the same relation). Thus, 9.1.1 and 9.1.2 together give a set of relators for the braid group.

One obvious invariant of isotopy of a braid is the permutation it induces on the order of the strands: given a braid $g$, the strands define a map $p(g)$ from the right set of endpoints to the left set of endpoints, which we interpret as a permutation of $\{1, \ldots, n\}$. In this way we get a homomorphism $p : B_n \to S_n$, where $S_n$ is the group of permutations of $\{1, \ldots, n\}$ (the symmetric group). The generator $t_i$ maps to the transposition $\tau_i = (i, i+1)$ of adjacent elements. The inverse $T_i = t_i^{-1}$ also maps to $\tau_i$, which has order two.

The connection between the braid group and the symmetric group is so intimate that it will be worthwhile to spend some time investigating the symmetric group more closely. We fix $A = \{t_1, \ldots, t_{n-1}\}$ as our set of generators for $S_n$, and we set $A' = \{t_1, \ldots, t_{n-1}, T_1, \ldots, T_{n-1}\}$.

Each permutation $\sigma$ gives rise to a total order relation $<_\sigma$ on $\{1, \ldots, n\}$, with $i <_\sigma j$ if $\sigma(i) < \sigma(j)$. Obviously, $\sigma$ is determined by the order $<_\sigma$, and therefore by the set of pairs $(i, j) \in \{1, \ldots, n\} \times \{1, \ldots, n\}$ such that $i < j$ and $j <_\sigma i$. We denote this set by $R_\sigma$. We have

**9.1.3.** $$R_{\sigma\tau} = \tau^{-1} R_\sigma \bigtriangleup R_\tau,$$

where $\triangle$ denotes symmetric difference and the image of a pair under a permutation is defined by taking the image of each component and reordering, if necessary, so the smaller number comes first. (Alternatively, we could define $R_\sigma$ using unordered pairs, but this complicates the bookkeeping elsewhere.)

If $e$ is the identity, we clearly have $R_e = \varnothing$. It follows that $R_{\sigma^{-1}} = \sigma R_\sigma$ and

**9.1.4.** $$R_{\sigma\tau^{-1}} = \tau R_\tau \triangle \tau R_\sigma = \tau(R_\tau \triangle R_\sigma).$$

**Lemma 9.1.5 (word length in symmetric group).** *The word length of an element $\sigma \in S_n$ is $|R_\sigma|$, the number of elements in $R_\sigma$.*

*Proof of 9.1.5:* Take $\sigma \neq e$ and look at a pair $(i, j) \in R_\sigma$ such that $\sigma(i) - \sigma(j)$ is minimal. It is easy to see that $\sigma(i) = \sigma(j) + 1$. Multiplying $\sigma$ on the left by $\tau_{\sigma(i)}$ removes one element from $R_\sigma$, namely $(i, j)$. By induction, we can reach the identity by multiplying $|R_\sigma|$ generators, so $|\sigma| \leq |R_\sigma|$.

Conversely, multiplying any permutation $\sigma$ on the right by any generator $\tau_i$ changes the cardinality of $R_\sigma$ by $\pm 1$, so $|\sigma| \geq |R_\sigma|$. 9.1.5

Given a set $R$ of pairs $(i, j)$ with $i < j$, does it equal $R_\sigma$ for some permutation $\sigma$? This is equivalent to asking the following question: we extend $R$ to an antisymmetric relation $R'$ by saying $(i, j) \in R'$ if and only if $i < j$ and $(i, j) \in R$ or $j < i$ and $(j, i) \notin R$; is $R'$ an order relation? We just have to check that this extension is transitive, and in terms of $R$ this translates into the following criterion:

**Lemma 9.1.6 (characterizing $R$).** *A set $R$ of pairs $(i, j)$, with $i < j$, comes from some permutation if and only if the following two conditions are satisfied:*

(a) *If $(i, j) \in R$ and $(j, k) \in R$, then $(i, k) \in R$.*
(b) *If $(i, k) \in R$, then $(i, j) \in R$ or $(j, k) \in R$ for every $j$ with $i < j < k$.*

9.1.6

Next, we define a partial order in $S_n$ by setting $\sigma \geq \tau$ if $R_\sigma \supset R_\tau$. We can express this order in terms of factorization into geodesic words:

**Lemma 9.1.7 (describing the order).** *Either of the following conditions is equivalent to $\tau \geq \sigma$:*

(a) *The concatenation of a geodesic from the identity to $\tau\sigma^{-1}$ with one from $\tau\sigma^{-1}$ to $\tau$ is a geodesic.*
(b) *The concatenation of a geodesic from the identity to $\sigma^{-1}$ with one from $\sigma^{-1}$ to $\tau^{-1}$ is a geodesic.*

*Proof of 9.1.7:* Condition (a) can be rewritten $|\tau| = |\tau\sigma^{-1}| + |\sigma|$. By Lemma 9.1.5 and Equation 9.1.4 this is equivalent to $|\tau(R_\tau \bigtriangleup R_\sigma)| = |R_\tau| - |R_\sigma|$, which happens if and only if $R_\sigma \subset R_\tau$.

Condition (b) can be rewritten $|\tau^{-1}| = |\sigma^{-1}| + |\sigma\tau^{-1}|$, which is equivalent to (a) because inverse elements have same word length. 9.1.7

The identity $e$ is the smallest element of $S_n$ with respect to $\geq$. The largest element is the permutation sending $(1, \ldots, n)$ to $(n, \ldots, 1)$, which we denote by $\omega$. Equation 9.1.3 shows that, if $\sigma$ is a permutation, $R_{\omega\sigma}$ is the complement of $R_\sigma$ in the set of pairs $(i, j)$ with $i < j$. Thus multiplication on the left by $\omega$ works as a complementation operator, which we also denote by $\neg$.

We now show that if $\sigma$ and $\tau$ are permutations, there is a unique largest permutation smaller than both $\sigma$ and $\tau$. To see this, we construct the set of reversals associated with this putative element, and show that this set satisfies Lemma 9.1.6 (characterizing $R$), and that it is maximal with this property among subsets of $R_\sigma \cap R_\tau$. The definition of the set is

$$R = \{(i, k) \in R_\sigma \cap R_\tau \mid (i, j) \in R \text{ or } (j, k) \in R \text{ for all } j \text{ with } i < j < k\}.$$

Although $R$ appears on the right-hand side, the definition makes sense by induction on $k - i$.

Condition (b) of Lemma 9.1.6 is satisfied by construction. To show that condition (a) is satisfied, we take $i < j < k$ with $(i, j) \in R$ and $(j, k) \in R$, and work by induction on $k - i$. Both $(i, j)$ and $(j, k)$ are in $R_\sigma \cap R_\tau$, so $(i, k) \in R_\sigma \cap R_\tau$. To show that $(i, k) \in R$, it is sufficient to show that $(i, j') \in R$ or $(j', k) \in R$ for any $j'$ with $i < j' < k$. We can assume $j' < j$; the case $j' > j$ is analogous and if $j' = j$ there is nothing to prove. Now since $i < j' < j$, we have $(i, j') \in R$ or $(j', j) \in R$ by the induction hypothesis. If $(i, j') \in R$, we are done. If $(j', j) \in R$, we use condition (b) of the lemma on $(j', j)$ and $(j, k)$, and conclude that $(j', k) \in R$, so again we are done. This completes the proof that $R$ satisfies Lemma 9.1.6.

Finally, $R$ is maximal because if $\rho$ is a permutation with $\sigma \geq \rho$ and $\tau \geq \rho$, condition (a) of the lemma applied to $\rho$ gives $R \supset R_\rho$. This shows the existence of a largest element $\sigma \wedge \tau$ smaller than $\sigma$ and $\tau$.

(Here is an example where $R_{\sigma\wedge\tau} \neq R_\sigma \cap R_\tau$: take $n = 4$, let $\sigma$ be the cycle (234) and let $\tau$ be the cycle (1423). Then $R_\sigma = \{(2, 4), (3, 4)\}$ and $R_\tau = \{(1, 2), (1, 3), (1, 4), (2, 3), (2, 4)\}$, so $R_\sigma \cap R_\tau = \{(2, 4)\}$, which does not satisfy the conditions in Lemma 9.1.6. Therefore $\sigma \wedge \tau$ is the identity.)

Using the complementation operator $\neg$ of left multiplication by $\omega$, we can now show that there is also a smallest element $\sigma \vee \tau$ larger than $\sigma$ and $\tau$; we just set $\sigma \vee \tau = \neg(\neg\sigma \wedge \neg\tau)$. We have just proved the following result:

**Proposition 9.1.8 (symmetric lattice).** *The partial order $\geq$ imposes a lattice structure on $S_n$, that is, given permutations $\sigma$ and $\tau$, there is a largest element $\sigma \wedge \tau$ smaller than $\sigma$ and $\tau$, and a smallest element $\sigma \vee \tau$ larger than $\sigma$ and $\tau$. There is a complementation operator $\neg$ on $S_n$ that reverses the order $\geq$.* [9.1.8]

The partial order $\geq$ is not very closely tied to the group structure: it is not invariant under left or right multiplication by a permutation, or under composition. It has, however, a weak form of group invariance.

**Proposition 9.1.9 (weak invariance).** *If $\rho$ is a permutation, the set of permutations bounded below by $\rho$ (and above by $\omega$) and the set of permutations bounded (below by the identity and) above by $\omega\rho^{-1}$ are sublattices of $S_n$, isomorphic under multiplication on the right by $\rho^{-1}$.*

*Proof of 9.1.9:* That the two sets are sublattices is immediate from the definition of a lattice. Take $\tau \geq \sigma$ in the first sublattice; then $\omega \geq \tau \geq \sigma \geq \rho$. By Lemma 9.1.7 (describing the order), the concatenation of geodesics from the identity to $\rho^{-1}$, from $\rho^{-1}$ to $\sigma^{-1}$, from $\sigma^{-1}$ to $\tau^{-1}$ and from $\tau^{-1}$ to $\omega^{-1}$ is still a geodesic. Multiplying on the left by *rho* the portion that goes from $\omega^{-1} = \omega$ to $\rho^{-1}$, we see that $\omega\rho^{-1} \geq \tau\rho^{-1} \geq \sigma\rho^{-1}$, so right multiplication by $\rho^{-1}$ maps the first sublattice to the second and preserves order. A similar argument shows that right multiplication by $\rho$ maps the second sublattice to the first and also preserves order. [9.1.9]

We now turn again to the connection between braids and permutations. Recall that our set of generators $A$ includes only positive crossings. An element of $A^*$ is called a *positive braid*. The *positive braid semigroup* $P = P_n$ is the quotient (Definition 6.1.1) of the free semigroup $A^*$ by the relations 9.1.1 and 9.1.2, interpreted as semigroup relations. Thus two positive braids represent the same element of $P_n$ if they are connected by a chain of positive braids related by the equalities 9.1.1 and 9.1.2. In particular, equivalent positive braids have the same *length* (number of crossings), because the two sides in each of the relations 9.1.1 and 9.1.2 have the same length.

It turns out that the map $P_n \to B_n$ induced from the inclusion $A^* \to A'^*$ is injective, a fact by no means obvious, and which can be rephrased by saying that positive braids that are equivalent as braids are equivalent as positive braids. This was proved by Garside [Gar69], and will be reproved below. Meanwhile we will distinguish carefully between the two types of equivalence.

The total (algebraic) number of crossings of two given strands in a braid is clearly an invariant of isotopy. Since a positive braid only has positive crossings, the absolute number of crossings of two strands in a *positive* braid is an invariant of isotopy. We call a positive braid *non-repeating* if any two of its strands cross at most once; any positive braid equivalent to a non-repeating braid is also non-repeating. We define $D = D_n \subset P_n$ as the set of classes of non-repeating braids.

In a non-repeating braid, the strands are overlaid back to front, without getting intertwined. To see this, consider the strand that starts at top right. At any crossing, it must have a positive slope, otherwise it would be crossing a strand that it had already crossed. Thus this strand is in front at all crossings, and can be lifted clear off the remaining ones. By induction, all strands can be lifted off, one by one; we can imagine that they lie on separate layers.

It follows that, using only the elementary moves corresponding to the relations 9.1.1 and 9.1.2, we can put a non-repeating braid in a canonical form that depends only on the permutation it induces. For example, one can shove all the crossings of the hindmost strand to the left of all other crossings, and continue with this process by induction. Thus two non-repeating braids that give rise to the same permutation are equivalent as positive braids (so the map $p : P \to S_n$ is injective when restricted to $D$), and in particular have the same length. Another consequence is that two non-repeating positive braids that represent the same element of $B$ represent the same element of $P$.

Likewise, $p : D \to S_n$ is surjective, that is, any permutation $\sigma \in S_n$ comes from a non-repeating braid. For we can write $\sigma$ as a geodesic string over $\tau_1, \ldots, \tau_{n-1}$, and replace each $\tau_i$ by $t_i$. The resulting string over $A$ must be a non-repeating braid: for if two strands crossed twice, we could erase the two crossings and get another (possibly inequivalent) braid mapping to $\sigma$. This would lead to a shorter word for $\sigma$ over $\tau_1, \ldots, \tau_{n-1}$ than the one we started from, which was supposedly a geodesic.

This argument shows that a positive braid that maps to a geodesic string over $\tau_1, \ldots, \tau_{n-1}$ is non-repeating, and that some non-repeating braid maps to any given geodesic string. But then every non-repeating braid maps to a geodesic string, for we already know that non-repeating braids representing the same permutation have the same length. The following lemma summarizes the preceding discussion (except for the last statement, which is obvious):

**Lemma 9.1.10 (permutation determines $D$ element).** *The homomorphism $p : P_n \to S_n$ restricts to a bijection $D \to S_n$. A positive braid $w$ is non-repeating if and only if the corresponding string over $\tau_1, \ldots, \tau_{n-1}$ is geodesic, that is, if and only if $|w| = |p(w)|$. If a non-repeating braid maps to a permutation $\sigma$, two strands $i$ and $j$ cross if and only if $(i, j) \in R_\sigma$ (where $i < j$ and strands are numbered by their positions at the beginning of the braid).* [9.1.10]

We will denote by $p^{-1} : S_n \to D$ the inverse bijection, so $p^{-1}(\sigma)$ will always mean an element of $D$. Of course, $p^{-1}$ is not a homomorphism, because the product of elements of $D$ is not always in $D$. But by considering geodesic representatives we see that if $R_\sigma \cap R_{\rho^{-1}} = \varnothing$, we do have $p^{-1}(\sigma\rho) = p^{-1}(\sigma)p^{-1}(\rho) \in D$. This proves the following lemma.

**Lemma 9.1.11 (product in $D$).** *If $a, b \in D$, we have $ab \in D$ if and only if $\neg(p(b)^{-1}) \geq p(a)$.* [9.1.11]

**Definition 9.1.12 (heads and tails).** We consider in the semigroup $P$ two partial orderings, defined as follows: if $ab = c$ for $a, b, c \in P$, we write $a \prec c$ and $c \succ b$, and say that $a$ is a *head* and $b$ is a *tail* of $c$. Watch out: $a \prec b$ and $b \succ a$ are not equivalent! (These two order relations can be associated with any semigroup. When the semigroup is a group, they are not very informative: any element is a head and tail of any other.)

We remark that any head or tail of an element of $D$ is also in $D$, because if a non-repeating braid is the concatenation of two positive subbraids, the two subbraids must also be non-repeating.

It will probably not come as a surprise that $p$ maps the order $\succ$ in $D$ to the order $\geq$ in $S_n$ (and also $\prec$ to $\leq$, provide $\leq$ is appropriately defined). Indeed, take $b, c \in D$ with $c \succ b$. Then there is $a \in D$ with $c = ab$, and by Lemma 9.1.10 we have $|p(c)| = |c| = |p(a)| + |p(b)|$, so that condition (a) in Lemma 9.1.7 (describing the order) is satisfied. Thus $p(c) \geq p(b)$. Conversely, if $\sigma \geq \rho \in S_n$, these two elements have geodesic representatives $s$ and $r$ over the generators $\tau_i$ with $r$ a suffix of $s$. Replacing $\tau_i$ by $t_i$ in $r$ and $s$, we get braids that represent the classes $p^{-1}(\rho)$ and $p^{-1}(\sigma)$ in $D$, again by Lemma 9.1.10. Clearly $p^{-1}(\sigma) \succ p^{-1}(\rho)$.

It follows that $D$ is a lattice with respect to $\succ$ and that $p : D \to S_n$ is a lattice isomorphism. We define a complementation operator $\neg$ in $D$ by pulling back the complementation operator of $S_n$. Although our main interest is in $D$, it is often convenient to switch back and forth between $D$ and $S_n$, because in $S_n$ we can take inverses. We must of course be careful when multiplying elements of $D$, as described in Lemma 9.1.11.

There is an important element $\Omega = \Omega_n \in P_n$, described physically by rotating the $n$ strands together 180° clockwise. We have $p(\Omega) = \omega$, where $\omega$ is the maximal element in $S_n$. We denote the image of $\Omega$ in $B_n$ by $\Omega$ as well; we will soon show that we are justified in doing so. One representative for $\Omega$, shown in Figure 9.2, is

**9.1.13.** $$(t_1 t_2 \ldots t_{n-1})(t_1 t_2 \ldots t_{n-2}) \ldots (t_1 t_2) t_1.$$

**Lemma 9.1.14 (characterizing $\Omega$ and $D$).** *A positive braid in the class $\Omega \in B$ is also in the class $\Omega \in P$. This happens if and only if the braid has length $|\omega| = \frac{1}{2}n(n-1)$ and maps to $\omega$ under $p$. A positive braid $a$ is in $D$ if and only if $\Omega \succ a$ (or, equivalently, $a \prec \Omega$) in $P$.*

*Proof of 9.1.14:* As already mentioned, the length and the permutation associated with two positive braids that map to the same element of $B$ are the same. Thus any positive braid representing $\Omega$ has same length and same permutation as 9.1.13.

Conversely, in a positive braid whose permutation is $\omega$, each pair of strands crosses by the last statement of Lemma 9.1.10; if the braid's length is $|\omega|$, the braid is non-repeating, and so represents $\Omega$ by the injectivity of $p : D \to S_n$ (Lemma 9.1.10).

We already know that a head or tail of $\Omega \in D$ is also in $D$ (see the remark following Definition 9.1.12). Conversely, given $a \in D$, we look at the permutation $p(a)$; Lemma 9.1.10 says that $|a| = |p(a)|$. There exists $b \in D$ such that $p(b) = p(a)^{-1}\omega$, by Lemma 9.1.10, and $|b| = |p(b)| = |\omega| - |p(a)|$. Thus $ab$ is a positive braid of the right length and giving the right permutation, so $ab = \Omega$ and $a \prec \Omega$. A similar argument shows that $\Omega \succ a$. 9.1.14

Using this, we can show that $D$ satisfies a right cancellation law: if $ab = a'b \in D$ with $a, a', b \in P$, we have $a = a'$. For $ab = a'b$ implies $p(ab) = p(a'b)$ and $p(a) = p(a')$, so $a = a'$ by Lemma 9.1.10 (we know that $a, a' \in D$ because they are heads of $ab \in D$). The same argument shows that $D$ satisfies a left cancellation law. In fact, we will see in the next section that both cancellation laws hold in all of $P$.

**Convention 9.1.15.** Because of right cancellation in $D$, if $b$ and $c$ are elements of $D$ with $c \succ b$, we can write $cb^{-1}$ for the unique element $a \in D$ such that $c = ab$.

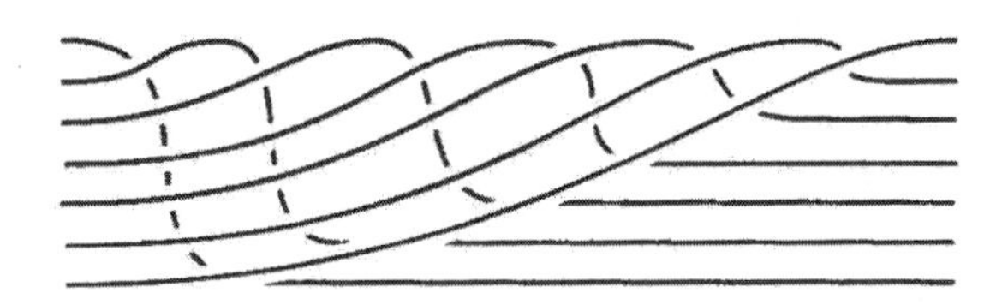

**Figure 9.2. The positive braid $\Omega_6$.** The braid is drawn in such a way as to correspond to Equation 9.1.13. A different drawing could give a different word in $H^*$.

To find out more about $\Omega$, we look at the semigroup automorphism of $A^*$ called the *flip*, which takes each generator $t_i$ to $\tilde{t}_i = t_{n-i}$. The name is justified because the image $\tilde{w}$ of a braid $w$ under this automorphism is indeed obtained by flipping $w$ around a horizontal axis. The flip is an involution, and since it preserves the relations 9.1.1 and 9.1.2, it defines an involution on $P$ (and also on $B$), also denoted by $b \mapsto \tilde{b}$.

**Proposition 9.1.16 (more about $\Omega$).** *For any $b \in P$, we have $\Omega b = \tilde{b}\Omega$ in $P$. The element $\Omega^2$ is central in $P$ and hence in the braid group. If $\Omega$ is a factor of $c$ in $P$, that is, if $c = a\Omega b$ with $a, b \in P$, then $c \succ \Omega$ and $\Omega \prec c$.*

*Proof of 9.1.16:* The last two statements follow immediately from the first. The first statement only needs to be proved for $b = t_i$. We look at the product $t_{n-i}(\Omega t_i^{-1})$ in $P$, where the notation is as in Convention 9.1.15. The image of this product in $S_n$ is $\tau_{n-i}\omega\tau_i^{-1}$, which is easily seen to equal $\omega$. Applying Lemma 9.1.14, we get $t_{n-i}(\Omega t_i^{-1}) = \Omega$, and therefore $t_{n-i}\Omega = \Omega t_i$. 9.1.16

**Proposition 9.1.17 (directed system).** *Given any element $a \in P$, there is an integer $n$ such that $\Omega^n \succ a$. In particular, the semigroup $P$ has the directed system property, that is, any two elements of $P$ are tails of some third element.*

*Proof of 9.1.17:* Let $w$ be a word representing $a$, and let it start with $t_i$. We multiply $w$ on the left by $\Omega(\Omega t_i^{-1})$, where the notation follows Convention 9.1.15. The result is a word $\Omega^2$ followed by a suffix of $w$. We move $\Omega^2$ to the right, and repeat the procedure for each character of $w$, to get an even power of $\Omega$. 9.1.17

## 9.2. Canonical Forms

We now construct a finite state automaton $M$ over $A$, having state set $D$. The interesting characteristic of $M$ is that, after a word $w$ is input, the state $d$ of $M$ is the maximal tail of $\overline{w} \in P$ that lies in $D$. Now *a priori* it is not obvious that such a maximal tail should exist—for example, it might be that applying $\vee$ to two elements of $D$, both of which are tails, gives something that is not a tail. But we will show that $M$ does behave as promised. Meanwhile, we use this defining property to find out what the initial state and the transitions of $M$ should be.

As a matter of notation, if $M$ is in state $a$ and reads a word $w \in A^*$, we will write $M(a, w)$ for the resulting state, or $M(w)$ if $a$ is the initial state. Taking $w = \varepsilon$, the nullstring, we see that the initial state should the class $e \in D$ of the trivial braid (the identity of $P$).

How about the transitions? Assume that $M$ is in state $a \in D$, and sees $t_i$ as the next character. Since $M$ has no record of the word $w$ that took it to state $a$, we must content ourselves with finding the maximal tail of $at_i$ that lies in $D$, and hope that this is also the maximal tail for $\overline{w}t_i$. There are two possibilities: if the two strands of $a$ that start (on the right) at positions $i$ and $i+1$ do not cross, $at_i$ is still in $D$. By Lemma 9.1.10 (permutation determines $D$ element), this happens if and only if $p(a) \not\geq \tau_i$, which can be rewritten $\neg\tau_i \geq p(a)$. If, on the contrary, the two strands cross, we set $M(a, t_i) = bt_i$, where $b$ is the maximal tail of $a$ in which the two strands do not cross (so that $bt_i \in D$). Now $b$ is given by $a \wedge \neg t_i = p^{-1}(p(a) \wedge \neg\tau_i)$, because we know that $p(a) \geq p(b)$ and $\neg\tau_i \geq p(b)$, and that $b$ is maximal with these properties. (Notice that we do not know yet that $M(a, t_i) = bt_i$ is a maximal tail of $at_i$; only that $b$ is a maximal tail of $a$ with $bt_i \in D$.)

To summarize, and observing that the second possibility above subsumes the first, we set

$$M(a, t_i) = p^{-1}(p(a) \wedge \neg\tau_i)t_i = (a \wedge \neg t_i)t_i.$$

We now extend this formula by induction. We first define $w^R$ as the reversal of a word $w \in A^*$; the map $w \mapsto w^R$ is an involutive anti-isomorphism of $A^*$ that respects the relations 9.1.1 and 9.1.2, and therefore defines an anti-automorphism of $P$. Physically, it corresponds to rotating a braid 180° around a vertical axis in the plane of the paper. Note that $p(\overline{w^R}) = p(\overline{w})^{-1}$, because $p(t_i) = \tau_i$ has order two for each $i$.

**Proposition 9.2.1 (action of $M$).** *If $M$ is in state $a$ and we input a word $w$ representing an element of $D$, its resulting state is*

$$M(a, w) = (a \wedge \neg\overline{w^R})\overline{w};$$

*the element $a \wedge \neg\overline{w^R} \in D$ is the maximal tail of $a$ that gives an element of $D$ when multiplied on the right by $\overline{w}$.*

*Proof of 9.2.1:* We first show that $b = a \wedge \neg\overline{w^R}$ satisfies the characterization given. Clearly $b$ is a tail of $a$. Thus $p(b) \leq \neg(p(\overline{w})^{-1})$, and $b\overline{w} \in D$ by Lemma 9.1.11. Let $b' \in D$ be another tail of $a$ with $b'\overline{w} \in D$. Then $\omega \geq p(b')p(\overline{w})$, and by Proposition 9.1.9 (weak invariance) we can multiply on the right to get $\neg p(\overline{w})^{-1} \geq p(b')$, or $\neg\overline{w^R} \succ b'$. Therefore $b = a \wedge \neg\overline{w^R} \succ b'$.

Next we show by induction that the formula for $M(a, w)$ is correct. If $|w| = 1$, this is just the definition of the transitions in $M$. So let $w = vt_i$; by the induction assumption,

$$M(a, w) = M(M(a, v), t_i) = ((a \wedge \neg\overline{v^R})\overline{v} \wedge \neg t_i)t_i.$$

Note that $t_i v^R = (vt_i)^R \in D$ and $\tau_i p(v)^{-1} \geq p(v)^{-1}$. Lemma 9.1.11 (product in $D$) now implies that $(\neg(t_i\overline{v^R}))\overline{v} \in D$. Applying $p$ and Lemma 9.1.10 (permutation determines $D$ element), we see that $\neg t_i = (\neg(t_i\overline{v^R}))\overline{v}$. By Proposition 9.1.9

$$M(a,w) = ((a \wedge \neg\overline{v^R} \wedge \neg(t_i\overline{v^R}))\overline{v})t_i = (a \wedge \neg(\overline{w^R}))\overline{w},$$

the term $\neg\overline{v^R}$ being absorbed by the one on its right. This completes the induction step. 9.2.1

Now remember that $t_i t_{i+1} t_i = t_{i+1} t_i t_{i+1}$ and $t_i t_j = t_j t_i$ for $j > i+1$ are elements of $D$. It follows that the state $M(a,w)$ depends only on $a$ and on the element $\overline{w}$ of $P$ represented by $w$, because $M$ satisfies the defining relations for $P$. In the light of this, we set $M(a,b) = M(a,w)$ for any $w$ representing $b \in P$, and we define $M(b)$ analogously.

We are now ready to show that $M$ does behave as promised at the start of this section: For any $b \in P$, the state $M(b)$ is the maximal tail of $b$ that lies in $D$. For certainly $M(b)$ is a tail for $b$. If $d \in D$ is another tail for $b$, with $b = ad$, say, we have $M(b) = M(M(a), d) \succ d$, where the last relation follows from Proposition 9.2.1 (action of $M$). Therefore $M(b)$ is maximal, as claimed.

The automaton $M$ can be enhanced slightly to become a *transducer*, or finite state automaton with output. When $M$ is in state $a$ and sees the character $t_i$, let it output the part of $a$ that is left behind in the transition, namely, $ab^{-1} \in D$, with $b = a \wedge \neg t_i$. We denote this output by $O(a,t_i)$; then $O(a,t_i)M(a,t_i) = at_i$. We define $O(a,w)$ by induction by setting $O(a,vt_i) = O(a,v)O(M(a,v),t_i)$; then $O(a,w)M(a,w) = a\overline{w}$. If $w$ represents an element of $D$, it is not hard to show by induction on the length of $w$ over the generators $t_i$ that $O(a,w)$ is still $ab^{-1} \in D$, with $b = (a \wedge \neg\overline{w^R})$. Another way of saying this is that, if $w \in D$, then $O(a,w)M(a,w) = a\overline{w}$ is a factorization of the right-hand side as the product of two elements of $D$, in such a way that $M(a,w)$ is maximal for $\succ$, or, equivalently, that $O(a,w)$ is minimal for $\prec$. Once again it follows that $O(a,w)$ depends only on the elements $a$ and $\overline{w}$ of $D$, and not on the representative $w \in A^*$. We set $O(a,b) = O(a,w)$ for any $w$ representing $b \in P$, and $O(b) = O(e,b)$, where $e$ is the identity. To summarize:

**Theorem 9.2.2 ($M$ finds maximal tail).** *Let $O(w) \in P$ and $M(w) \in D$ be the output and final state of the transducer $M$ when applied to a word $w$ over $A$. Then $O(w)M(w) = \overline{w}$ in $P$, and $O(w)$ and $M(w)$ depend only on the element $\overline{w}$ represented by $w$. Furthermore, $M(w)$ is the maximal tail of $\overline{w}$ that lies in $D$, and it is non-trivial (unless $w$ is the nullstring).* 9.2.2

Applying $M$ recursively to its own output, we obtain a factorization $\overline{w} = h_1 h_2 \ldots h_m$, where each $h_k$ is in $D$, and where $h_k$ is the maximal element of $D$ that satisfies $h_1 h_2 \ldots h_k \succ h_k$. In particular, $h_{k-1} \wedge \neg h_k^R = e$, that is, if two strands that are adjacent at the boundary of $h_{k-1}$ and $h_k$ cross in $h_{k-1}$, they also cross in $h_k$.

If we choose a particular braid representing each element of $D$, and replace each $h_k$ by its canonical representative $w_k$ in the expression above, we obtain the *right-greedy canonical form* of $\overline{w}$. This is illustrated in Figure 9.3.

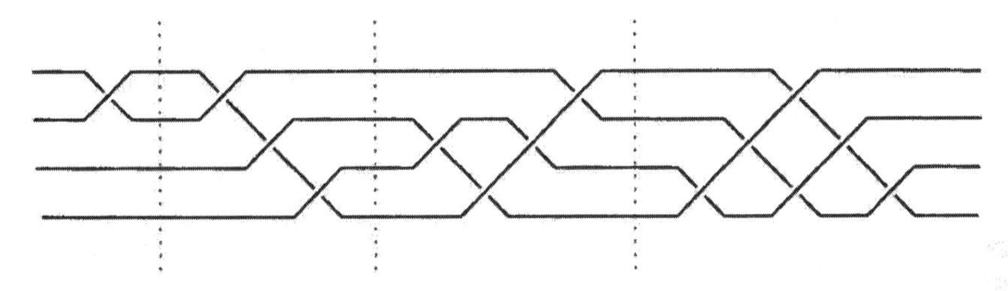

**Figure 9.3. Right-greedy canonical form.** Each chunk, starting from the right, is the maximal tail of the remaining positive braid.

A string over $A$ is in *left-greedy canonical form* if it is in right-greedy canonical form when read backward.

**Proposition 9.2.3 (characterizing canonical form).** *A string $w$ over $A$ is in right-greedy canonical form if and only if it has a decomposition*

$$w = w_1 w_2 \ldots w_m,$$

*where each $w_k$ is the canonical representative of $\overline{w_k} \in D$ and $\overline{w_{k-1}} \wedge \neg \overline{w_k^R} = e$, for $1 < k \leq m$ and $e$ the trivial braid. Geometrically, if two strands that are adjacent at the boundary of $w_{k-1}$ and $w_k$ cross in $w_{k-1}$, they also cross in $w_k$.*

*Likewise, $w$ is in left-greedy canonical form if it has a similar decomposition, but subject to the condition $\overline{w_k^R} \wedge \neg \overline{w_{k-1}} = e$ instead.*

*Proof of 9.2.3:* The characterizations for right- and left-greedy canonical form are clearly equivalent. We show the first one.

If $w$ is in right-greedy canonical form, it has a decomposition as claimed by definition: the condition $\overline{w_{k-1}} \wedge \neg \overline{w_k^R} = e$ follows from the maximality of $\overline{w_k}$, as observed above.

Conversely, let $w = w_1 w_2 \ldots w_m$ be a decomposition with $\overline{w_{k-1}} \wedge \neg \overline{w_k^R} = e$. We show that $\overline{w_m} = M(w)$, that is, that $w_m$ is the rightmost element of the right-greedy canonical decomposition of $\overline{w}$; by induction, it will follow that $w_1 w_2 \ldots w_m$ is the canonical decomposition of $\overline{w}$. To see that $\overline{w_m} = M(w)$, we also use induction: given that $\overline{w_{k-1}} = M(w_1 w_2 \ldots w_{k-1})$, we get $\overline{w_k} = M(w_1 w_2 \ldots w_k)$ by Proposition 9.2.1 (action of $M$) and the assumption $\overline{w_{k-1}} \wedge \neg \overline{w_k^R} = e$. 9.2.3

**Theorem 9.2.4 (canonical forms are regular).** *The set of positive braids that are in right-greedy canonical form is a regular language; similarly for left-greedy canonical forms.*

*Proof of 9.2.4:* We construct a finite state automaton $W$ that recognizes left-greedy canonical form—the result for right-greedy canonical form will then follow from Theorem 1.2.8 (reversal). $W$ must keep track of two chunks, $w_{k-1}$ and $w_k$, and check that they satisfy the condition $\overline{w_k^R} \wedge \neg \overline{w_{k-1}} = e$.

More precisely, the states of $W$ are pairs $(w_{\text{prev}}, w_{\text{cur}})$ of braids each with at most $\frac{1}{2}n(n-1)$ crossings. All pairs for which $\overline{w_{\text{cur}}^R} \wedge \neg \overline{w_{\text{prev}}}$ is not the identity are lumped together into a failure state. In the initial state, $w_{\text{prev}}$ and $w_{\text{cur}}$ are both the trivial braid; the accept states are those for which $w_{\text{cur}}$ is the canonical representative of its class.

When $W$ is in state $(w_{\text{prev}}, w_{\text{cur}})$ and sees a character $t_i$, it checks if $\overline{w_{\text{cur}}} \wedge t_i$ is the identity. If so, the two strands involved have not yet crossed within this chunk, and $W$ goes to state $(w_{\text{prev}}, w_{\text{cur}} t_i)$. Otherwise, $W$ goes to state $(w_{\text{cur}}, t_i)$ if $w_{\text{cur}}$ is the canonical representative of its class, and to the failure state otherwise. 9.2.4

The existence of canonical forms in $P$ has strong consequences. To begin with, we can show that left and right cancellation hold in $P$. It suffices to prove this for left and right multiplication by $\Omega$, since, by Proposition 9.1.17, every positive braid is a head and a tail of some power of $\Omega$. But cancellation of $\Omega$ is very easy to prove: if $b\Omega = c\Omega$, say, we write $b$ and $c$ in right-greedy canonical form as strings $w_b$ and $w_c$. Then $w_b\Omega$ and $w_c\Omega$ are still in right-greedy canonical form, and therefore are identical, showing that $w_b = w_c$. To prove left cancellation, use left-greedy canonical forms instead.

**Theorem 9.2.5 (Garside).** *$P$ embeds in $B$; that is, two positive braids that represent the same element in $B$ already coincide in $P$.*

*Proof of 9.2.5:* If two positive braids $w$ and $w'$ (that is, strings over $A$) represent the same element in $B$, there is some chain of possibly non-positive braids $w = w_0, w_1, \ldots, w_n = w'$ (that is, words over $A$) such that $w_k$ and $w_{k+1}$ differ in one of two ways: by the reduction or creation of a pair $t_i t_i^{-1}$ or $t_i^{-1} t_i$, or by one of the defining relations 9.1.1 and 9.1.2. This is an immediate consequence of the possibility of expressing $ab^{-1}$ in the form 2.2.2.

Let $N$ be the maximum number of occurrences of inverse generators $t_i^{-1}$ in all the $w_k$. We multiply each $w_k$ by $\Omega^{2N}$. Using the centrality of $\Omega^2$, we move an $\Omega^2$ to the place just before each inverse generator $t_i^{-1}$, which therefore gets replaced by a positive subbraid $\Omega(\Omega t_i^{-1})$. As a result, each $w_k$ is replaced by a positive braid; the braids corresponding to $w_k$ and $w_{k+1}$ differ either by one of the defining relations, or by the replacement of a pair $t_i\Omega(\Omega t_i^{-1})$ or $\Omega(\Omega t_i^{-1})t_i$ by $\Omega^2$. Except in the case of $t_i\Omega(\Omega t_i^{-1})$, it is immediate that this transformation is an equivalence in $P$. The case of $t_i\Omega(\Omega t_i^{-1})$ follows by right multiplication by $t_i$, the centrality of $\Omega^2$ and the cancellation property in $P$.

This shows that $\Omega^{2N}w = \Omega^{2N}w'$ in $P$, and therefore, by the cancellation property, that $w = w'$ in $P$. 9.2.5

## 9.3. The Braid Group Is Automatic

We extend the notion of canonical forms to non-positive braids. We know that left multiplication by $\Omega$ preserves left-greedy form, and likewise with "right" instead of "left." (Here we are using $\Omega$ to denote both an element of $P$ and its canonical representative.) We also know that any braid becomes positive when multiplied by a sufficiently large power of $\Omega$. To phrase this a different way, any braid is equivalent to the product, in either order, of a positive braid with a (possibly negative) power of $\Omega$.

We say that a word $w$ representing an element $b \in B$ is in *right-greedy form* if it has a decomposition of the form

$$w = w_1 w_2 \dots w_m \Omega^n,$$

where $n$ is positive or negative, $w_1 \dots w_m$ is a positive braid in right-greedy form, and $w_k \neq \Omega$ for all $k$. *Left-greedy form* is analogous, but the $\Omega$'s come at the beginning. The existence and uniqueness of left- and right-greedy forms for an element $b \in B$ follow from the previous paragraph and from the existence and uniqueness of the corresponding forms for positive braids. Theorem 9.2.4 (canonical forms are regular) clearly extends to canonical forms in $B$.

**Theorem 9.3.1 (right-greedy automation).** *The language of words in right-greedy form gives an automatic structure for the braid group.*

*Proof of 9.3.1:* By Theorem 2.3.5, we must show that the paths in the Cayley graph corresponding to the right-greedy forms of elements that differ by at most one generator are a bounded uniform distance apart. So suppose that $w = w_1 \dots w_m$ is a word in right-greedy form, and consider what happens when we try to put $wt_i$ in right-greedy form, where $t_i$ is a generator. If $w$

is not a positive word, it must end with one or more copies of $\Omega^{-1}$, which we can move past $t_i$ (possibly replacing $t_i$ by $\tilde{t}_i$). This gives a new word at a bounded uniform distance from $w$, because $\Omega^2$ commutes with $t_i$ by Proposition 9.1.16. From now on, then, we assume that $w$ is positive.

The final generator $t_i$ must be absorbed by the last chunk, $w_m$. This may force $w_m$ to give up some subchunk that must be absorbed by $w_{m-1}$, which as a result may spew up another subchunk, and so on, in a cascade. But the cascade effect is not very severe—the material exchanged between chunks is always less than $\Omega$! This is because, in the notation of Theorem 9.2.2 (see also the discussion leading to that theorem), we have $O(wd) = O(w)O(M(w), d)$, where $d$ is any element of $D$; here $d$ is the material coming into a chunk from the right, and $O(M(w), d) \in D$ is the material going out on the left.

It follows that the right-greedy forms for $w$ and $wt_i$, which we denote $w = w_1 \dots w_m$ and $w' = w'_1 \dots w'_{m'}$ (with $m'$ possibly differing from $m$ by 1), are uniformly close, because, for each $k$, the prefixes $w_1 \dots w_k$ and $w'_1 \dots w'_k$ differ in length by at most the length of $\Omega$, and their word difference is also bounded by $\Omega$. **9.3.1**

We have seen that right-greedy form changes only by a bounded amount under right multiplication by an element of $D$. How about left-greedy form? The language of words in left-greedy form is certainly regular, by Theorem 1.2.8 (reversal). As for the uniform distance condition, an analysis like the one in the previous proof would go more or less like this:

We start with a word $w = w_1 \dots w_m$ in left-greedy form, and multiply on the right by $t_i$. If $w_m$ cannot accept $t_i$ and still remain in $D$, the product is already left-greedy. Otherwise, $w_m$ can absorb $t_i$. Consequently, $w_{m-1}$ may now be able to accept a crossing from $w_m$: this will happen if the $i$-th and $(i+1)$-st strands were adjacent at the left edge of $w_m$, and if they had not crossed in $w_{m-1}$. Once $w_{m-1}$ accepts such a crossing, it can happen that it is now able to accept additional crossings from $w_m$, since new pairs of strands are now adjacent at the chunk boundary which may have crossed in $w_m$ but not in $w_{m-1}$... An entire cascade may be triggered, as crossings are handed down the line toward the earlier chunks, possibly allowing new crossings to flow into a certain chunk and freeing others which were jammed.

Despite this discouraging possibility, an uncontrolled landslide cannot actually happen, and only a bounded amount of material can flow past any chunk boundary. One way to see this is to consider right multiplication by $\Omega$ and the subsequent reorganization into left-greedy form, which transforms $w_1 \dots w_m \Omega$ into $\Omega \tilde{w}_1 \dots \tilde{w}_m$. What flows from the $(i+1)$-st to the $i$-th chunk is exactly $w_i^{-1}\Omega \in D$. It follows that only an element $g \prec w_i^{-1}\Omega$ can flow past any chunk boundary on right multiplication by any element of $D$, and in particular on right multiplication by a generator.

Another way to study the effect of right multiplication on left-greedy forms is based on a direct connection between the two canonical forms. By Proposition 9.2.3 (characterizing canonical form), right-greedy form can be identified by looking at chunk boundaries: if $w = w_1 \dots w_m$ is in right-greedy form, two strands of $w$ that are adjacent at the boundary between $w_i$ and $w_{i+1}$ and cross in $w_i$ also cross in $w_{i+1}$. Suppose we replace $w_i$ by the canonical representative $v_i$ of some element of $D$ such that strands adjacent at the right edge of $v_i$ cross in $v_i$ if and only they do not cross in $w_i$; and that we replace $w_{i+1}$ by $v_{i+1}$ such that strands adjacent at the left edge of $v_{i+1}$ cross in $v_{i+1}$ if and only they do not cross in $w_{i+1}$. Then the condition for *left-greedy* form will be satisfied at the boundary between $v_i$ and $v_{i+1}$. In terms of permutations, we want $p(\overline{v_i}) = \omega p(\overline{w_i})$ and $p(\overline{v_{i+1}}) = p(\overline{w_{i+1}})\omega$, which will be the case if we take

**9.3.2.** $$\overline{v_i} = (\overline{w_i}^{-1}\Omega)^R \quad \text{and} \quad \overline{v_{i+1}} = (\Omega\overline{w_{i+1}}^{-1})^R$$

(remember that reversal and inversion in $S_n$ are the same). Of course Equation 9.3.2 cannot hold for all values of $i$, but if we define $v_i$ by this equation for *even* values of $i$, so the left-greedy condition is satisfied at even-odd boundaries, it turns out that the left-greedy condition is satisfied at odd-even boundaries as well, because of identity $\Omega b = \tilde{b}\Omega$ (Proposition 9.1.16) and of the fact that canonical forms are invariant under flipping. We thus have the following lemma:

**Lemma 9.3.3 (greedy duality).** *Let $w = w_1 \dots w_m$ be a positive word in right-greedy form, and let $v = v_1 \dots v_m$ be defined by Equation 9.3.2 (for $i$ even). Then $v$ is a positive word in left-greedy form. Conversely, if $w$ is in left-greedy form, $v$ is in right-greedy form.* [9.3.3]

Now let $w$ and $w'$ be the left-greedy forms of elements that differ by right multiplication by one generator. By multiplying on the left by a large power of $\Omega$, we can assume that the elements are positive. If the number of chunks in $w = w_1 \dots w_m$ and $w' = w'_1 \dots w'_{m'}$ is different, we pad one of the words on the right with chunks equal to the trivial braid. Using Lemma 9.3.3, we transform $w$ and $w'$ into new words $v = v_1 \dots v_m$ and $v' = v'_1 \dots v'_m$ in right-greedy form. The elements represented by $v$ and $v'$ differ by right multiplication by an element of $D$, and therefore the amount of material that passes from one chunk to the next upon multiplication is bounded by $\Omega$, as explained in the proof of Theorem 9.3.1. But the process of adjusting $v$ into $v'$ is exactly dual to the process of adjusting $w$ into $w'$! This shows again the desired result that $w$ and $w'$ are uniformly close, and therefore that the language of words in left-greedy form is automatic.

**Theorem 9.3.4 (braid group is biautomatic).** *Words in left-greedy (or words in right-greedy form) constitute a language that gives an automatic structure for the braid group.*

*Proof of 9.3.4:* Theorem 9.3.1 (right-greedy automation) says that right-greedy form changes little under right multiplication by a generator. This is the same as saying that left-greedy form changes little under left multiplication by a generator. We have just seen that left-greedy form changes little under right multiplication by a generator. The result now follows from Lemma 2.5.5 (characterizing biautomatic). 9.3.4

In fact we have an even stronger result. Recall that a *symmetric automatic structure* is one for which the inverse of an accepted word is also accepted. We say that a word is in *mixed canonical form* if it is a word of minimal length of the form $u^{-1}v$, where $u$ and $v$ are both in left-greedy form. We will show that the set of words in mixed canonical form gives an automatic structure, which is clearly invariant under inversion. But first we must show that mixed canonical form is well defined:

**Lemma 9.3.5 (mixed canonical form).** *Every element of $B$ has exactly one representative satisfying the defining properties of mixed canonical form. If $u = u_1 \dots u_m$ and $v = v_1 \dots v_m$ are braids in left-greedy form, $u^{-1}v$ is in mixed canonical form if and only if $u_1^R \wedge v_1^R = e$, where $e$ is the identity.*

*Proof of 9.3.5:* First we prove the "only if" part of the last statement. Let $u$ and $v$ be as in the statement, and assume $u^{-1}v$ is in mixed canonical form. Since $u^{-1}v$ has minimal length, two strands adjacent at the boundary between $u^{-1}$ and $v$ cannot cross in both $u_1$ and $v_1$: otherwise we would be able to cancel the negative crossing in $u_1^{-1}$ with the positive crossing in $v_1$. In symbols, this means that $\overline{u_1}^R \wedge \overline{v_1}^R = e$.

We now prove uniqueness. Suppose that $w = u^{-1}v$ and $w = u'^{-1}v'$ satisfy the conditions for mixed canonical form, and that they represent the same element of $B$. We multiply $w$ and $w'$ on the left by a large power of $\Omega$. In $w$, we move one copy of $\Omega^2$ (which is central) to the boundary between each pair $u_{i+1}^{-1}$ and $u_i^{-1}$, for $i$ odd; we do the same for $w'$. The result is a positive word in left-greedy form: this follows from Proposition 9.2.3 (characterizing canonical form) by the argument leading to Lemma 9.3.3 (greedy duality), except that we have to check the left-greedy condition at the boundary between $u$ and $v$ as well. The latter is true as well: we have $\overline{u_1}^R = \neg(\Omega\overline{u_1}^{-1})$, so the previous paragraph implies $\neg(\Omega\overline{u_1}^{-1}) \wedge \overline{v_1}^R = e$, as required.

Since $w$ and $w'$ represent the same element of $B$, so do the words obtained in this way, and since they are both in left-greedy form the two words must be identical. But then $w$ and $w'$ are identical as well, since the process above is obviously reversible. This concludes the proof of uniqueness, and also shows the "if" part of the last statement of the lemma, because the argument did not use the minimality condition in the definition of mixed canonical form—it only used the left-greediness of $u, v, u', v'$ and the equations $u_1^R \wedge v_1^R = e$ and ${u'_1}^R \wedge {v'_1}^R = e$.

Existence of the mixed canonical form is also easy. Any element of $B$ can be expressed as a product $u^{-1}v$ with $u$ and $v$ positive and left-greedy (for example, we can take $u$ to be a power of $\Omega$); we just choose such an expression having minimal length. 9.3.5

**Theorem 9.3.6 (braid group has symmetric automation).** *Mixed canonical form gives a symmetric automation for the braid group.*

*Proof of 9.3.6:* We first show that the language of words in mixed canonical form is regular. In the previous section we constructed a finite state automaton $M$ that recognizes the largest element of $D$ that is a tail of a given positive braid (Theorem 9.2.2). There are also automata $L$ and $R$ that check whether a positive braid is in left-greedy or right-greedy form (Theorem 9.2.4). We call $M^{-1}$ and $R^{-1}$ the automata obtained by replacing the arrow labels of $M$ and $R$ by their inverses (we are *not* reversing the arrows).

There is also a finite-state automaton $Q$ that finds the longest prefix $\alpha$ of a positive braid $p$, such that $\overline{\alpha} \in D$. This is obvious, because this maximal prefix must be apparent after at most $|\Omega| + 1$ crossings have been examined. (We don't require that $Q$ find the maximal head in $D$, although there is an automaton that does that, too.)

To test a word $w$ for mixed canonical form, we run $M^{-1}$ and $R^{-1}$ on it so long as only negative generators are encountered. If $u^{-1}$ is this longest negative prefix of $w$, we check if $R^{-1}$ is in the accept state after $u^{-1}$ has been read. If not, $u$ is not in left-greedy form, so $w$ is not in mixed canonical form.

If $R^{-1}$ is in the accept state after reading $u^{-1}$, we save the state of $M^{-1}$. Next we apply $L$ and $Q$ to the remaining word $v$, keeping an eye for the reappearance of negative generators. If we ever find negative generators again, or if $L$ does not end up in the accept state, $w$ fails. If not, we compare the state of $L^{-1}$, which we saved, with the state of $Q$, to see if any cancellations are possible between the rightmost chunk of $u^{-1}$ and the leftmost chunk of $v$, in which case again the word is rejected. Otherwise, the word is accepted.

The fact that right multiplication by a generator changes mixed canonical form by a bounded amount follows by an argument similar to the one following Lemma 9.3.3 (greedy duality). As for symmetry, it is clear that the

inverse of a word in mixed canonical form is also in mixed canonical form.
9.3.6

**Corollary 9.3.7 (braid group is lattice).** *The braid group is a lattice with respect to the partial ordering defined by $b \succ c$ if and only if $bc^{-1}$ is a positive braid. The partial ordering $\succ$ and the lattice structure are invariant under multiplication on the right.*

Note that $\succ$ is an extension of the partial ordering of the same name on $P$, coming from the tail relation. However, $\succ$ is not the tail relation on the group $B$ considered as a semigroup (as already observed, this latter relation is trivial).

*Proof of 9.3.7:* Given two braids $b_1$ and $b_2$, let $u_1^{-1}u_2$ be the mixed canonical form of $b_1b_2^{-1}$. We show that $b_1 \vee b_2 = u_2b_2 = u_1b_1$ is the least braid that dominates $b_1$ and $b_2$. Clearly $b_1 \vee b_2 \succ b_1$ and $b_1 \vee b_2 \succ b_2$. Now suppose that $d \succ b_1$ and $d \succ b_2$, so that $d = q_1b_1 = q_2b_2$, where $q_1$ and $q_2$ are positive. Then $b_1b_2^{-1} = q_1^{-1}q_2$. We write $q_1$ and $q_2$ in left-greedy canonical form, and form the product $q_1^{-1}q_2$. By Lemma 9.3.5, this can be put in mixed canonical form after a finite number of cancellations at the boundary between $q_1^{-1}$ and $q_2$. In other words, there exist $p, r_1, r_2 \in P$ such that $q_1 = pr_1$, $q_2 = pr_2$ and $r_1^{-1}r_2$ is the mixed canonical form of $b_1b_2^{-1}$. But then $r_1 = u_1$ and $r_2 = u_2$, showing that $d = q_1b_1 \succ u_1b_1 = b_1 \vee b_2$.

To describe $b_1 \wedge b_2$, we notice that $a \succ b$ is equivalent to $b^- \succ a^-$, where $b^-$, the *negation* of $b$, is the braid obtained from $b$ by replacing every positive crossing by the corresponding negative crossing, and vice versa. If $u_1^{-1}u_2$ is the mixed canonical form of $(b_1b_2^{-1})^-$, it follows that $b_1 \wedge b_2 = u_1^-b_1 = u_2^-b_2$.
9.3.7

We now consider the connection between braid groups and mapping class groups. The braid group on $n$ strands acts as a group of diffeomorphisms of the $(n+1)$-punctured sphere, where the extra puncture is $\infty$. The kernel of this action is the centre of the braid group, generated by $\Omega^2$. The image has index $n+1$: it differs from the entire mapping class group only in that it fails to permute $\infty$ with the other points.

**Theorem 9.3.8 (mapping class group is automatic).** *The mapping class group for the $(n+1)$-punctured sphere is automatic.*

*Proof of 9.3.8:* The mixed canonical form for a word in the braid group is only dependent in a very simple way on the power of $\Omega$. One can represent every element in the quotient $B_n/\langle\Omega^2\rangle$ by words in mixed canonical form, but disallowing those where $\Omega$ or $\Omega^{-1}$ appears, unless the word is $\Omega$ or $\Omega^{-1}$.

This defines a regular collection of words, closed under inversion. (Alternatively, we could allow all words in mixed canonical form, if we don't mind having infinitely many representatives of a group element). There are several accepted words representing each element in the quotient. It is easy to check equality in the quotient group: one reads the two words from left to right; if they are equal, the chunk boundaries must match, and the word difference must be either $\Omega$ or the identity; at the end of the word, the difference must be the identity. The finite state automata for checking multiplication by generators are derived, in a similar way, from the algorithms for $B_n$. 9.3.8

**Open Question 9.3.9.** Does the mapping class group for the $(n+1)$-punctured sphere admit a symmetric automation?

**Open Question 9.3.10.** Do the current algorithms generalize to mapping class groups of other surfaces?

**Open Question 9.3.11.** Is there a Lorentz metric on the Teichmüller space for an $n$-punctured disk *rel* boundary such that a braid is positive if and only if it can be represented by a time-like path through the base point in the corresponding modular space (whose fundamental group is the braid group)?

Note that if black holes exist in our universe, the Lorentz metric on our space-time does not make it a lattice: for two people who have entered two distinct black holes, their possible futures have an empty intersection. In contrast, Teichmüller space shouldn't have black holes.

**Open Question 9.3.12.** Does the Teichmüller space for the $n$-punctured disk have a lattice structure extending the lattice structure on the braid group?

## 9.4. The Conjugacy Problem

When the left end of a braid is brought around in a circle and joined to the right end, it becomes a *closed braid*. The conjugacy class of the word representing a braid determines the isotopy class of the corresponding closed braid.

Since $\Omega^2$ is in the centre of the braid group, two elements $g$ and $h$ are conjugate if and only if $\Omega^{2k}g$ is conjugate to $\Omega^{2k}h$, for any $k$. Therefore, the conjugacy problem in the braid group can be reduced to the conjugacy problem for elements of $P$.

The question arises of whether an element of $P$ contains $\Omega$ up to conjugacy, that is, whether it is conjugate to an element of $P$ of the form $a\Omega b$. Recall from Proposition 9.1.16 that if an element of $P$ has an $\Omega$ anywhere it has an $\Omega$ as a tail and as a head.

**Theorem 9.4.1 (find $\Omega$ up to conjugacy).** *An element $g \in P$ is conjugate to an element that contains $\Omega$ if and only if there is a $d \in D$ such that $dg(d^{-1}\Omega) \succ \Omega^2$, or equivalently, if $M(O(d, g(d^{-1}\Omega))) = \Omega$ (see Section 9.2 for the notation).*

*There is a finite state automaton $M'$ that tells whether $g$ contains $\Omega$ up to conjugacy; if so, $M'$ produces an element $d \in D$ that conjugates $g$ to a word containing $\Omega$.*

*Proof of 9.4.1:* The conditions $dg(d^{-1}\Omega) \succ \Omega^2$ and $M(O(d, g(d^{-1}\Omega))) = \Omega$ are obviously equivalent: the second equation says that when $M$ is fed the word $dg(d^{-1}\Omega)$ twice, it finds an $\Omega$ at the end of the second pass, so therefore on the first pass as well.

If $dg(d^{-1}\Omega) \succ \Omega^2$, we have $dgd^{-1} \succ \Omega$, so $g$ contains an $\Omega$ up to conjugacy.

Conversely, suppose that $g$ contains an $\Omega$ up to conjugacy; explicitly, choose $X$ so that $XgX^{-1}$ is in $P$ and contains $\Omega$. We can multiply $X$ by a suitable power of $\Omega$ until $X$ is in $P$ and contains no $\Omega$'s. For if $XgX^{-1} = u\Omega$ for $u \in P$, we have $(\Omega^{\pm 1}X)g(\Omega^{\pm 1}X)^{-1} = \Omega^{\pm 1}u\Omega\Omega^{\mp 1} = \Omega u$.

Write $X = x_1 \dots x_m$ in right greedy form, so that

$$XgX^{-1} = x_1 \dots x_m g y_m \dots y_1 \Omega^{-m},$$

where $y_i = x_i^{-1}\Omega$ if $m-i$ is even or $\Omega x_i^{-1}$ if $m-i$ is odd. Note that $y_m = x_m^{-1}\Omega$.

We claim that the product $x_1 \dots x_m g$ must contain at least one $\Omega$. Indeed, each additional $y_i$ can promote at most one chunk of its right-greedy form to $\Omega$, so if there is no $\Omega$ in $x_1 \dots x_m g$, there are at most $m$ in $x_1 \dots x_m g y_m \dots y_1$, so there could have been no $\Omega$ in the entire word $XgX^{-1}$.

Therefore, $\Omega = M(x_1 \dots x_m g) = M(x_m, g)$, so $x_m g$ contains $\Omega$, say $x_m g = \Omega h$. We claim that $hx_m^{-1}\Omega$ must contain $\Omega$: if not, a machine like $M$, while reading the word $hy_m \dots y_1$ from right to left, would find no $\Omega$, since its final state is the same as if it had only read $hx_m^{-1}\Omega$ from right to left. The prefix $x_1 \dots x_{m-1}$ would then increase the number of $\Omega$'s by at most $m-1$, which is not enough.

This shows that $x_m g x_m^{-1}\Omega$ must contain $\Omega^2$, as desired.

The entire analysis of $g$ can easily be done in one pass, by a finite state automaton $M'$. It is constructed by hooking the transducer output of $M$ to another copy of $M$, and then running it with a "non-deterministic" initial state: that is, simultaneously check each state $S_d$ that the composite machine would be in, if it had started in state $d$. More formally, the state space of $M'$ is a function from $D$ to $D \times D$, and the start state is the function which takes $d$ to $(d, 1)$. The transition function on input $t_k$ takes the first coordinate to the same thing $M$ would have, and the second coordinate to the result of applying the transducer output of the first transition as input to the $M$ starting in the state specified by the second coordinate. 9.4.1

Garside's algorithm for the conjugacy problem is based on the following principle:

**Theorem 9.4.2 (conjugate within positive).** *The equivalence relation of conjugacy on $P$ is generated by conjugacy by elements of $D$.*

This would be obvious if we were not restricting to $P$. The statement says that if $p, q \in P$ are conjugate in $B$, then there is a sequence $p = p_0, p_1, \ldots, p_m = q$ such that there are elements $d_i \in D$ with $d_i p_i = p_{i+1} d_i$.

*Proof of 9.4.2:* This is similar to the previous proof.

If $p$ and $q$ are elements of $P$ which are conjugate in $B$, then there is $X \in P$ such that $XpX^{-1} = q$. As before, we write $X = x_1 \ldots x_m$, in right-greedy form. Then we have $q = x_1 \ldots x_m p y_m \ldots y_1 \Omega^{-m}$, where $y_i = x_i^{-1}\Omega$ or $\Omega x_i^{-1}$ depending on the parity of $m - i$. The product $x_1 \ldots x_m p x_m^{-1}\Omega$ must contain at least one $\Omega$, for otherwise the entire product excluding the negative power of $\Omega$ would contain at most $m - 1$ $\Omega$'s, which is not enough. Therefore, $x_m p x_m^{-1}\Omega$ contains an $\Omega$, so $p_1 = x_m p x_m^{-1}$ is positive. The construction proceeds recursively on $m$. 9.4.2

Using this theorem, all positive elements conjugate to $p$ can be algorithmically enumerated, by applying all possible conjugacies by elements of $D$ to all known elements conjugate to $p$ until no new elements are found.

The speed of this algorithm depends on how many elements of the positive braid semigroup are conjugate to a given element. (If it helps, one may restrict to $\Omega$-free positive braids.) The number of such elements grows exponentially at least if the number of strands is allowed to grow. As a simple example, consider a two-strand closed braid with $m$ twists, rearranged by wrapping it $k$ times around a circle to form a $2k$ strand braid. This can be expressed as

$$t_1^{m_1} t_3^{m_2} \ldots t_{2k-1}^{m_k} t_2 t_1 t_3 t_2 \ldots t_{2k-1} t_{2k-2},$$

where $\sum_i m_i = m$. For fixed $k$ and $m$, these elements of the positive braid semigroup are conjugate but distinct. This gives $C(m, k)$ different representatives.

This example can be modified slightly to make it seem less obviously trivial. For instance, there are $m^k$ conjugate elements

$$t_1^{m_1} t_3^{m_2} \ldots t_{2k-1}^{m_k} \quad t_2 t_4 \ldots t_{2k-2}^2 \quad t_1^{m-m_1} t_3^{m-m_2} \ldots t_{2k-1}^{m-m_k}$$

with $(k+1)m - 1$ crossings.

**Open Question 9.4.3.** Is the number of positive braids conjugate to a given positive braid bounded by a polynomial in the number of strands? We think the answer is yes; perhaps there is a proof based either on the geometry of Teichmüller space, or on the action of the braid group on the sphere of measured laminations.

There is one potential trap in an attempted simplification of the search for positive words conjugate to $p$: it is not sufficient to check only conjugacies of the form $gpg^{-1}$, where $g \in D$ is a tail of $p$. For instance, conjugating the entire braid by $\Omega$ is often not accessible by operations of the form above. The 3-strand braid $t_1^2 t_2^3$ illustrates this fact: conjugating by a tail in $D$ only accomplishes the cyclic permutations, while conjugation by $\Omega$ takes it to $t_2^2 t_1^3$.

This can be woven into more complicated examples: for instance, an $m$-strand example of this type can be wound in a neighbourhood of a strand of a $k$-strand example, to show that it is necessary to conjugate by more than $\Omega$ plus tails of the braid.

**Open Question 9.4.4.** Is there a polynomial-time algorithm in the spirit of this chapter for testing conjugacy of braids? (We think there is a polynomial time algorithm using the theory of pseudo-Anosov homeomorphisms of surfaces, but that is in a different spirit until the next question is answered.)

**Open Question 9.4.5.** Is there a connection between the finite state algorithms here, and the algorithms connected with the pseudo-Anosov theory?

## 9.5. Complexity Issues

The algorithm for the word problem developed by Artin [Art47] involves putting a braid in a canonical form whose length has only an exponential bound in terms of the length of the initial word: this is clearly unsatisfactory for practical purposes. Garside [Gar69] did not analyze his algorithms for the word problem and the conjugacy problem for difficulty, and at least superficially, they are exponentially slow. Birman's treatment [Bir75] of Garside's algorithms also does not discuss speed.

In the preceding sections we constructed finite state automata to put a braid in one of several canonical forms. These algorithms can be and have been implemented on computers in practical and reasonably efficient ways, for example by Bruce Ramsay.

Of course, any particular finite state automaton or transducer can be made to operate in linear time on the length of its input. The problem is to make algorithms which have a reasonable dependence on the number of strands. The obvious implementation of the automaton $M$ of Section 9.2, for example, is to create a table whose rows are permutations, and whose columns are generators. This rapidly grows impractical as the number of strands increase.

Instead, one can work with elements of $D$ as generators of the braid group, and represent them as permutations. A permutation can be described by a table of $n$ integers. With this notation, the group operations of multiplication and inversion can be performed quickly, in time proportional to $n$. An implementation of $M$ depends also on an implementation of the lattice operations.

We would like to thank Bernard Chazelle for pointing out the following result:

**Proposition 9.5.1 (timing of lattice operations in $S_n$).** *The lattice operations join and meet, or $\vee$ and $\wedge$, in the symmetric group $S_n$ can be implemented in time $O(n \log n)$. The negation operator can be implemented in time $O(n)$, and so can the product of two permutations.*

*Proof of 9.5.1:* The result about the timing for a product is obvious, if we hold a permutation as an array. Since negation is given by $\neg a = \omega a$, the timing of this operation is no greater than the timing of a product.

Let $\sigma \in S_n$ be a permutation, and let $J$ be a subset of $\{1, \ldots, n\}$ consisting of a block of consecutive numbers, $J = [L+1, L+k]$, where $0 \leq L < L+k \leq n$. We say that $J$ is *sorted according to* $\sigma$ if its elements are written as a $k$-tuple $(j_1, \ldots, j_k)$, where $\sigma(j_1) < \ldots < \sigma(j_k)$. Another way of looking at this is as follows. We define $\omega_J$ to be the permutation which keeps all elements outside $J$ fixed and which reverses the order of every element of $J$. Let $\rho = \omega_J \wedge \sigma$. Then $J$ is sorted according to $\sigma$ if and only if $j_i = \rho^{-1}(i + L - 1)$. So sorting $J$ according to $\sigma$ is equivalent to specifying $L$, $k$ and $\rho$.

The idea of the algorithm for the meet is to use a merge sort: given two permutations $a$ and $b$, divide the $n$ items into two lists of approximately half size, recursively sort these according to $a \wedge b$, and then merge the results. For the merging step, one repeatedly compares the lower elements of the two lists, and moves the lower of the two to the merged list.

In a conventional sorting situation, comparisons of two elements are easy. However, the linear ordering of $a \wedge b$ is defined recursively, and so the comparison is more difficult. We must do something to speed up the comparisons, or the time will end up being quadratic in $n$.

Let $X$ and $Y$ be adjacent disjoint blocks of consecutive integers in $\{1, \ldots, n\}$, each of which is sorted according to $a \wedge b$, and suppose that, if $x \in X$ and $y \in Y$, then $x < y$. To merge $X$ and $Y$, we first write down the elements of $X$ followed by the elements of $Y$, and then decide how to move elements of $X$ past elements of $Y$, while maintaining the order within $X$ and within $Y$. The criterion that $x \in X$ should be transposed with $y \in Y$, when sorting according to $a \wedge b$, is firstly that the infimum of the function $a$ among the items $x$ and rightward in $X$ is greater than the supremum of $a$ among the items $y$ and leftward in $Y$, and secondly that the same thing should be true for the function $b$. If this is so, the block consisting of $x$ and rightward in $X$ can be transposed with the block consisting of $y$ and leftward in $Y$.

We discussed above the meaning of "sorting according to $\sigma$" for a permutation $\sigma$, in terms of the lattice operations. In that interpretation, we have found $\rho_X = \omega_X \wedge a \wedge b$ and $\rho_Y = \omega_Y \wedge a \wedge b$. In order to find $\rho_{X \cup Y}$, we need to find the permutation $\alpha$, which fixes everything outside $X \cup Y$, preserves the order of elements of $X$, preserves the order of elements of $Y$, possibly interchanges some elements of $X$ with some elements of $Y$ and satisfies $\rho_{X \cup Y} = \alpha \rho_X \rho_Y$. (Note that $\rho_X$ and $\rho_Y$ commute.)

Starting from the right end of $X$, we make a pass toward the left, and, at each position, make a record of the infimum of $a$ at all positions to its right and similarly for $b$. Similarly for $Y$ we record the supremum of $a$ on all positions to the left of any given position and similarly for $b$. Once this is done, $a \wedge b$ comparisons of the two lists take constant time, so merging can proceed as for a conventional merge sort.

The time for merging is proportional to the total length of the two lists. The recursion involves $\log n$ levels, and the time spent for merges at each depth of recursion is proportional to $n$. Therefore, the running time is $O(n \log(n))$.

The procedure for $a \vee b$ is $\neg((\neg a) \wedge (\neg b))$. 9.5.1

**Corollary 9.5.2 (timing of $M$).** *The operation of the finite state automaton $M_n$, which recognizes the largest suffix of a positive braid with $n$ strands which is a prefix of $\Omega_n$, on input of a braid with word length $m$, can be simulated in time $O(mn \log n)$ (using either the conventional generators or the generating set $D$).*

Depending on the model of computation one chooses, it is also possible to construct another implementation of $M_n$ (using a table) in time $O(n^2 n! \log n)$ which will run in time which is linear in $m$ and independent of $n$; this would become completely impractical for say $n = 15$, since $15! = 1{,}307{,}674{,}368{,}000$. Note that it would take a similar amount of space to store the table.

**Corollary 9.5.3 (timing of canonical form).** *An $n$-strand braid of length $m$ can be put in any of its canonical forms, expressed in terms of the generating set $D$, in time $O(m^2 n \log n)$.*

*Proof of 9.5.3:* The right-greedy canonical form of a positive word can be constructed by iteratively applying $M_n$. This requires $O(m)$ passes, so the total time is quadratic in $m$.

For a word of mixed sign expressed in terms of the generating set $D$, any positive-negative pair in the product, $gh^{-1}$ can be replaced by a negative-positive pair $(g\Omega^{-1})(\Omega h^{-1})$, and vice-versa. Thus, all the negative terms can be arranged to come either at the beginning or at the end. Then the previous case with trivial variations can be applied, to put the word in any of the canonical forms. 9.5.3

There is another strategy for putting elements in canonical form which seems to work more efficiently in practice, but we don't know how to prove it is more efficient. Consider a positive word, to be put in right-greedy canonical form. It can be done by iteratively arranging initial subwords in canonical form. When multiplied on the right by an element of $D$, there is a simple cascade, which takes at most time $O(mn \log n)$, where $m$ is the length of the current subword. Thus, the entire process takes time at most $O(m^2 n \log n)$.

However, the cascade does not always propagate completely to the left, so potentially there may be a large savings. Experimentation with small numbers of strands indicates that for a random braid, the cascade propagates completely to the left only a small but definite proportion of the time. As the number of strands increases, if a word is chosen randomly with respect to conventional generators, the percentage of time in which the cascade propagates all the way decreases.

A further improvement of the algorithm seems to cut the average distance of propagation of the cascade to roughly a constant (something like five) for modest width random braids of long length. The trick is to stop the propagation whenever a $\Omega$ appears. Any $\Omega$ in the middle of the word is destined to drop all the way to the left — but the trail it leaves is completely predictable, it simply conjugates all the intermediate units by $\Omega$, which is a simple flip. The cascades can be simulated, stopping if it creates a $\Omega$ (becomes an avalanche) or decays to nothing. At least in the "random" situations we have tested, this method makes canonical form take time approximately linear in $m$. It is easy, at the end, to gather all the $\Omega$'s to the beginning where they belong.

It is certainly possible to defeat the above strategy, by feeding in mischievous sequences of generators. For instance, one can have a braid which is obtained from a six-strand braid by replacing one of its strands with a ten-strand braid. Any time a $\Omega$ appears for the secret ten-strand sub-braid (which can be carefully disguised), this will propagate all the way to the initial set of $\Omega$'s for the sub-braid.

Perhaps it is possible to construct a careful data-structure which will enable one to quickly predict the result of the propagation of a cascade over a long stretch, but it seems tricky.

**Proposition 9.5.4 (timing of products).** *The canonical form for the product of two group elements of lengths $m$ and $p$ in the $n$-strand braid group can be found in time $mpn \log n$.*

*Proof of 9.5.4:* This can be done by using the cascade process to incorporate generators of the smaller word into the canonical form of the larger word, as described above. [9.5.4]

Note that the expected running time may be much lower, but we don't know how to prove such a result.

**Corollary 9.5.5 (timing of lattice operations in $B_n$).** *If $w$ and $v$ are words of length $m$ and $p$ in generators $D$ in $B_n$, then the lattice products $w \wedge v$ and $w \vee v$ and the questions of inequalities $w \succ v$ and $v \prec w$ can be computed in time $O((m+p)^2 n \log n)$.*

*Proof of 9.5.5:* To find the meet, $w \wedge v$, compute the symmetric canonical form for $wv^{-1}$, and form the product of $v$ with its negative piece, as described in Corollary 9.3.7 (braid group is lattice). The join $w \vee v$ is similar. The discussion for $\prec$ is parallel to this. [9.5.5]

For small numbers of strands, there are ways to speed up the implementations. This hardly matters for the operation of putting braids in canonical form, because it is quick enough anyway; but for an attempt to use this approach for a practical classification of conjugacy classes, it could be important.

As an example, for braids having six strands, there are 15 pairs of strands. This number is less than 16, which is the size of a quickly accessed chunk of memory in many computers. If a permutation of six or fewer objects is represented by its set of order reversals, then this set can be encoded in the bits of a word. Computation of $\succ$ in $S_n$ is then extremely quick: it is a bitwise operation on words. Complementation is also quick. To compute the meet, as part of initialization, one can compute a table, for each of the 32,768 bit patterns of length 15, the result of bubble-sorting with these comparisons. (One could instead compute the entire $6! \times 6! = 518{,}400$ table for $\wedge$, but that is more time and space). Then $a \wedge b$ is obtained by forming the bitwise logical $\wedge$, then looking up the answer from the table.

Inversion in $S_n$ is also needed to implement the various algorithms. Since inversion is a unary operation, it can be described using a precomputed table of length 32,768.

The direct generalization of this method becomes quickly impractical as the size of the braid increases, since the number of bits grows as the square of the number of strands, and complete tables will not fit in memory. For instance, there are 28 pairs of strands in an 8-strand braid, and this is conveniently less than 32. However, $2^{28} = 268{,}435{,}456$, which is an impractical size for a table of the results of bubble-sorting.

A much harder problem is to find canonical forms of minimal length in the generators $t_i$, especially if one allows both the number of strings and the number of crossings to vary. Unless P=NP, such a form with nice properties cannot exist for braid groups. Paterson and Razborov [PR00] show that, if the number of strands is allowed to vary, the set of minimal braids is co-NP-complete, that is, the set

$$\{\langle n, b\rangle \mid b \text{ is a nonminimal braid in } B_n\}$$

is NP-complete. This implies that (again, unless P=NP) there is no polynomial algorithm to produce a minimal representation of a given braid.

**Open Question 9.5.6.** Is it an NP-complete question to find out whether a braid is minimal, if the number of strands is fixed? (This last condition means that the question is not the same as that resolved by Paterson and Razborov.)

# Chapter 10

# Higher-Dimensional Isoperimetric Inequalities

In this chapter we study isoperimetric inequalities for combable groups in dimension greater than two. The existence of such inequalities is due to Thurston, and the proofs presented here were worked out by Epstein. In Section 10.1 we set up combinatorial constructs that make it easier to carry out the proofs. Section 10.2 presents the inequalities proper. Section 10.3 extends the inequalities to more general chains, using a standard method from geometric measure theory, originally due to Federer and Fleming in [FF60]. Thanks are due to Brian White who explained this method to Epstein.

This machinery is used in Section 10.4 to prove that the special linear groups $\mathrm{SL}(n, \mathbf{Z})$ are not automatic for $n \geq 3$. This result is due to Thurston, and the details provided here have been worked out by Epstein. In contrast, $\mathrm{SL}(2, \mathbf{Z})$ has a free subgroup of index six, and therefore is automatic by Corollary 12.1.7 (free groups) and Theorem 4.1.4 (finite index).

A preliminary version of some of this material was presented in [ET88].

## 10.1. Cell Complexes and Lipschitz Maps

In this section we prove some proofs of properties of combable spaces which are combinatorial in nature. We will often need to argue by induction on dimension, and, within each dimension, by induction on the number of "chunks" into which the space is broken up. One convenient way to do this is through the use of CW complexes. We refer the reader to [Mun84] and [LW69] for elementary information on CW complexes.

First we specify the cell structures we are going to use on some standard spaces. When we talk of the *n-sphere*, we mean the CW complex with one cell in dimension 0 and one cell in dimension $n$. When we talk of the *n-disk*, we mean the standard CW complex with one cell in dimension 0, one cell in dimension $n-1$ and one cell in dimension $n$. We make $\mathbf{R}$ into a CW complex by placing a 0-cell at each integer point, with a 1-cell connecting each adjacent pair of integers. We make $[0,\infty)$ into a CW complex in a similar manner.

If $X$ is a subset of a CW complex $Y$, we denote by $|X|$ the number of open cells of $Y$ that intersect $X$.

**Definition 10.1.1 (cellular).** Let $X$ and $Y$ be CW complexes. Recall that a *cellular map* $f : X \to Y$ is a continuous map which sends the $i$-skeleton $X^i$ into the $i$-skeleton $Y^i$ for each $i$. Any continuous map $f$ between CW complexes is homotopic to a cellular map, and the homotopy $f_t$ can be chosen so that for each open cell $e$ of $X$ and for all $t$, the image $f_t(e)$ is contained in the smallest subcomplex of $Y$ containing $f(e)$.

Moreover, a cellular map is homotopic to a map with the property that for each open $i$-cell $e_Y$ of $Y$ and for each open $i$-cell $e_X$ of $X$, the intersection $e_X \cap f^{-1}(e_Y)$ is a finite union of open $i$-cells such that the closures are disjoint closed round euclidean $i$-disks, and the map on each of these closed round disks is a euclidean similarity with the standard unit disk $D^n$, followed by one of the characteristic maps for the CW structure of $Y$. (This is proved by induction: assuming the condition is satisfied on the $(i-1)$-skeleton, we make the map transverse regular to the centre of the $i$-cells of $Y$, without changing it on the $(i-1)$-skeleton. We perform a radial homotopy on each $i$-cell of $Y$.

When we talk of cellular maps, we will always mean maps satisfying this further restriction. We will also assume that all the attaching maps for all our CW complexes are cellular maps, as just described, of a sphere into some skeleton. This is different from the usual definition of a CW complex, where it is not required that each attaching map has a 0-cell in its image. The additional requirement makes no essential difference to the concept.

Suppose first that $X$ and $Y$ are finite dimensional CW complexes. We want a condition on maps from $X$ to $Y$ which will be analogous to the idea of a lipschitz map between metric spaces. To get some feeling for what the conditions should be like, consider a lipschitz map $S^2 \to \mathbf{R}$. If the image is a long interval, then the lipschitz constant is large. This gives us the second condition in Definition 10.1.2. Now consider a smooth map $f : S^2 \to S^2$. Let $v$ be the area form of $S^2$ and let $k$ be the lipschitz constant of $f$. Then $f^*v = gv$, where $g : S^2 \to \mathbf{R}$ is bounded by $k^2$. Therefore the degree of $f$ is bounded by $k^2$. So the degree gives an estimate of the lipschitz constant. This gives us the third condition in Definition 10.1.2.

**Definition 10.1.2 (CW-lipschitz).** A map $f : X \to Y$ is said to be $(k_1, k_2)$-*CW-lipschitz* if the following conditions are satisfied:

(1) $f$ is cellular.

(2) Let $e$ be any open cell of $X$. Let $L$ be the smallest subcomplex containing $e$, and let $L_0$ be the subcomplex of $L$ consisting of all cells except for $e$. Let $N$ and $N_0$ be the smallest subcomplexes of $Y$ containing $f(L)$ and $f(L_0)$ respectively. Then $N \setminus N_0$ is the union of at most $k_1$ open cells.

(3) For each open $i$-cell $e$ of $X$, the number of components of $e \cap f^{-1}(Y^i \setminus Y^{i-1})$ is at most $k_2$.

We say that a CW complex is *bounded* if, for each $n$, there is a constant $k_n$ such that every open $n$-cell is contained in a subcomplex with at most $k_n$ cells. All the CW complexes we work with will have this property, since each will either be a finite complex or a covering space of a finite complex.

We need to discuss lipschitz homotopies of maps. It is no good requiring the time parameter of the homotopy to vary over the unit interval, because then no space of infinite diameter could be lipschitz contractible. We need to move a point $x$ to the basepoint in a time which is of the same order as the distance of $x$ to the basepoint. So we parametrize the homotopy over $[0, \infty)$, and insist that locally the homotopy becomes stationary after a finite time, and that this finite time varies in a lipschitz way. We now formalize this idea in the CW context.

**Definition 10.1.3 (CW-lipschitz contraction).** Let $X$ be a CW subcomplex of the CW complex $Y$. Let $F : X \times [0, \infty) \to Y$ be a CW-lipschitz map, where the usual product CW structure is taken on the domain. We say that $F$ is a $(k_0, k_1, k_2)$-*CW-lipschitz contraction* of $X$ in $Y$ to a 0-cell $x_0 \in X$ if the following conditions are satisfied:

(1) $F$ is $(k_1, k_2)$-CW-lipschitz.

(2) $F_t$ fixes $x_0$ for each $t \in [0, \infty)$.

(3) $F(x, 0) = x_0$ for every $x \in X$.

(4) There is a (not necessarily continuous) function $h : Y \to [0, \infty)$, called the *duration function* of the homotopy, such that

 (i) $h(x_0) = 0$;

 (ii) $F(x, t) = x$ for $t \geq h(x)$.

 (iii) if $L$ is a connected finite subcomplex of $Y$, then $h(L)$ is contained in an interval with integral endpoints and with length $|L|\, k_0$.

(5) If $L$ is a connected finite subcomplex of $Y \times [0, \infty)$, then $F(X \times [0, \infty) \cap L)$ is contained in a connected subcomplex of $Y$ with at most $|L|\, k_1$ cells. (This is not implied by the condition that $F$ is $(k_1, k_2)$-CW-lipschitz, because it may be that $L$ is connected, but not $X \times [0, \infty) \cap L$.)

To understand Conditions 10.1.3(5) and (4)(iii), think of a path-metric space $Y$, with a subset $X$ which is connected by rectifiable paths. Our conditions are analogous to asking for a lipschitz contraction of $X$ in $Y$, where we use not the intrinsic path metric on $X$, but the metric induced from $Y$. Condition (4)(iii) corresponds to the term $|b_1 - b_2|$ in Definition 3.6.1.

**Lemma 10.1.4 (combing and contraction).** *A necessary and sufficient condition for a connected 1-dimensional CW complex $X$ to be combable is that the 0-skeleton of $X$ should be CW-lipschitz contractible in $X$.*

*Proof of 10.1.4:* As pointed out on page 85, in order to show that $X$ is combable, we need only construct an appropriate path from each vertex to the basepoint. A CW-contraction gives such paths.

Conversely, given a combing defined on the vertices, we can change each path so that it is cellular. We may assume that the path assigned to the basepoint is the constant path. We are given the duration function $h$ on the vertices of $X$. We extend it to the 1-cells of $X$ by taking the larger of the two values of $h$ at its endpoints. It is immediate to check that all the conditions for a CW-lipschitz contraction are satisfied. [10.1.4]

A *cellular i-chain* (or just a *chain*) of $X$ is a formal integral linear combination of cells of dimension $i$. The *CW-mass* of a (cellular) chain is defined to be the sum of the absolute values of the integer coefficients. We also apply the term "CW-mass" to any subset of $X$ or any cellular map which defines a cellular chain in some obvious way. Sometimes we specify the dimension by talking of the *CW-i-mass*. The *CW-diameter* of a subset $S$ of $X$ is the smallest integer $d$ such that any two points of $S$ lie in some connected subcomplex $L$ with at most $d$ cells. A set consisting of a single point has CW-diameter at least one. The CW-diameter of a cellular chain is defined by regarding it as the union of its cells. The boundary homomorphism is defined by thinking of a cellular chain as an element of $H_i(X^i, X^{i-1})$, and then using the usual homology boundary of the triple $(X^i, X^{i-1}, X^{i-2})$.

## 10.2. Estimates for Cell Complexes

In Theorem 3.6.6 we showed that any combable group satisfies a quadratic isoperimetric inequality; in the present context this means that the filled Cayley graph of the group (page 154) satisfies a quadratic isoperimetric inequality relating the length of a curve to the minimum area of a disk bounded by the curve, the length and area being measured by the number of cells. In this section we prove isoperimetric inequalities in higher dimensions, both statements and proofs using the notions of mass and diameter which we defined for CW complexes on page 214.

**Theorem 10.2.1 (mass times diameter estimate).** *Let $G$ be a combable group and let $X$ be a locally compact CW-complex with basepoint on which $G$ acts cellularly and properly discontinuously with compact quotient. We suppose that $X$ is $k$-connected for some $k \geq 0$. Then:*

(1) *There is a CW-lipschitz contraction of the $k$-skeleton $X^k$ to the basepoint, and the contraction can be assumed to extend a given CW-lipschitz contraction on the $(k-1)$-skeleton.*

(2) *If $k \geq 1$, there is a constant $a_k > 0$ such that any cellular map $f : S^k \to X$ extends to a cellular map $f'$ of the $(k+1)$-disk, with $\text{mass}_{k+1}(f') \leq a_k \operatorname{diam}(f(S^k)) \text{mass}_k(f)$.*

(3) *If $k \geq 1$, there is a constant $b_k > 0$, such that for any cellular $k$-cycle $z$, there is a cellular $(k+1)$-chain $c$, such that $\partial c = z$ and $\text{mass}_{k+1}(c) \leq b_k \operatorname{diam}(z) \text{mass}_k(z)$.*

*Proof of 10.2.1:* The proof of this theorem is by induction on $k$, starting with $k = 0$. In Theorem 3.3.6 (pseudoisometry of Cayley graph) we showed that $X$ is pseudoisometric to the Cayley graph. It follows from Theorem 3.6.4 (combable is invariant) that the one-skeleton $X^1$ is combable. Lemma 10.1.4 (combing and contraction) starts the induction with a contraction of $X^0$, which fixes the basepoint.

We now have to prove the theorem for $k > 0$, assuming it is true for smaller values of $k$. We first prove the following technical lemma.

**Lemma 10.2.2 (lipschitz bounds).** *Let $G$ be a combable group and let $X$ be a locally compact CW-complex with basepoint on which $G$ acts cellularly and properly discontinuously with compact quotient. We suppose that $X$ is $(k-1)$-connected for some $k \geq 1$. Given integers $\alpha_1 > 0$ and $\alpha_2 > 0$, there exist integers $\beta_1 > 0$ and $\beta_2 > 0$, with the following property. Let $f, g : (S^k, *) \to (X^k, *)$ be $(\alpha_1, \alpha_2)$-CW-lipschitz maps. If $k = 1$, $f$ and $g$ are assumed to be equal in $\pi_1(X^1)$. If $k > 1$, $f$ and $g$ are assumed to define equal cellular $k$-chains. Then there is a $(\beta_1, \beta_2)$-CW-lipschitz map $F : (S^k \times I, * \times I) \to (X^k, *)$ which is a homotopy between $f$ and $g$.*

*Proof of 10.2.2:* We choose a finite subcomplex $K$ of $X$ such that the closure of any cell of $X$ can be moved by some element of $G$ into the interior of $K$. It follows that there is number $b > 0$ such that any open cell of $X$ is in the closure of at most $b$ other cells.

Let $L$ be the union of all connected subcomplexes of $X^k$ containing the basepoint, and containing at most $\alpha_1$ open cells. By induction on $\alpha_1$, this means that $|L| \leq b^{\alpha_1}$. In particular, $L$ is a finite complex.

If $X$ is a CW complex, let $CX$ denote the cone on $X$, with the obvious CW structure. We have the maps

$$L \xrightarrow{j} L/L^{k-1} = S^k \vee \cdots \vee S^k \xrightarrow{w} L \cup CL^{k-1} \xrightarrow{u} X^k$$

where $w$ is an inverse homotopy equivalence to the projection $p : L \cup CL^{k-1} \to L/L^{k-1}$, $j = p|_L$, $u|_L$ is the inclusion, while $u|_C L^{k-1}$ is induced by the CW-lipschitz contraction of the $(k-1)$-skeleton to a point (assumed to exist by induction using the first part of Theorem 10.2.1). Let $H$ be a homotopy between $uwj$ and the inclusion $i : L \to X^k$. In the commutative diagram

$$\begin{array}{ccccc} \pi_k L & \longrightarrow & H_k L & \xrightarrow{j*} & H_k(L, L^{k-1}) \\ \downarrow & & \downarrow & & \downarrow \\ \pi_k X^k & \longrightarrow & H_k X^k & \longrightarrow & H_k(X^k, X^{k-1}) \end{array}$$

the maps between homology groups are all injective.

Let $f', g' : S^k \to L$ be defined by $if' = f$ and $ig' = g$. From the commutative diagram above, the maps $f'$ and $g'$ have the same image in $H_k(L, L^{k-1})$. If $k > 1$, it follows from the Hurewicz isomorphism theorem that $\pi_k(L/L^{k-1}) \cong H_k(L/L^{k-1})$. Therefore, if $k > 1$,

$$jf' \simeq jg' : (S^k, *) \to (S^k \vee \cdots \vee S^k, *).$$

If $k = 1$, the same result is true. In fact, in that case, $f' \simeq g'$, as we see by lifting to maps of the unit interval into the universal cover of $X^1$. The maps $jf'$ and $jg'$ are both $(\alpha_1, \alpha_2)$-CW-lipschitz maps and they both have a standard geometric form, namely there are at most $\alpha_2$ disjoint closed round $k$-disks in the domain $S^k$, the complement of these disks is mapped to the basepoint, and each of the disks is mapped by a euclidean similarity to the unit disk $D^k$, followed by the standard identification $D^k \to D^k/S^{k-1} = S^k$ with one of the $k$-spheres in the range. It follows that there is an $(\alpha_1, \alpha_2)$-CW-lipschitz map

$$Q : (S^k \times I, * \times I) \to (S^k \vee \cdots \vee S^k, *)$$

which is a homotopy between $jf'$ and $jg'$.

We choose $\gamma_1$ and $\gamma_2$ so that $H$ and $uw$ are $(\gamma_1, \gamma_2)$-CW-lipschitz. Since these depend on $L$, $\gamma_1$ and $\gamma_2$ depend on $\alpha_1$. We get a homotopy between $f = if'$ and $g = ig'$ by combining the homotopy $Hf'$ between $if'$ and $uwjf'$, the homotopy $uwQ$ between $uwjf'$ and $uwjg'$ and the homotopy $Hg'$ between $uwjg'$ and $ig'$. Each of the three homotopies is $(\alpha_1\gamma_1, \alpha_2\gamma_2)$-CW-lipschitz. The composition of these homotopies is therefore $(3\alpha_1\gamma_1, 3\alpha_2\gamma_2)$-CW-lipschitz. We set $\beta_1 = 3\alpha_1\gamma_1$ and $\beta_2 = 3\alpha_2\gamma_2$. 10.2.2

The next lemma implies the statement of 10.2.1(2) showing that a sphere is the boundary of a disk with controlled mass.

**Lemma 10.2.3 (lipschitz extension).** *Let $G$ be a combable group and let $X$ be a locally compact CW-complex with basepoint on which $G$ acts cellularly and properly discontinuously with compact quotient. We suppose that $X$ is $k$-connected, where $k \geq 1$. Then there is a constant $p$ with the following property. Let $f : (S^k, *) \to (X^k, *)$ be a CW-lipschitz map with CW-$k$-mass $v$ and CW-diameter $d$. Then there is an extension to a CW-lipschitz map $f : D^{k+1} \to X^{k+1}$ with $(k+1)$-mass bounded by $pvd$. Given $\alpha_1$ and $\alpha_2$, there exist $\gamma_1$ and $\gamma_2$ such that, if $f|_{S^k}$ is $(\alpha_1, \alpha_2)$-CW-lipschitz, then $f|_{D^{k+1}}$ is $(\gamma_1, \gamma_2)$-CW-lipschitz.*

*Proof of 10.2.3:* We assume by induction using Theorem 10.2.1 (mass times diameter estimate) that, for some $q > 0$, we have a $(q, q, q)$-CW-lipschitz contraction $F$ of $X^{k-1}$ in $X$. We will also assume that $q$ is chosen big enough so that all attaching maps for $X$ are $(q, q)$-CW-lipschitz. Note that $q$ depends only on $X$. Given any point $x \in f(S^k) \cap X^{k-1}$, we can choose a connected subcomplex $L$ of $X$ containing both the basepoint and $x$, such that $|L| \leq d$. By Definition 10.1.3 (CW-lipschitz contraction), the duration function of $F$ is bounded by $qd$ on $f(S^k) \cap X^{k-1}$.

We will describe a basepoint preserving homotopy of maps $H_t : (S^k, *) \to (X^{k+1}, *)$, which takes place over the interval $[0, qd]$, starts with the constant map of $S^k$ to the basepoint of $X^{k+1}$ and ends with the given map $f$. This homotopy enables us to extend $f$ to $D^{k+1}$, by using the identification

**10.2.4.** $$D^{k+1} \cong S^k \times I / S^k \times 0.$$

We will discuss the associated constants below.

According to Definition 10.1.1 (cellular), the inverse image under $f$ of the open $k$-cells of $X^k$ consists of $v$ disjoint closed round $k$-disks in $S^k$, and each disk is mapped in by a euclidean similarity with $D^k$ followed by a characteristic map. On the complement $Y$ in $S^k$ of the interiors of these round disks, the homotopy we seek is the map $f$ followed by the given contraction $F : X^{k-1} \times [0, qd] \to X^k$.

Let $D$ be the closure of one of the small round $k$-disk components of $S^k \backslash Y$ and let $S$ be its boundary. We give $D$ a temporary CW structure as an $k$-cell with the usual three open cells of dimensions 0, $k-1$ and $k$, to enable us to use the induction hypothesis. Now $f|_S$ is $(q, q)$-CW lipschitz by assumption (because it is an attaching map for $X$) and $F$ is a $(q, q, q)$-CW-lipschitz contraction. Therefore the composition $F \circ (f|_S \times \mathrm{id})$ is $(q^2, q^2)$-CW-lipschitz; in particular, for each integer $i$ $(0 \leq i < qd)$, $F_i \circ f|_S$ is a $(q^2, q^2)$-CW-lipschitz map. We are assuming the truth of Lemma 10.2.3 (lipschitz extension) in dimension $k-1$. This gives us constants $\alpha_3$ and $\alpha_4$, depending only on $q$, and hence only on $X$, such that $F_i \circ f|_S$ can be extended to a $(\alpha_3, \alpha_4)$-lipschitz map of $D$ into $X^k$. We use $F$, together with these extensions, to define the desired map $H$ on the $k$-skeleton of $S^k \times [0, qd]$.

Now we need to define $H$ on the top-dimensional cells of $S^k \times [0, qd]$. We have

$$(F|_{X^{k-1}\times[i,i+1]}) \circ (f|_S \times \text{id})$$

for $0 \leq i < qd$, which is a $(q^2, q^2)$-CW-lipschitz map. We want to define $H$ on $D \times [i, i+1]$. So far, we have defined $H$ on the boundary sphere $\Sigma_i$ of $D \times [i, i+1]$ and this sphere has dimension $k$.

We wish to apply Lemma 10.2.2 (lipschitz bounds) to the sphere $\Sigma_i$; this requires us to change the CW-structure on $\Sigma_i$ so that it has two cells, instead of eight. After this change, $H|_{\Sigma_i}$ is $(\alpha_5, \alpha_6)$-CW-lipschitz for some $\alpha_5$ and $\alpha_6$ depending only on $X$. It is easy to see from Lemma 10.2.2 (lipschitz bounds) that there are only a finite number of homotopy classes of $(\alpha_5, \alpha_6)$-CW-lipschitz maps of a $k$-sphere into $X^k$, once we fix the image of the basepoint. Moreover, we only have to consider a finite number of different possible images of the basepoint, because the quotient by $G$ is compact. We fix a finite set $F$ of representatives of $(\alpha_5, \alpha_6)$-CW-lipschitz maps of a $k$-sphere into $X^k$, which, together with their translates under $G$, give a complete set of homotopy classes. Since $X$ is $k$-connected, and there are only a finite number of elements in $F$, we can find constants $\beta_1$ and $\beta_2$ depending only on $X$, and a $(\beta_1, \beta_2)$-CW-lipschitz extension of each element of $F$ to a $(k+1)$-disk. Using Lemma 10.2.2, we choose $\beta_1$ and $\beta_2$ large enough so that any $(\alpha_5, \alpha_6)$-CW-lipschitz map $S^k \to X$ is $(\beta_1, \beta_2)$-CW-lipschitz homotopic to an element of $F$.

We have a $(\beta_1, \beta_2)$-CW-lipschitz map $\Sigma_i \times I \to X^k$, which starts with $H|_{\Sigma_i}$ on $\Sigma_i \times 0 = \Sigma_i$ and ends with an element $s \in F$. We have a CW structure on the $(k+1)$-disk, obtained by gluing a smaller $(k+1)$-disk $E$ to $\Sigma_i \times I$ at one end (see 10.2.4). This CW complex has seven cells, and we have a $(2\beta_1, 2\beta_2)$-CW-lipschitz map of it to $X$. We then change the cell structure to the product structure on $D \times [i, i+1]$, which has nine cells (see Figure 10.1), and we find constants $\beta_5$ and $\beta_6$, depending only on $X$, such that we can extend $H|_{\Sigma_i}$ to a $(\beta_5, \beta_6)$-CW-lipschitz map $D \times [i, i+1] \to X^{k+1}$.

The part of Lemma 10.2.3 (lipschitz extension) referring to the mass of the extension to the $(k+1)$-disk now follows on setting $p = q\beta_6$. We still need to discuss the constants $\gamma_1$ and $\gamma_2$ in the statement of Lemma 10.2.3 (lipschitz extension). First we work out CW-lipschitz constants $(\sigma_1, \sigma_2)$ for $H : S^k \times [0, qd] \to X$, when the obvious cell structure with $2(2qd+1)$ cells

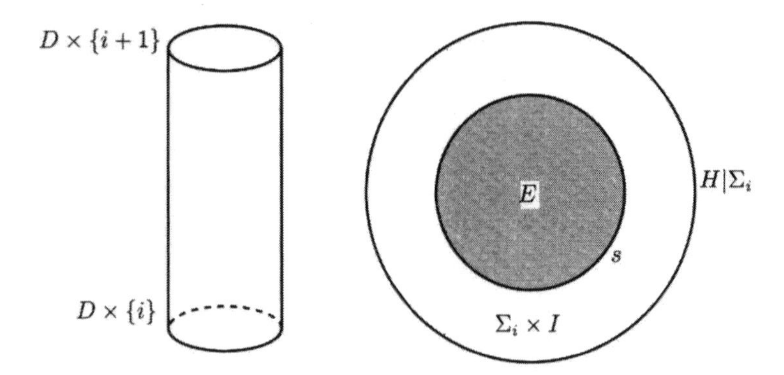


**Figure 10.1. Two cell structures on the $(k+1)$-disk.** On the left we illustrate the product cell structure $D \times I$ with $3 \times 3 = 9$ cells. On the right we have the product cell structure on $\Sigma_i \times I$ with a $(k+1)$-cell $E$ attached, which has $(2 \times 3) + 1 = 7$ cells. The labels $s$ and $H|_{\Sigma_i}$ on the right denote the respective restrictions to $\Sigma_i \times \{1\}$ and to $\Sigma_i \times \{0\}$ of the extension we need.

is used on the domain. We can take $\sigma_1 = \alpha_1 q + \beta_5$, where the first term $\alpha_1 q$ comes from the subspace $Y$ of $S^k$ defined at the beginning of this proof, and the second term $\beta_5$ comes from disks like $D$ discussed above. We can take $\sigma_2 = q + \alpha_2\beta_6$, where the first term $q$ comes from the 1-cells of the form $x \times [i, i+1]$, where $x$ is a vertex, and the second term $\alpha_2\beta_6$ comes either from $(k+1)$-cells of the form $e \times [i, i+1]$, where $e$ is the unique $k$-cell of $S^k$, or $k$-cells of the form $e \times i$.

If we now think of $H$ as providing an extension of $f|_{S^k}$ to $D^{k+1}$ and we take the usual cell structure on $D^{k+1}$ with three cells, we may take $\gamma_1 = (2qd+1)\sigma_1$ and $\gamma_2 = qd\sigma_2$. Clearly $d \le \alpha_1 + 1$ and so $\gamma_1$ and $\gamma_2$ may be chosen so as to depend only on $X$ and on $\alpha_1$ and $\alpha_2$. 10.2.3

The next lemma is part of the statement of Theorem 10.2.1 (mass times diameter estimate).

**Lemma 10.2.5 (contraction of skeleton).** *Let $G$ be a combable group and let $X$ be a locally compact CW-complex with basepoint on which $G$ acts cellularly and properly discontinuously with compact quotient. We suppose that $X$ is $k$-connected for some $k \ge 0$. Then there is a CW-lipschitz contraction of the $k$-skeleton $X^k$ to the basepoint, and the contraction can be assumed to extend a given CW-lipschitz contraction on the $(k-1)$-skeleton.*

*Proof of 10.2.5:* The contraction, which we denote by $F$, is defined by induction on $k$. For $k = 0$, the contraction is equivalent to the fact that $G$ is combable, by Lemma 10.1.4 (combing and contraction). Suppose by induction that we have $q > 0$ and a $(q,q,q)$-CW-lipschitz contraction of $X^{k-1}$ to a point in $X$. We can find $q_1$ such that each attaching map $S^{k-1} \to X^{k-1}$ is $(q_1,q_1)$-CW-lipschitz. The composite $S^{k-1} \times [0,\infty) \to X^k$ is $(qq_1,qq_1)$-CW-lipschitz. Choose $q_2 > qq_1$ such that, for each integer $i \geq 0$, we can apply Lemma 10.2.3 (lipschitz extension) to extend the map on $S^{k-1} \times \{i\}$ to a $(q_2,q_2)$-CW-lipschitz map $D^k \times \{i\} \to X^k$, extending by a constant map if possible.

Now take $q_3 > q_2$ such that we can extend further to a $(q_3,q_3)$-CW-lipschitz map $D^k \times [i,i+1] \to X^{k+1}$, by once again applying Lemma 10.2.3 (lipschitz extension) (which necessitates changing the CW structure on the boundary). If possible, we define the extension to be the constant map. Putting these extensions together, we obtain a homotopy $D^k \times [0,\infty) \to X^{k+1}$, which in turn gives us a $(q,q_3,q_3)$-CW-lipschitz contraction of $X^k$ in $X$. 10.2.5

To prove 10.2.1(3), we use the action of $G$ to assume without loss of generality that the basepoint is contained in the support of $z$. On the support of $z$, the duration function is bounded by a constant times the diameter of $z$. The contraction of $X^k$ gives the required chain $c$. 10.2.1

**Theorem 10.2.6 (combable group has reasonable $K(G,1)$).** *If $G$ is a combable group, there is an Eilenberg-Maclane space $K(G,1)$ having a finite number of cells in each dimension.*

Alonso [Alo89] has proved independently that a combable group has a free $G$-resolution which is finitely generated in each dimension. Related results have been proved by K.S. Brown [Bro89], D.J. Anick [Ani86] and C.G. Squier [Squ87].

**Open Question 10.2.7.** If $G$ is a combable group, is there a contractible space on which $G$ acts properly discontinuously with compact quotient? Does it make it easier if we assume that $G$ is automatic?

*Proof of 10.2.6:* The Cayley graph $\Gamma$ of $G$ with respect to the chosen generators can be identified with the 1-skeleton of the universal cover of a $K(G,1)$. Suppose that we have constructed the $k$-skeleton $X$ of the universal cover of a $K(G,1)$, where $k > 0$. Then $X$ is $(k-1)$-connected. We need to attach a finite number of $(k+1)$-cells to $X/G$, so as to kill the $k$-th homotopy group.

In order to see which cells to attach, we proceed in the same way as in the proof of Lemma 10.2.3 (lipschitz extension), starting with an arbitrary cellular map $S^k \to X$. The only place in that proof where we used the fact that $X$ is $k$-connected was to extend each member of a finite set $F$ of $(\alpha_5, \alpha_6)$-CW-lipschitz maps $S^k \to X$ to $D^{k+1}$. The numbers $\alpha_5$ and $\alpha_6$ did not depend on our arbitrary cellular map $S^k \to X$. Instead of such extensions being given beforehand, we now bring them into existence by using each element of $F$ to attach a $(k+1)$-cell to $X/G$. This corresponds to attaching to $X$ one $(k+1)$-cell for each element of $F \times G$. The new CW complex is $k$-connected and has a free cellular action by $G$, with compact quotient. We use each element of $F$ to attach a $(k+1)$-cell to $X/G$. Note that the universal cover of this space is acted on freely by $G$ and is $k$-connected, as we see by following the proof of Lemma 10.2.3 (lipschitz extension) to extend the arbitrary cellular map $S^k \to X$ to $D^{k+1}$. 10.2.6

## 10.3. Combable Groups and Riemannian Manifolds

In Section 10.2 we proved various isoperimetric inequalities of a *combinatorial* nature, using the cell structure of a CW complex to define notions such as mass and diameter. A more familiar point of view is to use the definitions of mass (sometimes called "volume" for differentiable chains) and diameter from riemannian geometry.

A *lipschitz k-chain* in a connected riemannian manifold $M$ is, by definition, a singular $k$-chain, that is a formal finite sum $\sum \pm f_i$ of maps $f_i : \Delta^k \to M$, such that each $f_i$ is lipschitz. Rademacher's Theorem states that a lipschitz map is differentiable almost everywhere. (There is an accessible proof in [Sim84].) If $f : \Delta^k \to M$ is a lipschitz map, we define its *mass* as follows. If the derivative $D_x f$ exists at a point $x \in \Delta^k$, then it sends an orthonormal basis at $x$ to a $k$-tuple of vectors tangent to $M$ at $fx$. This $k$-tuple defines a parallelopiped in the tangent space $T_x M$, which has a $k$-volume $V(x) \geq 0$ because of the riemannian structure on $M$. $V(x)$ is called the *k-dimensional Jacobian* of $f$ at $x$. The mass of $f$ is the integral of the function $V$ over $\Delta^k$. $V$ is defined almost everywhere and it is measurable. It is bounded because the norm of $D_x f$ is bounded by the lipschitz constant of $f$; therefore the mass is finite. The *k-mass* or *mass* of a lipschitz $k$-chain is defined to be the sum of the masses of the $f_i$. The *diameter* of a chain is the diameter of $\bigcup f_i \Delta^k$ as a subset of the metric space $M$.

The way we have explained things, it appears that the mass depends on the fact that we have chosen the domain to be the regular euclidean $k$-simplex with side length one. However, any riemannian metric on $\Delta^k$ would give the same mass, as is seen by using the change of variable formula. There are several other equivalent definitions of the mass, which we will not discuss. (See [Sim84].)

We plan to prove certain isoperimetric inequalities for lipschitz chains. It is straightforward to prove the same inequalities for more general chains, of the kind which are discussed in geometric measure theory (see [Sim84, Fed69]). However, the present context requires less background knowledge of the reader and is general enough for the applications we need. Sometimes in mathematics choosing an appropriate level of generality makes results easier to prove, because the objects naturally occurring in the course of the proof remain within the domain of discourse. For example, the processes we want to apply would change a smooth chain to a non-smooth one, unless very special precautions are taken, necessitating a great deal of complication. But the process changes a lipschitz chain to a lipschitz chain.

**Theorem 10.3.1 (triangulation).** *Let $M$ be a riemannian manifold of dimension $m$, and let $G$ be a group acting properly discontinuously on $M$ by isometries. Then there exists a $C^1$ triangulation for $M$ which is $G$-equivariant.*

This should be a standard result in the literature. Unfortunately, the only references we could find proved much more general and somewhat different results, which made them unsuitable for our purposes. It would be a worthwhile project for someone to write a complete proof along the lines of the sketch we are about to present.

*Proof of 10.3.1:* Given $x \in M$, let $G_x$ be its stabilizer in $G$. Let $U_x$ be a small $G_x$-invariant disk neighbourhood of $x$—much smaller than the injectivity radius at that point. The action of $G_x$ on $U_x$ is orthogonal with respect to coordinates which come from the tangent space at $x$ via the exponential map of the riemannian structure. We may assume that if $gU_x \cap U_x \neq \varnothing$, then $g \in G_x$. Given any locally finite subset $X$ of $M$, there is a standard construction that assigns to each point $x$ the set of points $C_x$ of $M$ which are nearer to $x$ than to any other point of $X$. It is easy to make $X$ $G$-invariant and to choose $X$ so that each $C_x$ is contained in some $U_y$.

If $X$ is chosen reasonably densely, the local geometry is almost euclidean. To show that the $C_x$ form a nice cell complex, $X$ needs to be chosen so that all intersections are generic, and this will happen for most arrangements of $X$—the proportion in the sense of measure theory of the bad choices can be made to decrease as more and more points are added to $X$. The $C_x$ give a $G$-invariant cell structure to $M$ in which the cells are finite intersections of the $C_x$. If $C = C_x$ is a cell, let its stabilizer be $G_C$. Since $C \subset U_y$ for some $y$, we have $G_C \subset G_y$. If $U_y$ is sufficiently small, then, as in the euclidean case, we have at least one point $p_C$ in the interior of $C$ which is fixed by $G_C$. The points $p_C$ are chosen in a $G$-equivariant manner. We construct the triangulation by induction on the dimension of $C$. If the boundary of $C$ is triangulated, then the cone from $p_C$ triangulates $C$.

An alternative approach is to repeat the arguments in [Mun66], when there is a group action. 10.3.1

For the remainder of this section, we fix $M$ and a $G$-invariant $C^1$ triangulation $\tau$ of $M$.

**Lemma 10.3.2 (cellular implies comparable).** *There is a constant $c > 0$, depending on $\tau$, with the following property. Let $T$ be a differentiable $k$-chain $\sum f_i$ such that each $f_i : \Delta^k \to M$ is cellular with respect to the natural CW structure on $\Delta^k$. Then we have the following inequalities:*

$$\begin{array}{rcccl} \text{mass}(T)/c & \leq & \text{CW-mass}(T) & \leq & c\,\text{mass}(T) \\ \text{diam}(T)/c & \leq & \text{CW-diam}(T) & \leq & c(\text{diam}(T)+1). \end{array}$$

*Proof of 10.3.2:* If we change the riemannian metric on the simplices of the $C^1$-triangulation, the riemannian mass and diameter change in a lipschitz way. Therefore we may change the metric on each $i$-simplex to be the standard euclidean metric on the regular euclidean $i$-simplex with side length one.

We then have $c_k$ CW-mass$(T) =$ mass$(T)$, where $c_k$ is the volume of the standard euclidean $k$-simplex. Since each simplex of $\tau$ has diameter one, we have diam $T \leq$ CW-diam$(T)$.

We now show that CW-diam$(T) \leq c(\text{diam}\, T + 1)$ for some $c > 0$. First note that, if $i > 1$, there are two metrics on the boundary of a regular euclidean $i$-simplex with side length one. The first is induced using the euclidean length of straight lines in the $i$-simplex. The second is induced from the length of piecewise paths in the boundary of the $i$-simplex. These two metrics are lipschitz related, by a constant $c_i > 1$. Given a path $\sigma$ in the $i$-skeleton of $\tau$, we can therefore push it into the $(i-1)$-skeleton, at the cost of increasing its length by at most a factor $c_i$. Repeating this process, we eventually get a path $\sigma'$ in the 1-skeleton, whose length is at most $c'\,|\sigma|$, where $c' = c_2 \dots c_m$. Therefore $\sigma'$ contains no more than $(c'\,|\sigma| + 1)$ distinct vertices of $\tau$. The union of all simplices of $\tau$ containing such a vertex is a connected subcomplex of $\tau$ with at most $c(|\sigma| + 1)$ simplices, for some constant $c > 1$. This proves the desired inequality. 10.3.2

We will prove isoperimetric inequalities for differentiable chains by means of an analogue of the deformation theorem of geometric measure theory. This was first proved in [FF60]. Interested readers should also consult [Sim84].

**Theorem 10.3.3 (deformation theorem).** *There is a $c > 0$, depending only on $\tau$, with the following property. Let $T$ be a lipschitz $k$-cycle in $M$.*

*Then $T = Q + \partial R$, where $R$ is a lipschitz $(k+1)$-chain, $Q$ is a smooth $k$-cycle all of whose simplices are simplicial maps $\Delta^k \to \tau$, and moreover* $\mathrm{mass}_k(Q) \leq c\, \mathrm{mass}_k(T)$, $\mathrm{mass}_{k+1}(R) \leq c\, \mathrm{mass}_k(Q)$, *and $Q$ and $R$ are contained in the smallest subcomplex of $\tau$ containing $T$.*

*Proof of 10.3.3:* As in Lemma 10.3.2 (cellular implies comparable), we may assume that each simplex of $\tau$ is a regular euclidean simplex with side length one. The first thing we have to do is to push $T$ into the $k$-skeleton of $\tau$. This is done in the time-honoured way, by induction, starting by pushing $T$ into the $(m-1)$-skeleton, where $m$ is the dimension of $M$, then into the $(m-2)$-skeleton, and so on. To do the induction step, we suppose that $T$ is contained in the $i$-skeleton, where $i > k$, and we need to push it into the $(i-1)$-skeleton. Since the support of $T$ is compact and has $i$-measure zero, we can find a point in each $i$-simplex which is not in the support of $T$, and project radially from that point. The radial lines provide us with a homotopy of $T$, and therefore a $(k+1)$-chain $R$.

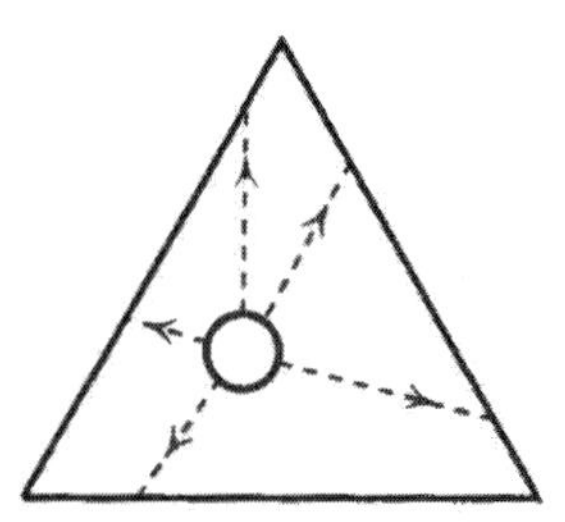

**Figure 10.2. Projecting from the wrong centre.** Suppose $T$ is a small circle in a triangle. If we project from the centre of the circle, the mass of $T$ increases by a huge factor.

This construction certainly gives us lipschitz chains as required, but it has a serious defect, namely that there is no control over the masses of the chains produced. Suppose for example, $i = 2$ and $k = 1$, and $T$ is a very small circle surrounding the centre of projection (see Figure 10.2). Then the ratio of the length of the projection of $T$ to the length of $T$ itself is arbitrarily large. Instead, our strategy will be to project from all possible choices of centre, and to average the masses of the corresponding chains. We find that the average mass is bounded by a constant times the mass of $T$, and therefore there must be plenty of centres from which it is permissible to project. Now we give the details of the argument.

Let $4r_i$ be the radius of the largest ball which can fit inside $\Delta^i$, the regular euclidean $i$-simplex with side length one. Let $B = B(x_0, r_i)$ be the ball with centre equal to the barycentre of $\Delta^i$ and with radius $r_i$. If $u \in B$, then $B(u, 2r_i)$, the ball with centre $u$ and radius $2r_i$, is contained in $\Delta^i$ and contains $B$. Let $\pi_u : B(u, 2r_i) \setminus \{u\} \to \partial B(u, 2r_i)$ be projection onto the boundary sphere from the centre $u$, and extend $\pi_u$ to be the identity map outside $B(u, 2r_i)$.

Since $T = \sum \pm f_j$ is a cycle, the singular simplices which occur in the chains $\partial f_j$ cancel out in pairs. We can therefore glue together all the $k$-simplices in the domains of the $f_j$, one standard $k$-simplex for each $j$, along their $(k-1)$-dimensional faces, and obtain a compact polyhedron $P$, which is an oriented manifold when the $(k-2)$-skeleton is removed. Moreover $P$ has an induced path metric and $P \setminus P^{k-2}$ is a (non-complete) euclidean manifold with this metric. We may regard $T$ as a lipschitz map $T : P \to M$. (If $P$ is not connected, we can place the components a large distance apart. Or, more naturally, we can place them an infinite distance apart and extend the concept of a metric space to allow distances to be infinite.)

We focus attention on one particular open $i$-simplex $E$ of $\tau$ and denote by $S$ the restriction of $T$ to $T^{-1}E$. We have

$$\mathrm{mass}_k(\pi_u S) \leq \int_{S^{-1}B(u,2r)} \frac{V(x)(2r)^k}{|Sx - u|^k} dx + \mathrm{mass}_k(S),$$

where $r = r_i$ and $V$ is the $k$-dimensional Jacobian defined on page 221. Let $v > 0$ and let $\mu$ be Lebesgue $i$-dimensional measure on $E$. We set

$$\alpha(v) = \mu\{u \in B : \mathrm{mass}_k(\pi_u(S)) > v \ \mathrm{mass}_k(S)\}.$$

We claim that there is a constant $c$, depending only on $i$ and $k$, such that $\alpha(v) < c/v$. If $\mathrm{mass}_k(S) = 0$, then, since $\pi_u$ is lipschitz on the image of $S$, $\mathrm{mass}_k(\pi_u S) = 0$, and so $\alpha(v) = 0$. We may therefore assume that $\mathrm{mass}_k(S) > 0$. We have

$$\begin{aligned}
&\alpha(v)v \ \mathrm{mass}_k(S) \leq \int_B \mathrm{mass}_k(\pi_u S) du \\
&\leq (2r)^k \int_{S^{-1}B(u,2r)} \left( V(x) \int_B |u - Sx|^{-k} du \right) dx + \mathrm{mass}_i(B) \, \mathrm{mass}_k(S) \\
&< (2r)^k \int_{S^{-1}B(u,2r)} \left( V(x) \int_{B(0,3r)} |u|^{-k} du \right) dx + \mathrm{mass}_i(B) \, \mathrm{mass}_k(S) \\
&\leq c \ \mathrm{mass}_k(S)
\end{aligned}$$

for some $c$ depending only on $i$ and $k$. Since we are assuming $\mathrm{mass}_k(S) > 0$, we deduce that $\alpha(v) < c/v$, as claimed.

We choose a point $u$ in the interior of each $i$-simplex of $\tau$ from which to project, and let $\pi$ be the map whose restriction to an $i$-simplex is the corresponding $\pi_u$. We have seen that for most choices of the $u$'s, $\text{mass}_k(\pi T) \leq v\ \text{mass}_k(T)$, provided $v$ is large. (The necessary size of $v$ is independent of $T$.)

We now look at the mass of the $(k+1)$-chain $R$, such that $\partial R = T - \pi T$. $R$ is given by the homotopy $H : P \times I \to M$, where

$$H(x,t) = tSx + (1-t)\pi_u Sx$$

for $Sx \in B(u, 2r)$. Here the addition takes place in $E$, which is a regular $i$-simplex of side length one. For other values of $x$, $H(x,t)$ is independent of $t$. The contribution to the mass of $R$ from the piece contained in a single $i$-simplex of $\tau$ is bounded by

$$(2r)^{k+1} \int_{S^{-1}B(u,2r)} V(x)|u - Sx|^{-k} dx.$$

We integrate this with respect to $u$ over $B$, and the same argument as before shows that for most $u \in B$, $\text{mass}_{k+1}(R) \leq v\ \text{mass}_k(T)$.

We have shown that there is no loss of generality in supposing that $T$ is contained in the $i$-skeleton and, in each $i$-simplex $\sigma$ of $\tau$, is disjoint from the ball $B_\sigma = B(x_0, r)$ of radius $r$ centred at the barycentre of $\sigma$. Radial projection from the barycentre of $\sigma$, mapping $\sigma \setminus B_\sigma$ to the boundary of $\sigma$ is lipschitz, with lipschitz constant depending only on $i$. It is therefore straightforward to construct a homotopy $P \times I \to M$, which starts with $T$ and ends with image in the $(i-1)$-skeleton. Furthermore the $(k+1)$-mass of the homotopy and the $k$-mass of the end of the homotopy are bounded by a constant times $\text{mass}_k(T)$.

Proceeding in this way, we see that we may assume without loss of generality that $T$ is contained in the $k$-skeleton of $M$. We next change $T$ so that it maps $P^{k-2}$ into the $(k-2)$-skeleton of $\tau$. The procedure will be to change $T|_{P^j}$ by upward induction on $j$, using the homotopy extension property and radial pushing, so that $T(P^j) \subset \tau^j$. So suppose that $H_0 = T$, and we have constructed $H_t : P^{j-1} \to M$, so that the map $H : P^{j-1} \times I \to M$ is lipschitz and $H_1(P^i) \subset \tau^i$ for $0 \leq i < j$. We also assume that $H_t$ is independent of $t$ for $t > 1 - 2^{-j}$. We extend $H$ to a $j$-dimensional simplex $\Delta$ of $P$ as follows. Let $x_0$ be the barycentre of $\Delta$. Let $\rho : \Delta \times [0,1] \to \Delta \times \{0\} \cup \partial\Delta \times [0,1]$ be radial projection from $(x_0, 2)$ (see Figure 10.3). On $\Delta \times [0, 1-2^{-j}]$, we define $H$ to be $H\rho$. On $\Delta \times [1-2^{-j}, 1-2^{-(j+1)}]$, we apply the process described at the beginning of this proof to provide a lipschitz homotopy moving the image of $\Delta$ into the $j$-skeleton. On $\Delta \times [1-2^{-(j+1)}, 1]$ $H_t$ is defined to be independent of $t$.

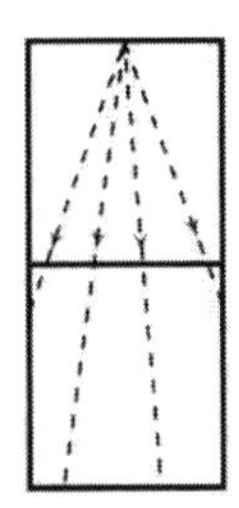

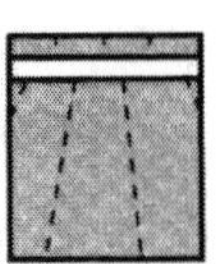

**Figure 10.3. Extending the homotopy.** We start with a map defined on $\Delta \times \{0\} \cup \partial\Delta \times I$. On the left, we show how radial projection from the barycentre of $\Delta$ at height two gives an extension to $\Delta \times I$. On the right we illustrate the process we actually need. In the large shaded piece at the bottom, we use the radial projection, as on the left. In the small shaded piece at the top, we indicate a constant homotopy. In the thin, unshaded sliver, we use a more complicated homotopy as in the beginning of the proof, which moves $\Delta$ into $P^j$, as we go from bottom to top of the unshaded piece.

In this way, we define a lipschitz map $H : P^{k-2} \times I \to M$. The extension of $H$ to $P \times I$ is obtained by composing with a lipschitz retraction $P \times I \to P \times \{0\} \cup P^{k-2} \times I$. Since the $k$-mass of $P^{k-2} \times I$ is zero, it follows that $H_1$ has the same $k$-mass as $T$. Since the image of $H$ is contained in $\tau^k$, $\text{mass}_{k+1}(H) = 0$.

We have shown that there is no loss of generality in assuming that $T(P^{k-2}) \subset \tau^{k-2}$ and that $T(P) \subset \tau^k$. Let $\Delta$ be a fixed $k$-simplex of $\tau$ with interior $\mathring{\Delta}$. Let $X$ be a component of $T^{-1}\mathring{\Delta} \subset P$. The map $T|_X : X \to \mathring{\Delta}$ is proper; let its degree be denoted $d_X$ (whose sign is fixed once we fix orientations for $X$ and $\Delta$).

**Lemma 10.3.4 (degree).** *Let $V_k$ be the volume of the regular euclidean $k$-simplex with side length one. Then* $\text{mass}_k(T|_X) \geq V_k |d_X|$.

*Proof of 10.3.4:* Given $\delta > 0$, we choose a small closed neighbourhood $U$ of $\tau^{k-1}$ in $M$ with $\text{vol}_k(\Delta \setminus U) = V_k - \delta$ as illustrated in Figure 10.4. Let $Y = X \cap T^{-1}(\Delta \setminus U)$. Then $Y$ is open and has compact closure in $X$.

Let $\phi : \mathbf{R}^k \to [0, \infty)$ be a smooth function with support in the unit disk. We assume that $\phi$ is symmetric under the orthogonal group $O(k)$ and that $\int \phi = 1$. For $\varepsilon > 0$, we define $\phi_\varepsilon(x) = \phi(x/\varepsilon)/\varepsilon^k$. Then $\phi_\varepsilon$ converges to a delta function as $\varepsilon$ tends to zero. Since $X$ has a euclidean structure and each $k$-simplex of $\tau$ is regarded as a regular euclidean $k$-simplex with side length one, the convolution $T_\varepsilon(x) = T * \phi_\varepsilon(x)$ is well-defined for $x \in Y$, provided $\varepsilon$ is small enough. We piece this together with $T|_X \setminus Y$ by using a function

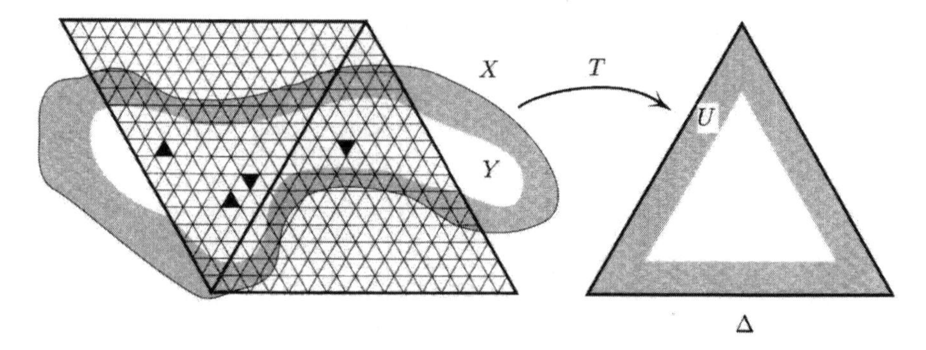


**Figure 10.4. Making $T$ cellular.** We show two $k$-simplices of $P$. These are subdivided using planes parallel to the faces. The map which is finally constructed is a similarity on each of the black simplices. These are the simplices called $\sigma_1, \ldots, \sigma_d$ in the proof.

$\psi : P \to \mathbf{R}$ which is equal to zero on a compact neighbourhood of $Y$ in $X$ and to zero outside some larger compact neighbourhood.

Note that $T$ is absolutely continuous because it is lipschitz, and therefore integration by parts, using the derivative of $T$, is permissible. It follows that the derivative of the smooth function $T_\varepsilon$ is equal to the convolution of the derivative of $T$. Suppose $f \in L^1$, where $f : \mathbf{R}^k \to \mathbf{R}$. Then $f_\varepsilon = f * \phi_\varepsilon$ converges to $f$ in $L^1$ as $\varepsilon$ tends to zero. It follows that $f_\varepsilon$ converges to $f$ in measure. From all this it follows that the derivative $DT_\varepsilon$ converges to $DT$ in measure. Therefore the mass of $\hat{T} = \psi T + (1-\psi)T_\varepsilon$ converges to the mass of $T$ as $\varepsilon$ tends to zero. So we may assume that $\mathrm{mass}_k(\hat{T}|_X) \leq \mathrm{mass}_k(T|_X) + \delta$.

Let $Z = (\hat{T})^{-1}(\Delta \setminus U)$. Then $T|_Z : Z \to \Delta \setminus U$ is a proper map, with degree $d_X$. Let $\omega$ be the volume form on $\Delta$. Then

$$|d_X|(V_k - \delta) \leq \left| \int_Z (\hat{T})^* \omega \right| \leq \mathrm{mass}_k(\hat{T}|_Z) \leq \mathrm{mass}_k(\hat{T}|_X) \leq \mathrm{mass}_k(T|_X) + \delta.$$

Since $\delta > 0$ is arbitrary, the lemma is proved. 10.3.4

We can now complete the proof of Theorem 10.3.3. We find $d = d_X$ disjoint regular euclidean $k$-dimensional simplices $\sigma_1, \ldots, \sigma_d$ in $X$ and a lipschitz map $T' : X \to \Delta$, which is equal to $T$ outside some compact subset of $X$, maps each $\sigma_i$ by a similarity to $\Delta$, and maps $X \setminus \cup_{i=1}^d \sigma_i$ into a small neighbourhood of $\partial\Delta$. To show that $T'$ exists, we can use standard obstruction theory, or an independent proof can be given using standard techniques of differentiable topology as in [Mil65]. To get going with these differentiable techniques, we can approximate $T$ by a smooth map as explained in the proof of Lemma 10.3.4 (degree). Another approach is to follow Brouwer's original treatment of degree and use simplicial approximation on subdivisions of the triangulations. It is convenient to assume that each $\sigma_i$ is a simplex in some subdivision of some $k$-simplex of $P$—for example, a subdivision by planes parallel to the faces, as illustrated in Figure 10.4.

There is a lipschitz homotopy between $T$ and $T'$ which sends $(x,t)$ to $(1-t)T(x)+tT'(x)$. Since $T = T'$ outside a compact subset of $X$, the homotopy is fixed there. Proceeding in the same way for each component like $X$, we obtain a lipschitz homotopy of maps of $P$ into $\tau^k$, and the homotopy is constant on a neighbourhood of $P^{k-2}$. We use the same notation $T'$ to denote the extension to $P \to \tau^k$. Let $C$ be the $k$-skeleton of $\tau$, with the barycentre of each $k$-simplex removed. Let $\rho : C \to \tau^{k-1}$ be radial projection from the barycentre in each $k$-simplex. Let $T'' = \rho T'$ on each subset of $P$ of the form $X \setminus \cup_{i=1}^{d} \sigma_i$, and let $T'' = T'$ on each subset of the form $\sigma_i$. Since $\rho$ is homotopic to the identity map, $T''$ is homotopic to $T'$ and we can make the homotopy between $T'$ and $T''$ lipschitz.

We have $\mathrm{mass}_k(T'') = V_k \sum |d_X|$, where the sum is over all subsets of the form $X$. This sum makes sense, because $d_X = 0$ except for a finite number of components like $X$. By Lemma 10.3.4 (degree), $\mathrm{mass}_k(T'') \leq \mathrm{mass}_k(T)$. The homotopy between $T$ and $T''$ takes place in $\tau^k$ and so its $(k+1)$-mass is zero.

We have shown that there is no loss of generality in assuming that there are a finite number of regular $k$-simplices contained in $P \setminus P^{k-2}$, that $T$ maps each of them by a similarity to some $k$-simplex of $\tau$, and that $T$ maps the complement of these simplices into $\tau^{k-1}$. These $k$-simplices in $P \setminus P^{k-2}$ are simplices in a subdivision of $P$.

There is a standard chain homotopy between simplicial chain complexes induced by subdivision. This allows us to write $T - T' = \partial R$, where $T'$ (not the same $T'$ as above) is a sum of the simplices in the subdivision and $R$ is a $(k+1)$-chain with mass zero and $\mathrm{mass}_k(T') = \mathrm{mass}_k(T)$. Most of the terms of $T'$ are contained in the $(k-1)$-skeleton—let the sum of these be $T_1$. The other singular simplices are similarities between $k$-simplices—let the sum of these be $T_2$. It is easy to see that $T_2$ is a cycle—for example, by looking at the map $H_k(\tau^k, \tau^{k-1}) \to H_{k-1}(\tau_{k-1}, \tau_{k-2})$.

It follows that $T_1$ is a $k$-cycle in $\tau^{k-1}$. In order to complete the proof, we need only show that $T_1 = \partial R_1$, where $R_1$ is a lipschitz $(k+1)$-chain, with zero mass. One way to see this is to note that homology based on lipschitz chains is equal to simplicial homology on finite simplicial complexes, for example because it satisfies the requisite axioms. Since the $k$-dimensional homology of a $(k-1)$-dimensional simplicial complex is zero, the desired result follows.

10.3.3

We can now prove the main result of this section.

**Theorem 10.3.5 (riemannian isoperimetric inequality).** *Let $G$ be a combable group of isometries acting properly discontinuously with compact quotient on a $k$-connected riemannian manifold $M$. Then there is a constant $c$ with the following property. Let $z$ be a lipschitz $k$-cycle in $M$. Then there exists a lipschitz $(k+1)$-chain $u$, such that $\partial u = z$ and* $\mathrm{mass}_{k+1}(u) \leq c\,\mathrm{diam}(z)\,\mathrm{mass}_k(z)$.

*Proof of 10.3.5:* There is no loss of generality in assuming that the riemannian metric is invariant under $G$ and that there is an $\varepsilon > 0$ such that, for each $x \in M$, the stabilizer of $x$ in $G$ acts approximately linearly in a ball of radius $\varepsilon$. If the diameter of $z$ is less than $\varepsilon$, we can construct $u$ by contracting $z$ to a point along geodesics. So we may assume that $\mathrm{diam}(z) \geq \varepsilon$. By Theorem 10.3.3 (deformation theorem), we can find a lipschitz $(k+1)$-chain $u_1$ and a $k$-cycle $z_1$, which is a sum of simplices of $\tau$, such that $\partial u_1 = z - z_1$, $\mathrm{mass}_{k+1}(u_1) \leq c_1\,\mathrm{mass}_k(z)$, $\mathrm{mass}_k(z_1) \leq c_1\,\mathrm{mass}_k(z)$ and $\mathrm{diam}(z_1) \leq \mathrm{diam}(z) + c_1$, for some constant $c_1$. We can apply Theorem 10.2.1 (mass times diameter estimate) and Lemma 10.3.2 (cellular implies comparable) to prove the existence of a cellular $(k+1)$-chain $u_2$ and a constant $c_2$, with $\partial u_2 = z_1$ and

$$\begin{aligned}\mathrm{mass}_{k+1}(u_2) &\leq c_2\,\mathrm{mass}_k(z_1)(\mathrm{diam}(z_1)+1)\\ &\leq c_3\,\mathrm{mass}_k(z)(\mathrm{diam}(z)+c_1+1) \leq c_4\,\mathrm{mass}_k(z)\,\mathrm{diam}(z)\end{aligned}$$

for suitable constants $c_3$ and $c_4$. 10.3.5

The following theorem is proved by the same methods.

**Theorem 10.3.6.** *Let $G$ be a combable group of isometries acting properly discontinuously with compact quotient on a $k$-connected riemannian manifold $M$. Then there is a constant $c$ with the following property. Let $z : S^k \to M$ be a lipschitz map. Then there exists a lipschitz map $u : D^{k+1} \to M$ extending $z$, such that* $\mathrm{mass}_{k+1}(u) \leq c\,\mathrm{diam}(z)\,\mathrm{mass}_k(z)$. 10.3.6

## 10.4. The Special Linear Groups

In this section we show that the special linear groups $\mathrm{SL}(n, \mathbf{Z})$ are not automatic for $n \geq 3$. The method of proof is to find a suitable contractible manifold on which $\mathrm{SL}(n, \mathbf{Z})$ operates properly discontinuously with compact quotient, and to prove that a higher dimensional isoperimetric inequality is false in that space. Applying Theorem 10.3.5 (riemannian isoperimetric inequality) then proves that $\mathrm{SL}(n, \mathbf{Z})$ is not combable and hence not automatic.

The standard contractible space on which $\mathrm{SL}(n, \mathbf{Z})$ operates properly discontinuously is the symmetric space $E_n = \mathrm{SO}(n)\backslash\mathrm{SL}(n, \mathbf{R})$, the quotient of $\mathrm{SL}(n, \mathbf{R})$ by $\mathrm{SO}(n)$ acting on the left. $\mathrm{SL}(n, \mathbf{Z})$ acts on the right on this space. However, the quotient is not compact, so there is no immediate connection between the truth or falsity of the higher dimensional isoperimetric inequalities in $E_n$ and the question of whether or not $\mathrm{SL}(n, \mathbf{R})$ is combable or automatic. Nonetheless, $E_n$ is a rich source of interesting geometry, and we therefore start by investigating it in some detail.

Let $\mathbf{R}^n$ have the standard inner product. A *lattice* in $\mathbf{R}^n$ is a discrete cocompact subgroup of $\mathbf{R}^n$, which is free abelian of rank $n$. In other words, the generators of the subgroup form a basis of the vector space $\mathbf{R}^n$ over $\mathbf{R}$. The (group theoretic) quotient of $\mathbf{R}^n$ by the lattice is a topological torus, with local euclidean geometry. By a *euclidean torus* we mean a riemannian manifold which is isometric to such a torus. We will usually restrict our attention to euclidean tori of volume one.

We can specify a lattice by means of a non-singular $n \times n$ matrix, by making the generators of the lattice equal to the columns of the matrix. The condition that the torus have volume one is equivalent to demanding that the matrix have determinant $\pm 1$. We will consider only matrices of determinant one, that is, elements of $\mathrm{SL}(n, \mathbf{R})$.

An element $A \in \mathrm{SL}(n, \mathbf{R})$ can be thought of as a lattice $L$ together with an ordered set of $n$ generators, whose orientation is equal to the standard orientation on $\mathbf{R}^n$ and which span a parallelepiped of volume one. We write $A = (a_1, \ldots, a_n)$, where the column vectors $a_i \in \mathbf{R}^n$ represent the $n$ generators. An orientation-preserving change of generators for the lattice is equivalent to multiplying $A$ on the right by an element of $\mathrm{SL}(n, \mathbf{Z})$.

We may also think of $A \in \mathrm{SL}(n, \mathbf{R})$ as the oriented euclidean torus $\mathbf{R}^n/L$ of volume one, together with a preferred ordered set of $n$ generators for its fundamental group and a field of orthonormal frames invariant under translation by any element of the torus. Both the ordered set of generators of the fundamental group and the orthonormal frames are oriented consistently with the orientation of the torus. The orthonormal frame field comes from the standard basis $(e_1, \ldots, e_n)$ of $\mathbf{R}^n$.

Given a euclidean torus of volume one, a field $(e_1, \ldots, e_n)$ of orthonormal frames invariant under translation by elements of the torus, and a preferred ordered set of $n$ generators of the fundamental group, we can recover the matrix $A$ of determinant one as follows. We fix a basepoint and represent each of the preferred set of generators of the fundamental group as a closed geodesic. The tangent vectors to these geodesics at the basepoint give us an ordered set $(t_1, \ldots, t_n)$ of tangent vectors at the basepoint, with the length of $t_i$ equal to the length of the corresponding closed geodesic. We set $a_{ij} = \langle e_i, t_j \rangle$.

The action on the left by an element $U \in \mathrm{SO}(n)$ on the matrix $A$ corresponds to operating on the orthonormal frame $(e_1, \ldots, e_n)$ without changing the underlying torus or preferred ordered set of generators of the fundamental group. The $j$-th element of the new orthonormal frame is $\sum_r u_{jr} e_r$.

An element of $E_n = \mathrm{SO}(n) \backslash \mathrm{SL}(n, \mathbf{R})$ can therefore be thought of as an oriented euclidean torus of volume one, with preferred ordered set of $n$ generators for the fundamental group. An element of $\mathrm{SL}(n, \mathbf{R})/\mathrm{SL}(n, \mathbf{Z})$ can be thought of as an oriented euclidean torus of volume one, with an equivariant field of orthonormal frames. The space $X_n = \mathrm{SO}(n) \backslash \mathrm{SL}(n, \mathbf{R})/\mathrm{SL}(n, \mathbf{Z})$ is the space of all euclidean tori of dimension $n$ and volume one, without specifying generators for the fundamental group. $\mathrm{SL}(n, \mathbf{Z})$ acts properly discontinuously on the right of $E_n$.

For an riemannian manifold acted on by a transitive group of isometries, the injectivity radius at a point of the manifold does not depend on the point. The injectivity radius of the torus corresponding to a lattice $L$ is equal to half the distance in $\mathbf{R}^n$ from the origin to the nearest non-zero point of $L$. The injectivity radius is a real valued strictly positive continuous function on $X_n$. We will also apply the term "injectivity radius" to a lattice, meaning the injectivity radius of the corresponding quotient torus.

$X_n$ is a not compact. For example, the sequence of matrices

$$\begin{pmatrix} 1/n & 0 \\ 0 & n \end{pmatrix}$$

defines a sequence of tori with injectivity radius tending to zero, and hence with no convergent subsequence.

**Lemma 10.4.1 (space of lattices compact).** *For any $\varepsilon > 0$, the space of all lattices in $\mathbf{R}^n$ (or, equivalently, all euclidean tori) with volume one and injectivity radius greater than or equal to $\varepsilon$ is compact.*

*Proof of 10.4.1:* We denote the lattice by $\Gamma$. Let $\gamma_1, \ldots, \gamma_n \in \Gamma$ be linearly independent elements chosen inductively so that $\gamma_i$ is a nearest element to the real vector subspace $V_i$ of $\mathbf{R}^n$ spanned by $\gamma_1, \ldots, \gamma_{i-1}$. It is easy to see that $\{\gamma_1, \ldots, \gamma_n\}$ generates $\Gamma$ as an abelian group.

Let $d_i$ be the distance of $\gamma_i$ to $V_i$ (see Figure 10.5). Then

**10.4.2.** $$d_{i+1} \geq \sqrt{3} d_i/2, \quad d_1 \geq 2\varepsilon \text{ and } d_1 d_2 \ldots d_n = 1.$$

These equations bound each of the $d_i$ above and below, away from zero, as we vary over all lattices with injectivity radius greater than or equal to $\varepsilon$. This enables us to choose a convergent sequence of lattices from any sequence of lattices with volume one. 10.4.1

**Figure 10.5. The nearest point construction.** The radius of the circle is $d_1$ and the distance between the two horizontal lines is $d_2$. No point on the upper line may appear inside the circle. So the closest the upper line may come is $(\sqrt{3}/2)d_1$. To show that $d_{i+1} \geq (\sqrt{3}/2)d_i$, we use the same diagram, applied to $\mathbf{R}^n/V_{i-1}$.

Let $E_n(\varepsilon)$ be the subspace of $E_n$ corresponding to tori of volume one and injectivity radius at least $\varepsilon$. In particular, $E_n(0) = E_n$.

**Lemma 10.4.3.** *$E_n$ deformation retracts to $E_n(\varepsilon)$ for any $\varepsilon \leq \frac{1}{2}$, and the deformation can be chosen to be* $\mathrm{SL}(n, \mathbf{Z})$*-equivariant.*

*Proof of 10.4.3:* Let $\Gamma$ be a lattice corresponding to an element of $E_n$. A shortest non-zero vector in $\Gamma$ has length $2\alpha$, where $\alpha$ is the injectivity radius of the corresponding torus. Let $V \subset \mathbf{R}^n$ be the vector subspace spanned by all elements of $\Gamma$ of length $2\alpha$. We do not change $\Gamma$ as a subset of $\mathbf{R}^n$. Instead, we gradually change the inner product on $\mathbf{R}^n$, making each vector in $V$ longer by a uniform factor, and each vector in $V^\perp$ shorter by a uniform factor, in such a way that the volume in the quotient torus is preserved.

At first the set of shortest vectors remains unaltered, but at a certain instant, one or more additional vectors in $\Gamma$ outside $V$ also become shortest. We then increase the size of $V$ so that it continues to be the vector subspace spanned by the set of shortest non-zero vectors in $\Gamma$. After a finite number of steps, the set of shortest vectors spans $\mathbf{R}^n$. Let the common length of these vectors be $d$.

We use the notation of the proof of Lemma 10.4.1 (space of lattices compact), which we apply to the lattice in which the set of shortest vectors spans $\mathbf{R}^n$. Clearly $d_i \leq d$ for each $i$. Hence $d^n \geq 1$, and so $d \geq 1$. It follows that the injectivity radius of a corresponding torus is at least $\frac{1}{2}$. This proves the theorem, bearing in mind that we can stop the process of changing the inner product as soon as the length of the shortest vectors has increased to $2\varepsilon$, provided $\varepsilon \leq \frac{1}{2}$. [10.4.3]

From the above results we immediately deduce the following theorem.

**Theorem 10.4.4.** *If $\varepsilon < \frac{1}{2}$, then $E_n(\varepsilon)$ is a contractible space on which $\mathrm{SL}(n, \mathbf{Z})$ acts on the right, properly discontinuously with compact quotient.* 10.4.4

Let $G_n$ be the subgroup of $\mathrm{SL}(n, \mathbf{R})$ consisting of matrices of the form $\left(\begin{smallmatrix} B & b \\ 0 & 1 \end{smallmatrix}\right)$, where $B$ is an $(n-1) \times (n-1)$ matrix of determinant one, and $b$ is an $(n-1) \times 1$ column vector. For our purposes, this is easier to work with than $\mathrm{SL}(n, \mathbf{R})$, but it still contains enough information for us to use it in showing that a higher dimensional isoperimetric inequality does not hold.

Let $A = (a_1, \ldots, a_n) \in E_n$ be an oriented frame of $\mathbf{R}^n$ of volume one, defined up to rotation (action on the left) by an element of $\mathrm{SO}(n)$. Let $V$ be the oriented hyperplane defined by the first $(n-1)$ vectors, and let $d$ be the signed distance of $a_n$ from $V$. Since all our frames are oriented, we have $d > 0$. The orbit under $G_n$ acting on the right on $A$ consists of all oriented frames of volume one such that the signed distance from the last vector to the subspace spanned by the first $(n-1)$ vectors is $d$. This means that $E_n/G_n$ is diffeomorphic to $(0, \infty)$. We define $T(d)$ to be the set of frames (defined up to action on the left by elements of $\mathrm{SO}(n)$) on which the above signed distance function has value $d$, and we define $T = T(1)$. For each $d > 0$, $G_n$ acts transitively on $T(d)$.

Note that $E_n$ is diffeomorphic to $T \times (0, \infty)$. The diffeomorphism takes a frame and changes scale uniformly in the space $V$ spanned by the first $(n-1)$ vectors and changes scale in $V^\perp$, in order to make $d = 1$, while keeping the volume equal to one. In $T$, the first $(n-1)$ vectors give an $(n-1)$-frame of volume one. We therefore have a fibre bundle $T \to E_{n-1}$, with fibre $\mathbf{R}^{n-1}$.

Elements of the symmetric space $E_n$ are oriented frames defined up to action on the left by $\mathrm{SO}(n)$. There is exactly one element of $E_n$ represented by an orthonormal frame, namely, the coset of the identity. The stabilizer of this element in $G_n$, acting on the right, is $\mathrm{SO}(n-1)$. We identify $T$ with $\mathrm{SO}(n-1)\backslash G_n$.

The fibre bundle $T \to E_{n-1}$ is a product in a natural way. To see this, we define a map $\sigma : G_n \to \mathbf{R}^{n-1}$ by

**10.4.5.**
$$\sigma\begin{pmatrix} B & b \\ 0 & 1 \end{pmatrix} = B^{-1}b.$$

Then $\sigma$ sends each orbit under $\mathrm{SO}(n-1)$, acting on the left, to a point. Therefore it defines a map, also denoted by $\sigma$, from $T$ to $\mathbf{R}^{n-1}$, which gives a product structure to the fibre bundle $T \to E_{n-1}$.

There is a standard way to define a metric on $E_n$, as follows: The tangent space to $\mathrm{SL}(n, \mathbf{R})$ at the identity can be identified with the space of real matrices of trace zero. This can be made into an inner product space by defining

**10.4.6.** $$\langle X, Y\rangle = 2\,\mathrm{trace}(X^T Y),$$

where $X^T$ denotes the transpose of $X$. This means that $\langle X, X\rangle$ is twice the sum of the squares of the entries. The inner product is invariant under conjugation by elements of $\mathrm{SO}(n)$. The subspace tangent to $\mathrm{SO}(n)$ is the space of skew-symmetric matrices, and the orthogonal complement is the space of symmetric matrices. The inner product on the space of symmetric matrices gives a well-defined positive definite bilinear form on the tangent space to $E_n$ at its basepoint. The action of $\mathrm{SL}(n, \mathbf{R})$ on the right gives a bilinear form on the tangent space at each point of $E_n$. This is the desired riemannian metric.

We have an induced riemannian metric on $T = \mathrm{SO}(n-1)\backslash G_n$. The group $F$ of matrices of the form $\left(\begin{smallmatrix} I & b \\ 0 & 1 \end{smallmatrix}\right)$ operates freely on $T$ on the right, and the next lemma shows that the orbits are the fibres of the fibre bundle $T \to E_{n-1}$.

**Lemma 10.4.7 (euclidean metric).** *A typical fibre of $T \to E_{n-1}$ is the set of cosets of the form $\mathrm{SO}(n-1)\left(\begin{smallmatrix} B & b \\ 0 & 1 \end{smallmatrix}\right)$, where $B \in \mathrm{SL}(n-1, \mathbf{R})$ is fixed and $b \in \mathbf{R}^{n-1}$ varies. The embedding $f_B : \mathbf{R}^{n-1} \to T$ defined by*

$$f_B(b) = \mathrm{SO}(n-1)\begin{pmatrix} B & b \\ 0 & 1 \end{pmatrix}$$

*preserves the riemannian metric.*

*Proof of 10.4.7:* Consider the curve

$$\mathrm{SO}(n-1)\begin{pmatrix} B & b+ty \\ 0 & 1 \end{pmatrix},$$

parametrized by $t$. To find the norm of its derivative at $t = 0$, we need to right translate to the identity coset. We get

$$\begin{aligned} \mathrm{SO}(n-1)\begin{pmatrix} B & b+ty \\ 0 & 1 \end{pmatrix}\begin{pmatrix} B^{-1} & -B^{-1}b \\ 0 & 1 \end{pmatrix} &= \mathrm{SO}(n-1)\begin{pmatrix} I & ty \\ 0 & 1 \end{pmatrix} \\ &= \mathrm{SO}(n-1)\exp\begin{pmatrix} 0 & ty \\ 0 & 0 \end{pmatrix}. \end{aligned}$$

We have

$$\begin{pmatrix} 0 & y \\ 0 & 0 \end{pmatrix} = \begin{pmatrix} 0 & y/2 \\ y^T/2 & 0 \end{pmatrix} + \begin{pmatrix} 0 & y/2 \\ -y^T/2 & 0 \end{pmatrix}.$$

In the metric on the symmetric space, we ignore the second summand when computing the norm. The norms in the symmetric space and in $\mathbf{R}^{n-1}$ are equal—see Equation 10.4.6. $\boxed{10.4.7}$

Let $\omega$ be the $(n-1)$-form on $T$ which is the pullback of the volume form on $\mathbf{R}^{n-1}$ under the map $\sigma : T \to \mathbf{R}^{n-1}$.

**Lemma 10.4.8 (properties of $\omega$).** *The form $\omega$ is closed. Its restriction to any fibre is the volume form. Its pullback to $G_n$ is preserved by conjugation by any element of* $\mathrm{SL}(n-1, \mathbf{R})$.

*Proof of 10.4.8:* The form is closed because it is the pullback of a closed form. To compute the restriction of $\omega$ to a fibre, note that $\sigma f_B : \mathbf{R}^{n-1} \to \mathbf{R}^{n-1}$ is volume preserving. Hence $f_B^* \omega$ is the volume form on $\mathbf{R}^{n-1}$. Since $f_B$ is an isometry by Lemma 10.4.7 (euclidean metric), this shows that $\sigma$ is the volume form on the fibre.

Next we prove that $\omega$ pulled back to $G_n$ is preserved by conjugation by an element of the form $\left(\begin{smallmatrix} P & 0 \\ 0 & 1 \end{smallmatrix}\right)$. Let $\gamma_P : G_n \to G_n$ be conjugation by this element. Then $P \circ \sigma = \sigma \circ \gamma_P : G_n \to \mathbf{R}^{n-1}$. Since $P : \mathbf{R}^{n-1} \to \mathbf{R}^{n-1}$ preserves the volume form, the required result follows. 10.4.8

Since the symmetric space $E_n$ is diffeomorphic to $T \times (0, \infty)$, we can extend $\omega$ to a closed $(n-1)$-form on $E_n$, by pulling back using the projection $E_n \to T$. The norm of $\omega$ on $T$ is one. On $T(d)$, the volume of the $(n-1)$-frame spanned by the first $(n-1)$ vectors is $1/d$. The change of scale in this vector space, needed when mapping $T(d)$ to $T$, is $d^{1/n-1}$. It follows that the norm of $\omega$ on $T(d)$ is $d$.

In order to prove that the special linear groups are not combable, we will prove that the isoperimetric inequalities are not satisfied. To carry out the proof, we need to find $n-1$ elements of $\mathrm{SL}(n-1, \mathbf{Z})$ which commute and which are as independent as possible, in a sense that we will make precise. To find such matrices, we turn to algebraic number theory.

Recall that an algebraic number field is said to be *totally real* if it and all its Galois conjugates are subfields of the field of real numbers.

**Lemma 10.4.9.** *For each $n \geq 2$, there is a monic polynomial of degree $n$ with integer coefficients, which is irreducible over the integers, and such that each root is real and positive.*

*Proof of 10.4.9:* Let $q$ be a prime. The polynomial

$$p_n(x) = (x-2q)(x-4q)\dots(x-2nq) \; + \; q$$

has the desired properties. It is irreducible by Eisenstein's criterion. The sign of $p_n((2j+1)q)$ is equal to $(-1)^{n+j}$, for $0 \leq j \leq n$. Therefore $p_n$ has $n$ positive roots. 10.4.9

Let $M$ be the ring of algebraic integers defined by $p_{n-1}$. Then $M$ is a free abelian group of rank $n-1$. If we fix a basis for $M$, multiplication by an element $x \in M$ gives us an $(n-1)\times(n-1)$ integer matrix. If $m$ is a unit, the matrix is invertible, so that its determinant is $\pm 1$. Multiplication by a root of $p_{n-1}$ has a minimum polynomial equal to $p_{n-1}$, and it is diagonalizable over the reals, with eigenvalues equal to the $n-1$ distinct roots of $p_{n-1}$. Since the multiplication in $M$ is commutative, we can simultaneously diagonalize multiplication by any element of $M$. As a ring, $M \otimes \mathbf{R}$ is the direct sum of $n-1$ copies of $\mathbf{R}$.

The units of $M$ form an abelian group under multiplication. According to the Dirichlet units theorem, this abelian group has rank $n-2$. Each unit is diagonalizable over $\mathbf{R}$. We can choose $n-2$ independent units, $u_1, \ldots, u_{n-2}$ such that each of the eigenvalues for each of the units is positive. (If necessary, we replace a unit by its square.) It follows that the determinant of $u_i$, regarded as an $(n-1)\times(n-1)$ matrix, is 1 rather than $-1$. Part of the Dirichlet units theorem tells us that if we take the logarithms of the diagonal entries for the diagonalized form of the $u_i$, we get a basis for the vector space of dimension $n-2$ consisting of elements $x \in \mathbf{R}^{n-1}$ such that $\sum x_i = 0$. This condition is the logarithm of the condition that the determinant should be 1.

Consider the subgroup of $H_n \subset G_n$ of matrices of the form $\left(\begin{smallmatrix}\Lambda & h\\ 0 & 1\end{smallmatrix}\right)$, where $\Lambda$ is an $(n-1)\times(n-1)$ real diagonal matrix of determinant one, with all entries strictly positive, and $h \in \mathbf{R}^{n-1}$. The fibre bundle $T \to E_{n-1}$ defined on page 234 restricts to a fibre bundle $H_n \to D_{n-1}$, where $D_{n-1}$ is the group of diagonal matrices with strictly positive diagonal entries and determinant one. (Note that $D_{n-1} \cap \mathrm{SO}(n-1)$ is trivial, so we need not use cosets to represent points in the symmetric spaces, in contrast to the situation with $E_{n-1}$ and $T$.)

**Proposition 10.4.10 (conjugating into the thick part).** *There is an $\varepsilon > 0$ and $C \in \mathrm{SL}(n,\mathbf{R})$, such that the image in the symmetric space $E_n$ of $C^{-1}H_nC$ is contained in $E_n(\varepsilon)$. Moreover, we can choose $C = \left(\begin{smallmatrix}P & 0\\ 0 & 1\end{smallmatrix}\right)$, where $P$ is an $(n-1)\times(n-1)$ matrix.*

*Proof of 10.4.10:* Let $K$ be the abelian subgroup of $\mathrm{SL}(n-1,\mathbf{Z})$ of rank $n-2$, chosen as above by using the units in the ring of algebraic integers of the splitting field of $p_{n-1}$. Then $K$ is diagonalizable over $\mathbf{R}$. Let $P \in \mathrm{SL}(n-1,\mathbf{R})$ be chosen so that $PKP^{-1}$ is diagonal. Taking logarithms, we have a lattice in a vector space of dimension $n-2$.

The quotient of $H_n$ by the group of matrices

$$\begin{pmatrix} P & 0\\ 0 & 1\end{pmatrix}\begin{pmatrix} K & \mathbf{Z}^{n-1}\\ 0 & 1\end{pmatrix}\begin{pmatrix} P^{-1} & 0\\ 0 & 1\end{pmatrix}$$

is compact. It follows that there is a compact subset $X \subset H_n$ such that

$$X\begin{pmatrix} P & 0 \\ 0 & 1 \end{pmatrix}\begin{pmatrix} K & Z^{n-1} \\ 0 & 1 \end{pmatrix}\begin{pmatrix} P^{-1} & 0 \\ 0 & 1 \end{pmatrix} = H_n.$$

Therefore

$$C^{-1}H_nC = C^{-1}XC\begin{pmatrix} K & \mathbf{Z}^{n-1} \\ 0 & 1 \end{pmatrix}.$$

We take $\varepsilon > 0$ such that $\mathrm{SO}(n)C^{-1}XC \subset E_n(\varepsilon)$. Multiplying on the right by an element of $\mathrm{SL}(n,\mathbf{Z})$ does not change the lattice defined by a matrix, only its preferred set of generators. Therefore $\mathrm{SO}(n)C^{-1}H_nC \subset E_n(\varepsilon)$.

10.4.10

**Corollary 10.4.11 ($\omega$ bounded).** *With $C$ as in Proposition 10.4.10 (conjugating into the thick part), the norm of the $(n-1)$-form $\omega$ is bounded on $C^{-1}H_nC$.*

*Proof of 10.4.11:* By 10.4.2, it is easy to see that, for a given $\varepsilon$, $T(d) \cap E_n(\varepsilon) = \varnothing$ if $d$ is large enough. The result follows since the norm of $\omega$ on $T(d)$ is $d$.

10.4.11

We recall from page 237 the definition of the group $H_n$ and that it is embedded in the symmetric space $E_n$. The next lemma reduces the problem of showing that $\mathrm{SL}(n,\mathbf{Z})$ is not combable to a question about the geometry of $H_n$.

**Lemma 10.4.12.** *Suppose that, for each positive integer $m$ we can find in $H_n$ an $(n-2)$-cycle $z_m$, depending on a positive integer $m$, whose mass and diameter are bounded above by a polynomial in $m$. Suppose that, for each $m$, there exists some $(n-1)$-chain $c_m$ in $H_n$, such that $\partial(c_m) = z_m$ and $|\omega(c_m)|$ is bounded below by a function which increases exponentially with $m$. Then $\mathrm{SL}(n,\mathbf{Z})$ is not combable.*

*Proof of 10.4.12:* Let $P \in \mathrm{SL}(n-1,\mathbf{R})$, $C \in \mathrm{SL}(n,\mathbf{R})$ and $\varepsilon > 0$ be chosen as in Proposition 10.4.10 (conjugating into the thick part). By Lemma 10.4.8 (properties of $\omega$), we know that $\omega$ is invariant under conjugation by $C$. Therefore the image $y_m$ in $C^{-1}H_nC$ of $z_m$ is a cycle, which is the boundary of an $(n-1)$-chain $d_m$, such that $|\omega(d_m)|$ is bounded below by a function which is exponential in $m$. Since $\omega$ is closed and $E_n$ is contractible, the value of $\omega$ on each $(n-1)$-chain with boundary $y_m$ is the same. This is true in particular for any such chain which lies in $E_n(\varepsilon)$.

Conjugation by $C$ induces a bilipschitz diffeomorphism between the riemannian manifolds $H_n$ and $C^{-1}H_nC$ (with the riemannian structure as submanifolds of the symmetric space). To see this note that each has a riemannian metric which is right invariant with respect to its own group structure.

Therefore the bilipschitz property need only be checked at the identity element. But here we can use the standard fact that every linear isomorphism between two normed vector spaces is bilipschitz. It follows that the mass and diameter of $y_m$ are bounded above by polynomials in $m$.

We prove by contradiction that $\mathrm{SL}(n, \mathbf{Z})$ is not combable. So suppose it is combable. Then Theorem 10.3.5 (riemannian isoperimetric inequality) shows that $y_m$ is the boundary of an $(n-1)$-chain $d_m$ in $E_n(\varepsilon)$, whose mass is bounded above by a polynomial in $m$. By Corollary 10.4.11 ($\omega$ bounded), $\|\omega\| \leq c_0$ on $E_n(\varepsilon)$, for some constant $c_0$. Now $|\omega(d_m)| \leq \|\omega\| \operatorname{mass}(d_m) \leq c_0 \operatorname{mass}(d_m)$. Therefore the $\operatorname{mass}(d_m)$ is bounded below by a function which is exponential in $m$. This contradiction proves the lemma. 10.4.12

Our $(n-2)$-cycle $z_m$ is combinatorially equivalent to the outward pointing normal bundle $Z$ to the boundary of the $(n-1)$-cube $[-1,1]^{n-1}$ in $\mathbf{R}^{n-1}$. The outward pointing normal vector at a point in the interior of an $(n-2)$-dimensional face of the cube is the standard unit vector. We need to spell out what we mean by a normal vector at a boundary point of one of these faces. Let $x \in \partial([-1,1]^{n-1})$. Let $I$ and $J$ be disjoint subsets of $\{1, \ldots, n-1\}$ chosen so that $x_i = 1$ if and only if $i \in I$ and $x_i = -1$ if and only if $i \in J$. Then $I \cup J$ is not empty.

Any convex linear combination of the various $e_i$ for $i \in I$ and $-e_j$ for $j \in J$ is defined to be a normal vector at $x$. The normal vectors based at $x$ form a simplex with vertices the points $e_i$ and $-e_j$. The corresponding cell of $Z$ is the product of a simplex of dimension $|I| + |J| - 1$ and a cube of dimension $n - 1 - |I| - |J|$, so it has dimension $n - 2$. Pictures are shown for $n = 3$ in Figure 10.6 and for $n = 4$ in Figure 10.7.

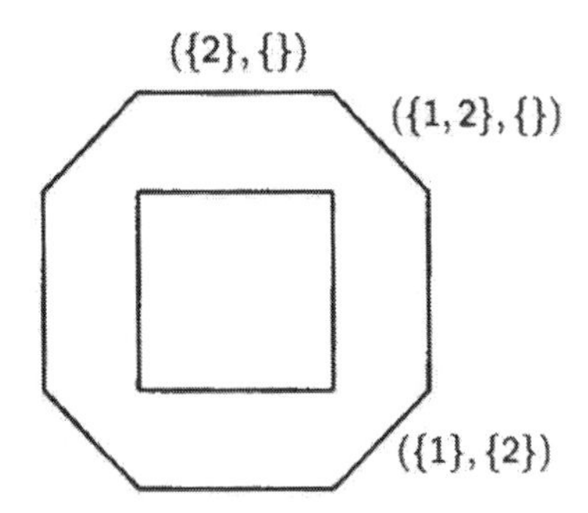


**Figure 10.6. The space $Z$ when $n = 3$.** $Z$ is a combinatorial circle. Some of the edges have been labelled with the pair $(I, J)$ as explained in the text.

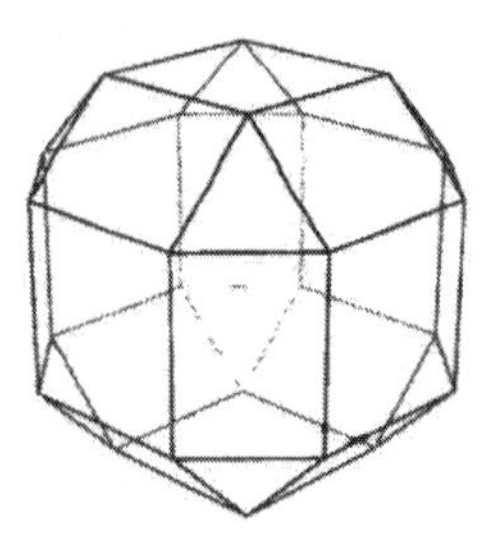

**Figure 10.7. The space $Z$ when $n = 4$.** $Z$ is a combinatorial two-sphere.

Another description of $Z$ is given as follows. Given disjoint subsets $I$ and $J$ as above, consider the linear function

$$\lambda_{I,J}(x) = \sum_{i \in I} x_i - \sum_{j \in J} x_j - 1 - |I| - |J|.$$

The combination of the inequalities $\lambda_{I,J}(x) \leq 0$, as $I$ and $J$ vary, defines a convex body in $\mathbf{R}^{n-1}$, whose boundary is $Z$.

We now embed $Z$ in $H_n$ to give the cycle $z_m$. We do this by embedding each face separately. We recall that $H_n$ has a product structure, making it diffeomorphic to $S \times \mathbf{R}^{n-1}$, where the first factor $S$ is the $(n-2)$-dimensional vector subspace of $\mathbf{R}^{n-1}$, consisting of the logarithms of the entries of the diagonal part of $H_n$ whose sum is zero, and the second factor is given by $\sigma$ (see Equation 10.4.5). The image $z_{I,J,m}$ of the $(I, J)$ face of $z_m$ in $H_n$ corresponds to the product of a simplex in $S$ with a cube in $\mathbf{R}^{n-1}$. The simplex in $S$ is a simplex with one vertex $v_k$ for each element $k \in I \cup J$. We define $v_k$ to be the point with $k$th coordinate equal to $-m(n-2)$ and all other coordinates equal to $m$. The cube in $\mathbf{R}^{n-1}$ consists of all points $x$ with $x_i = e^m$ if $i \in I$, $x_j = -e^m$ if $j \in J$ and $-e^m \leq x_k \leq e^m$ if $k \notin I \cup J$.

**Lemma 10.4.13 (metric properties of cycle).** *As a riemannian submanifold of the symmetric space, $z_{I,J,m}$ is the cartesian product of a euclidean simplex with a euclidean cube, where the product structure is as described above. In riemannian terms, the cube is a standard cube of sidelength 2 and the simplex is a standard simplex when it has log coordinates as explained above. In particular* $\operatorname{mass}(z_m)$ *is bounded by a polynomial in $m$, and* $\operatorname{diam}(z_m)$ *is bounded by a linear function in $m$.*

*Proof of 10.4.13:* We express $z_{I,J,m}$ in terms of the matrices, that it, as elements of $H_n$. An element of $H_n$ has the form $\left(\begin{smallmatrix} \Lambda & h \\ 0 & 1 \end{smallmatrix}\right)$, where $\Lambda$ is a diagonal

$(n-1)\times(n-1)$ matrix with determinant one and $h \in \mathbf{R}^{n-1}$. Let $e^{x_i}$ denote the $i$th diagonal entry of $\Lambda$ and $h_i$ the $i$th entry of $h$. A typical point of $z_{I,J,m}$ has $x_i = m$ and $-1 \le h_i \le 1$ if $i \notin I \cup J$. If $i \in I$, then $e^{-x_i}h_i = e^m$ and if $i \in J$, then $e^{-x_i}h_i = -e^m$. We further require that $x = (x_1, \ldots, x_{n-1})$ be a convex combination of the vertices $v_k$ of the simplex defined above.

Now let $x_i$ and $h_i$ vary smoothly with a parameter $t$, all the while representing a point of $z_{I,J,m}$. We work out the norm of the derivative at $t = 0$ by right translation to the identity element. Right translation to the identity gives us a curve

$$\begin{pmatrix} e^{x_1(t)-x_0(0)} & 0 & \cdot & \cdot & 0 & h_1(t) - e^{x_1(t)-x_1(0)}h_1(0) \\ 0 & \cdot & & & & \cdot \\ \cdot & & \cdot & & & \cdot \\ \cdot & & & \cdot & & \cdot \\ \cdot & & & & e^{x_{n-1}(t)-x_{n-1}(0)} & h_{n-1}(t) - e^{x_{n-1}(t)-x_{n-1}(0)}h_{n-1}(0) \\ 0 & \cdot & \cdot & \cdot & 0 & 1 \end{pmatrix}.$$

whose derivative is

$$\begin{pmatrix} x_1'(0) & 0 & \cdot & \cdot & 0 & h_1'(0) - x_1'(0)h_1(0) \\ 0 & \cdot & & & & \cdot \\ \cdot & & \cdot & & & \cdot \\ \cdot & & & \cdot & & \cdot \\ \cdot & & & & x_{n-1}'(0) & h_{n-1}'(0) - x_{n-1}'(0)h_{n-1}(0) \\ 0 & \cdot & \cdot & \cdot & 0 & 0 \end{pmatrix}.$$

The definition of $z_{I,J,m}$ tells us that if $i \notin I \cup J$, then $x_i' = 0$ and if $i \in I \cup J$, then $h_i'(0) - x_i'(0)h_i(0) = 0$. It follows that the $i$th row of the derivative has at most one non-zero entry, and this is in the diagonal position and equal to $x_i'(0)$ if $i \in I \cup J$ and in the righthand column and equal to $h_i'(0)$ if $i \notin I \cup J$. **10.4.13**

To complete the proof that $\mathrm{SL}(n, \mathbf{Z})$ is not combable, we need to find an $(n-1)$-chain $c_m$ whose boundary is $z_m$, and then to compute $\omega(c_m)$. The first step is to move the cycle $z_m$ into a single fibre of the fibre bundle $H_n \to S$. We will move it into the fibre over $0 \in S$.

We recall that the image of the face $z_{I,J,m}$ in $S$ is a simplex with vertices $v_k$. We move each of these vertices linearly to $0 \in S$ during a unit of time. This gives a homotopy of $z_{I,J,m}$ in $H_n$, defined by keeping the coordinates under the map $\sigma : H_n \to \mathbf{R}^{n-1}$ constant (see page 234 for the definition of $\sigma$). The homotopy defines an $(n-1)$-chain $c_{I,J,m}$. Since $\sigma(c_{I,J,m})$ is a cube of dimension $n - 1 - |I| - |J|$ and $I \cup J$ is not empty, $\omega(c_{I,J,m}) = 0$.

The homotopy just described therefore gives us a map $h : Z \times I \to H_n$, such that $h^*\omega = 0$, $h_0 = z_m$ and $h_1$ is a chain which represents the boundary of a cube of sidelength $2\exp(m)$. (Most of the $z_{I,J,m}$ are sent by $\sigma$ to degenerate chains. Those which are not degenerate are the $2(n-1)$ faces where $I \cup J$ is a singleton.) To complete the $(n-1)$-chain to a chain $c_m$ whose boundary is $z_m$, we fill in the cube. It follows that $\omega(c_m) = 2^{n-1}e^{(n-1)m}$.

This completes the proof of the following theorem:

**Theorem 10.4.14 (SL$(n, \mathbf{Z})$ not combable).** *For $n \geq 3$, SL$(n, \mathbf{Z})$ is not combable.* 10.4.14

# Chapter 11

# Geometrically Finite Groups

In this chapter we prove Epstein's result that the fundamental group of a geometrically finite hyperbolic group is automatic. Thanks are due to Brian Bowditch for helpful conversations.

If the manifold is not compact, it turns out to be convenient to place a different basepoint in each cusp. This leads naturally to the necessity for dealing with automatic groupoids, which we do in Section 11.1 (Groupoids). It is also convenient to deal with generators of different lengths, which is done in Section 11.2 (Generators of Differing Lengths). The main part of the proof consists of exploring the nature of quasigeodesics in the space obtained from hyperbolic space by cutting away disjoint horoballs. We show that, under suitable hypotheses, quasigeodesics with respect to the subspace's path metric, give, by a direct construction, quasigeodesics with respect to the hyperbolic metric. This enables us to use methods very similar to those used to analyze the cocompact case (see the proof of Theorem 3.4.1 (negative curvature)).

## 11.1. Groupoids

The definition of an automatic group can be extended to groupoids. First we give the definition of a groupoid.

**Definition 11.1.1 (groupoid).** A *groupoid* is a category such that there is a morphism between any two objects and such that each morphism is invertible. A groupoid is said to be *generated* by a set $S$ of morphisms, if every morphism is a composition of morphisms in $S$ or in $S^{-1}$. We say the groupoid is *finitely generated* if there is a finite set of generators. When considering automatic groups, we always required our generators to generate as a semigroup (see the beginning of Section 2.1 (Groups As Languages)). Similarly here we will require that generators for the groupoid generate the set of morphisms without the use of inverses—that is, they generate as a category.

A group is a groupoid with exactly one object. It is immediate that a finitely generated groupoid has only a finite number of objects. For any object $x$ in a groupoid $G$, the set of morphisms from $x$ to itself forms a group, called the *vertex group* which we denote by $G_x$. An object of $G$ will be called a *vertex of the groupoid*. We use the same notation $G$ for a groupoid and the set of its morphisms (in fact in some treatments of category theory, a category *is* equal to its set of morphisms). As $x$ varies, the different vertex groups $G_x$ are all isomorphic to each other. One natural example of a groupoid is to take as the set of vertices the set of points of a path-connected topological space $X$ and as morphisms from $x \in X$ to $y \in X$ the set of homotopy classes of paths from $x$ to $y$. This is called the *fundamental groupoid* of $X$. Another important example comes with a finite connected simplicial complex. The objects are the vertices of the complex and the morphisms are homotopy classes of edge paths.

It is often convenient to use the dual of this construction. Given an $n$-dimensional manifold, we suppose it is divided into closed $n$-cells to give a cell complex. We suppose that this cell complex satisfies some suitable conditions, which we will not bother to make explicit—a triangulation would be more than sufficient. We have an object for each $n$-cell, and a generating morphism for each $(n-1)$-cell which is a common face of two $n$-cells. We take formal composites of such generators, and define a morphism to be an equivalence class of formal composites, where the equivalence is defined as follows. We place a vertex in each $n$-cell, and we join two vertices by an edge if and only if the two $n$-cells have an $(n-1)$-dimensional face in common. A formal composite can be thought of as an edge path, using the edges just defined. Two formal composites will be considered to be equivalent if and only if the corresponding edge paths are homotopic in the manifold, keeping the basepoints fixed.

The advantage of these constructions over the fundamental group is that no favouritism is shown in respect of choice of basepoint. This can make a difference if one wants to use an automatic structure to make a drawing—the density of points is more likely to be fairly distributed if we use one of these groupoids instead of the fundamental group. Moreover the algebraic apparatus is much closer to the geometry than if we work with a single basepoint and the fundamental group. This can make associated computer programs easier to write and understand.

A group can be defined by specifying generators and relators. The same is true for a groupoid. Let $K$ be a connected 1-dimensional complex (that is, a connected graph) in which the two endpoints of a single edge are allowed to be equal. We define the *free groupoid* $F(K)$ on $K$. The vertices of the groupoid are the vertices of $K$. The morphisms from a vertex $x$ to a vertex $y$ is the set of homotopy classes of paths in $K$ from $x$ to $y$. If $K$ has only one vertex, the free groupoid is the free group generated by the edges.

We can think of each direction Alternatively, the morphisms can be defined as the set of edge paths from $x$ to $y$ such that an edge is never immediately followed by its inverse, and composition is defined by concatenation and cancellation.

The groupoid $F(K)$ can also be specified by its universal properties. Suppose we have a map $f$ of $K$ into a groupoid $G$, which sends each vertex to a vertex of $G$ and each directed edge to a morphism of $G$. If $e$ is a directed edge of $K$ going from $x$ to $y$ then we insist that $fe : fx \to fy$. If $e^{-1}$ is the same edge in the reverse direction, we insist that $(fe)^{-1} = f(e^{-1})$. Then $f$ extends uniquely to a functor with domain category the free groupoid generated by $K$ and with range category $G$.

We label the directed edges of $K$ with distinct symbols. To each symbol is associated a domain vertex and a range vertex. We can think of a directed edge as an arrow (see page 9). Then any path of arrows represents a morphism of the free groupoid generated by $K$. The morphisms of the free monoid are in one-to-one correspondence with the set of paths of arrows such that an arrow is never immediately followed by its inverse.

The role played by the generators in group theory is played here by $K$. The generators of a group give rise to a free group. Here, $K$ gives rise to the free groupoid $F(K)$. A group given by generators and relators is defined by taking the quotient of the free group by the normal subgroup defined by the relators. In the case of a group, a relator is an arbitrary element of the free group. In a groupoid, a *relator* is a morphism in $F(K)$ with domain and range equal to each other. Given $K$ and a set $R$ of relators, the groupoid $\langle K : R\rangle$ is defined as follows. Its vertices are the vertices of $K$. A morphism from $x$ to $y$ is an equivalence class of paths of arrows, where the equivalence relation is generated by $urv \sim uv$ for each relator $r \in R$. The effect on a vertex group is as follows. Let $F_z$ be the vertex group at $z$ of $F(K)$. Let $r$ be a relator. Then $r$ is based at some vertex $x$. We choose a path of arrows $u(r)$ from $z$ to $x$. The vertex group $\langle K : R\rangle_z$ is equal to $F_z$ modulo the normal subgroup generated by all elements of the form $u(r)ru(r)^{-1}$. We call $K$ the *generating graph* for the groupoid $\langle K : R\rangle$.

$\langle K : R\rangle$ can be specified by its universal properties. Let $G$ be any groupoid and let $f : K \to G$ be any map satisfying the conditions above and sending each relator $r$ to an identity morphism. Then $f$ extends uniquely to a functor from $\langle K : R\rangle$ to $G$.

One could also define *relations* in $F(K)$. Let $R$ be a set of pairs $(r, s)$, where $r$ and $s$ are morphisms in $F(K)$ with the same domain as each other and the same range as each other. Once again we can form $\langle K : R \rangle$ in the obvious way, using universal properties. Equivalently, given a relation $(r, s)$, we can form the relator $s^{-1}r$, reducing the problem to the one considered above.

Just as every group can be given by generators and relators, so every groupoid can be given by means of a generating graph and relators. Given a set $A$ of generators of a groupoid $G$, we obtain the generating graph $K$ as follows. The vertices of $K$ are the objects of the category $G$. There is also a one-to-one correspondence between $A$ and the set of edges of $K$. Of course, in this work we are mainly interested in *finitely presented groupoids*, given by a finite set of generators and a finite set of relations.

**Definition 11.1.2 (Cayley graph of a groupoid).** The Cayley graph of a groupoid with respect to a set of generators can be formed in the same way as the Cayley graph of a group. We pick as a basepoint $x_0$, some vertex of the groupoid. A vertex of the Cayley graph is a morphism with source $x_0$. An arrow from one vertex to another is a triple $(f_1, a, f_2)$, where $f_1 : x_0 \to x_1$ and $f_2 : x_0 \to x_2$ are vertices, $a : x_1 \to x_2$ is a generator, and $f_1 a = f_2$, where we are writing morphisms on the right.

When dealing with groups, the Cayley graph is a covering space of a 1-complex with a single vertex. The subgroup corresponding to the covering is the subgroup of relators. In the case of groupoids, the Cayley graph of $\langle K : R \rangle$ is once again a covering space, in fact a regular covering of $K$ (that is, the group of covering translations acts transitively on the fibre over any fixed point in $K$). The fundamental group of $K$ based at $x_0$ is the vertex group of $F(K)$ at $x_0$ and the covering corresponds to the subgroup of relators in the vertex group.

Geometrically, for each relator $r$, we attach to $K$ a 2-cell, by mapping the boundary circle according to the edge path given by $r$. The Cayley graph of the groupoid is the 1-skeleton of the universal cover of the resulting 2-complex.

The theory of automatic groups extends in a natural manner to the more general context of automatic groupoids. The very minor changes necessary in the definitions and statements of results are left to the reader, and these changes will be assumed to have been made as necessary. We will confine ourselves to a few comments.

We can define a finite state automaton $K_A$ whose states are the vertices of the groupoid and whose arrows are labelled by the generators. We give each of these generators distinct names. (Later on, we will drop this assumption, but it can always be restored if desired, by adding the name of the source and the target vertex to the label on each arrow.) All states of $K_A$ are accept states and all states are start states. The language $L(K_A)$ is just the set of labels of paths of arrows. The language of each other automaton connected with the study of this groupoid and its generators then has to be intersected with $L(K_A)$ in order to get meaningful results. In particular, there is no longer a map $A^* \to G$; instead the functions of this map in the case of groups are taken over in the context of groupoids by $\pi : L(K_A) \to G$. Results like Theorem 2.3.5 (characterizing synchronous) continue to hold. Another important result whose proof is left to the reader is Theorem 2.4.1 (changing generators) in the case of groupoids. The proofs are virtually identical to the corresponding proofs for groups.

The following proposition shows that there is no essential difference between automatic groups and automatic groupoids.

**Proposition 11.1.3 (basepoint and automatic groupoids).** *The following conditions are equivalent.*

(1) *For some choice of basepoint in the Cayley graph, $G$ is an automatic groupoid.*

(2) *For each choice of basepoint in the Cayley graph, $G$ is an automatic groupoid.*

(3) *For some choice $x$ of basepoint in $K$, $G_x$ is an automatic group.*

(4) *For each vertex $x$ of the groupoid, $G_x$ is an automatic group.*

*Proof of 11.1.3:* Suppose $G$ is automatic for one choice of basepoint in the Cayley graph, say $b_1$. Let $b_2$ be another choice of basepoint. Let $L_1$ be the language of accepted strings starting at $b_1$, which gives an automatic structure for $G$. We have to define a regular language $L_2$ of strings starting at $b_2$, which will also give an automatic structure. We fix a path of generators $\omega$ from $b_2$ to $b_1$ and set $L_2 = \omega L_1$. Using this language, it is easy to see that the first condition in the statement implies the second.

We now show that, for any choice of basepoint $b$, the group $G_b$ is automatic, given that $G$ is automatic. Using Theorem 2.4.1 (changing generators) (in the form that applies to groupoids rather than to groups), we may assume that each inverse of a generator of $G$ is also one of the generators of $G$. Note that the generators of $G$ do not in general lie in $G_b$. We proceed as follows. Let $T$ be a maximal tree in the one-complex $K$. Each directed edge in $K$ now represents a definite element of $G_b$, defined by first running along a path in $T$ from the basepoint, then using the arrow, and then returning inside $T$ to the basepoint. We regard each arrow which lies in $T$ as equal to the identity element in $G_b$. However we retain the name of the arrow, so that these different identity elements can be distinguished. So we have the same ordered set of generators for $G_b$ as for $G$. However, a symbol for a generator of $G$ means something somewhat different when it represents a generator of $G_b$.

Let $L$ be the given language of accepted strings starting at $b$. Let $L_b$ be the sublanguage of $L$ consisting of words which end at $b$. We claim that $L_b$ gives an automatic structure for $G_b$. To see this, we need only check the effect of right multiplying by a generator of $G_b$. Now a generator for $G_b$ corresponds to product of generators of $G$ (namely a path along $T$ followed by a generator of $G$ followed by another path along $T$). It follows using Theorem 1.4.6 (predicate calculus) that, for each generator $x$ of $G_b$, there is a finite state automaton in two variables which can verify whether a pair $(w_1, w_2)$ of elements of $L_b$ satisfy $\overline{w_1 x} = \overline{w_2}$. This shows that $G_b$ is automatic as a group.

Now we need to show that if $G_b$ is automatic, then so is $G$. We may assume that the set of generators is the same, and that they correspond as explained above. This means that each generator $x$ of $G_b$ corresponds to a loop in $K$, consisting first of a path in $T$ starting at the basepoint, then of the generator $x$ of $G$, and then a path in $T$ back to the basepoint. In addition we may assume that each of these loops has the same length, by including identity elements as generators, and padding out loops with identity elements based at $b$. We call this length $\lambda$.

Let $L_b$ be the language of accepted strings for $G_b$. We define $L$ to be the set of all strings which result from concatenating a string in $L_b$ with a path of arrows in $T$ without repeated vertices. Since there are only a finite number of such paths of arrows in $T$, $L$ is a regular language. We need only prove the conditions for Theorem 2.3.5 (characterizing synchronous) for $L$. There are two ways of thinking of a word $w$ in $L$ as a path in the Cayley graph of $G$. Firstly we can think of each letter of $w$ as a single edge in the Cayley graph, and secondly we can think of each letter of $w$ as a loop in the Cayley graph of length $\lambda$. For the purposes of this proof, we hop from one point of view to the other. We require $L$ to satisfy the conditions of Theorem 2.3.5 (characterizing synchronous) from the first point of view. This can be deduced from the same conditions, with the second point of view. But these in turn can be deduced from the same conditions for $L_b$. 11.1.3

**Definition 11.1.4 (biautomatic groupoid).** We can also talk of a groupoid $G$ being *biautomatic*. This means that there is a regular language $L$ over a finite set $A$ of generators and a constant $k \geq 1$, with the following properties:

(1) $L \subset L(K_A)$.

(2) The map $\pi : L \to G$ is onto. (Recall that we are using $G$ to denote its own set of morphisms.)

(3) If $w_1, w_2 \in L$ label paths from $v_1'$ to $v_1''$ and from $v_2'$ to $v_2''$ respectively, then the uniform distance between the paths is bounded by $k(d(v_1', v_2') + d(v_1'', v_2'') + 1)$, where the distance $d$ is measured in $\Gamma(G, A)$.

The question of basepoints does not arise in the case of a biautomatic structure.

**Proposition 11.1.5 (biautomatic groupoids: equivalent conditions).** *The following conditions are equivalent.*

(1) *$G$ is a biautomatic groupoid.*

(2) *For some $x$, $G_x$ is a biautomatic group.*

(3) *For each vertex $x$ of the groupoid, $G_x$ is a biautomatic group.*

*Proof of 11.1.5:* The second and third conditions are clearly equivalent, because all vertex groups are isomorphic. We will prove only that the second condition implies the first. We fix a maximal tree $T$ in the generating graph $K$. In the Cayley graph, the inverse image of $T$ consists of disjoint components $T_i$, each of them a copy of $T$. Each copy of $T_i$ will contain exactly one vertex, which we call $x_i$, mapping to $x \in K$.

Given an automatic structure $(A, L)$ for $G_x$, we obtain an automatic structure for $G$, by adding to the beginning of each element of $L$ paths (without repeated vertices) in $T$ joining each vertex of $K$ to $X$, and similarly at the end of each element of $L$. 11.1.5

## 11.2. Generators of Differing Lengths

In the definition of an automatic structure, the multiplier automata are required to read in a pair of strings, and the two strings are required to be read in at the same speed. We now give an equivalent formulation, where this requirement is relaxed a little. (If the requirement is relaxed too much, we obtain an asynchronous automatic structure, which is *not* an equivalent concept—see Definition 7.1.1 (asynchronous automaton).)

**Definition 11.2.1 (two-variable bounded difference machine).** We will define a two-variable bounded difference machine as a special kind of asychronous machine and we use the notation introduced in Definition 7.1.1. Suppose that for each letter $x \in A$ we have been given a positive integral *weight* $n(x)$ and a number $\lambda$ such that $\lambda \geq n(x)$ for each $x$. We also have an end-of-string symbol \$, to which no weight is attached. We have a function

$\delta : S_L \cup S_R \to [-\lambda, \lambda] \subset \mathbf{Z}$, called the *difference function*, such that $\delta(s) > 0$ if $s \in S_R$ and $\delta(s) \leq 0$ if $s \in S_L$. If $s \in S_L$, then $\delta(sx) = \delta(s) + n(x)$, and, if $s \in S_R$, then $\delta(sx) = \delta(s) - n(x)$. The start state is in $S_L$ and is sent to 0 by $\delta$. The effect is that $\delta$ records the difference in weighted length of the two strings being read, and the automaton ensures that one does not get too far out of step with the other. A two-variable bounded difference machine is not essentially different from a conventional two-variable machine.

Suppose we are given a group (or groupoid) $G$ with a set of weighted generators $A$, a regular language $L \subset L(K_A)$, such that $\pi : L \to G$ is surjective, and, for each generator $x$ a two variable bounded difference machine $M_x$ as just described, such that $M_x$ accepts $(w_1, w_2)$ if and only if $\overline{w_1 x} = \overline{w_2}$ and $w_1, w_2 \in L$. We can then find an automation of $G$ as follows. For each symbol $y \in A$, we adjoin a new padding symbol, which we call $\$_y$. Then we change $L$ by introducing rules that say that each $y$ must be immediately followed by exactly $(n(y) - 1)$ $\$_y$ symbols, each representing the identity element of $G$. Otherwise $L$ is unchanged. Changing each $M_x$ in a corresponding way, it is not hard to see that we are dealing with a concept which is equivalent to the usual description of an automatic structure. In the case of a groupoid, the symbol $\$_y$ can be thought of as an identity morphism on the target of $y$ (we are writing morphisms on the right). Another way of thinking about the situation is to divide each edge labelled $y$ into $n(y)$ pieces, each of length one. That is, we introduce $n(y) - 1$ new basepoints. (This has the possibly undesired consequence that, if we start with a group, we end with a groupoid.)

In this way we see that the concept of a two-variable bounded difference machine gives rise to a new concept of (synchronous) automatic group which is equivalent to the old concept. That is, a group or groupoid with weighted generators is automatic according to this new definition if and only if it is automatic with respect to the old definition (with all weights made equal).

We now come to a substantial generalization of Theorem 3.2.1 (geodesic automaton 1) and Theorem 3.2.2 (geodesic automaton 2). Let $G$ be a groupoid with a finite ordered set $A$ of weighted generators, and let $\Gamma(G, A)$ be its Cayley graph. Let $B$ be another finite ordered set of weighted generators for $G$, with corresponding Cayley graph $\Gamma(G, B)$. Then the vertices of $\Gamma(G, A)$ are in one-to-one correspondence with the vertices of $\Gamma(G, B)$, since a vertex is defined independently of choice of generators (see Definition 11.1.2). This makes the two Cayley graphs into metric spaces in the obvious way. It is easy to see that the identity map on vertices is a pseudoisometry, which therefore induces a pseudoisometry between the Cayley graphs themselves.

Let $\Theta$ be a connected subgraph of $\Gamma(G, B)$ which contains all the vertices. Then $\Theta$ also has a natural metric, which is not the induced metric from $\Gamma(G, B)$, but the intrinsic path metric. We fix a basepoint $b_0$ in $\Theta$.

**Theorem 11.2.2 (geodesic automaton 3).** *Let $\Theta$ be as above. We suppose that the identity map on the vertices is a pseudoisometry between the path metric of $\Theta$ and the path metric of $\Gamma(G, B)$ (or, equivalently, of $\Gamma(G, A)$). Let $V$ be a finite state automaton over $B$ and let $L(V)$ be the language accepted by $V$. Let $L(V)$ be prefix closed and consist of certain strings starting at $b_0$ and lying entirely within $\Theta$. Let $L \subset L(V)$ be the set of strings representing paths which are geodesic for the path metric of $\Theta$. We suppose that every vertex is the endpoint of some path in $L$. We also suppose that there is a number $k$ with the following property:*

> *Let $w_1$ and $w_2$ be two elements of $L$ with corresponding paths $\widehat{w_1}$ and $\widehat{w_2}$ in the Cayley graph starting at $b_0$. Suppose these paths end within a distance $d$ of each other in the path metric of $\Theta$. Then, in the uniform metric induced by the path metric of $\Theta$, the distance between the paths $\widehat{w_1}$ and $\widehat{w_2}$ is at most $k(d+1)$.*

*Under the above hypotheses, $(L, B)$ is an automatic structure on $G$.*

Since distances between vertices are bounded below by one, and we are using the identity map, pseudoisometry is the same as lipschitz equivalence on the vertices. In the above statement we could therefore use the path metric of $\Gamma(G, B)$ to measure the distance and the uniform metric; the effect of this is to increase the constant $k$. Note that we can deduce the condition for all $d \geq 0$, if we know it for $d \leq \lambda$. The reason is that the condition on the uniform distance between $\widehat{w_1}$ and $\widehat{w_2}$ is equivalent to the condition that there is a constant $k'$ such that, if $d = 0$, then the uniform distance is bounded by $k'$, and if $d \geq 1$, the uniform distance is bounded by $k'd$. Another way of relaxing the condition is to replace the uniform metric by the hausdorff metric. Arguing as in Lemma 3.2.3 (hausdorff implies uniform), with geodesics ending less than a distance $\lambda$ apart in the path metric of $\Theta$, we can deduce the uniform metric condition from the hausdorff metric condition.

*Proof of 11.2.2:* Let each element of $B$ have weight at most $\lambda$. We will start by describing $M$, the *word difference machine* based on words of length $k\lambda$. $M$ is a two-variable bounded difference partial deterministic automaton. Each state of $M$ is an element $g$ of $L(K_B)$ of (weighted) length at most $k\lambda$, together with the difference variable $i$ satisfying $-\lambda \leq i \leq \lambda$ or $i = \pm\infty$. In addition we have a success state $s^\$$. We have $S_L = \{(g, i) | -\lambda \leq i \leq 0\}$, $S_R = \{(g, i) | 0 < i \leq \lambda\}$, $S_L^\$ = \{g, -\infty\}$ and $S_R^\$ = \{(g, \infty)\}$.

Suppose $b \in B$ and we have a state $(g : x \to y, i)$ of $M$. Suppose $i \leq 0$. Then we have an arrow labelled $b$, read from the left tape, from the state $(g, i)$ to the state $(b^{-1}g, i + n(b)$ if and only if $b^{-1}g$ is defined (that is, the source of $b$ is $x$) and can be represented by a word over $B$ of length at most $k\lambda$. (Note that we are using a convention for composing morphisms which corresponds to writing maps on the right.) Suppose $i > 0$. Then we have an arrow labelled $b$, read from the right tape, from the state $(g, i)$ to the state $(gb, i - n(b))$ if and only $gb$ is defined (that is the source of $b$ is $y$) and can be represented by a word over $B$ of length at most $k\lambda$. The action of the end-of-string symbol \$ is to send $(g, i)$ to $(g, \infty)$ if $-\lambda \leq i \leq 0$ and to $(g, -\infty)$ if $0 < i \leq \lambda$, and is otherwise not defined. If $i = \pm\infty$ and $g$ is an This means that an end-of-string symbol can be read at most once. The start state of $M$ is the identity morphism at the basepoint. The accept states are the identity morphisms at any object of $G$.

The language of pairs $(u, v)$ of strings over $B$, with $|u| > |v|$ (computing the lengths using the weightings on the letters of $B$) is a regular language accepted by a two-variable bounded difference machine—we need only keep track of the length difference while it is between $-\lambda$ and $\lambda$.

By Theorem 1.4.6 (predicate calculus)

$$L_1 = \{u \mid u \in L(V) \wedge (\exists v)(|u| > |v| \wedge (u, v) \in L(M) \wedge v \in L(V))\}$$

is a regular language consisting of strings which are definitely not $\Theta$-geodesics. We claim that the set $L$ of $\Theta$-geodesics in $L(V)$ is equal to $L_2 = L(V) \backslash L_1 B^*$.

To prove this claim, note first that $\varepsilon \in L \subset L_2$. Next note that, since $L(V)$ is prefix closed, $L_2$ also is. We now prove a contradiction from the assumption that there is an element $w \in L_2$L, that is, $w$ is not a $\Theta$-geodesic. We may suppose that $w$ is a shortest such element. Let $w = ux$, where $u$ is a string and $x \in B$. By our choice of $w$, $u \in L$. Let $v \in L$ represent $\overline{ux}$; then $|v| < |u| + 1 = |w|$. By our assumptions, $\widehat{w}$ and $\widehat{v}$ are within a uniform distance in $\Theta$ and hence in $\Gamma(G, B)$ of at most $k\lambda$, since the length of $x$ is less than $\lambda$. Hence $(ux, v) \in L(M)$. But this means that $ux \in L_1$, which is impossible, since $v$ is a $\Theta$-geodesic.

We form the multiplier automaton for $x \in B$ as follows. Let $M^x$ have the same states, arrows and start state as $M$, but let the accept state be the morphism $x$. Then our multiplier automaton is the two-variable machine which accepts $(L \times L) \cap L(M^x)$. 11.2.2

There is also a version of Theorem 11.2.2 which works for biautomatic structures.

**Theorem 11.2.3 (geodesic automaton 4).** *Let $\Theta$ be as above. We suppose that the identity map on the vertices is a pseudoisometry between the* *metric induced from the path metric of* $\Theta$ *and the path metric of* $\Gamma(G, B)$ *(or, equivalently, of* $\Gamma(G, A)$*). Let* $V$ *be a finite state automaton over* $B$ *and let* $L(V)$ *be the language accepted by* $V$*. Let* $L(V)$ *be prefix closed and consist of certain strings which can be traced out entirely within* $\Theta$*; these strings will not in general be labels on paths all starting at the same point. Let* $L \subset L(V)$ *be the set of strings representing paths which are geodesic for the path metric of* $\Theta$*. We suppose that, for every pair* $(v', v'')$ *of vertices of* $\Gamma(G, B)$*, there is a path from* $v'$ *to* $v''$ *labelled by an element of* $L$*. We also suppose that there is a number* $k$ *with the following property:*

> *Let* $w_1$ *and* $w_2$ *be two paths in* $\Theta$ *labelled by elements of* $L$*. Let their initial points be the vertices* $w_1'$ *and* $w_2'$ *and their final points be the vertices* $w_1''$ *and* $w_2''$*. Then, in the uniform metric induced by the path metric of* $\Theta$*, their distance apart is* $k(d(w_1', w_2') + d(w_1'', w_2'') + 1)$ *at the most.*

*Under the above hypotheses,* $(L, B)$ *is a biautomatic structure on* $G$*.*

*Proof of 11.2.3:* The proof is the same as the proof of Theorem 11.2.2 (geodesic automaton 3). The automaton $M$ is slightly different, in that this time there is no basepoint, and paths can start at any vertex. $M$ has a set of start states, namely the set of identity morphism at any vertex of the groupoid, and these are also the accept states. 11.2.3

Here is a further indication that the change from groups to groupoids and the weighting of generators are technical conveniences rather than an essential generalizations.

**Proposition 11.2.4 (Cayley graph of group and groupoid).** *If we change the weights on the generators of a groupoid, the Cayley graph of the groupoid changes by a pseudoisometry. (The identity map is a pseudoisometry.) Let* $G$ *be a groupoid, with vertex group* $G_x$*, with generators chosen as in the proof of Proposition 11.1.3 (basepoint and automatic groupoids). Then the obvious embedding of the Cayley graph of* $G_x$ *in the Cayley graph of* $G$ *is a pseudoisometry.*

*Proof of 11.2.4:* The first statement is trivial. To prove the second we apply Lemma 3.3.2 (subspace). 11.2.4

## 11.3. Geodesics and Horoballs

Let $S$ be any manifold obtained from hyperbolic space $\mathbf{H}^n$ by removing the interiors of a disjoint set of horoballs. The typical way in which such a manifold will arise from our work is as the universal cover of a geometrical finite hyperbolic manifold, from which the cusp regions have been cut out. Then $S$ is a manifold with boundary. We give $S$ the path metric, which is not the same as the induced metric from $\mathbf{H}^n$ unless the set of horoballs is empty. Recall that each horosphere has a path metric which is euclidean. Any two points in $S$ are joined by a geodesic in $S$ and such a geodesic coincides with a hyperbolic geodesic in the interior of $S$ and with a euclidean geodesic on the boundary of $S$. A geodesic in $S$ meets the boundary of $S$ tangentially, except possibly at an endpoint of the geodesic.

**Theorem 11.3.1 (quasigeodesics outside horoballs).** *Let $k \geq 1$ and $\varepsilon \geq 0$ be fixed real constants. Then there is a positive real number $l$, depending only on $k$ and on $\varepsilon$, with the following property. Let $r > 3l$. Let $X$ be a union of disjoint horoballs in $\mathbf{H}^n$, such that any two components of $X$ are a distance at least $r$ apart and let $S$ be the complement of the interior of $X$. Let $\alpha : [a,b] \to S$ be path in $S$ which is a $(k,\varepsilon)$-quasigeodesic for the path metric of $S$. Let $\phi$ be the hyperbolic geodesic from $\alpha(a)$ to $\alpha(b)$. Then the union of the $l$-neighbourhood of $X$ and the $l$-neighbourhood of $\phi$ contains the image of $\alpha$.*

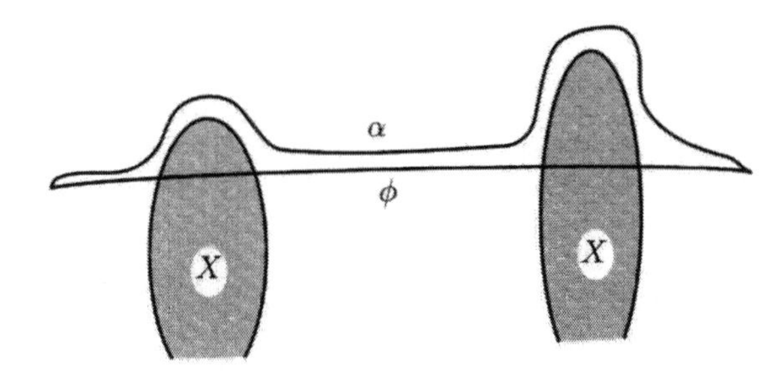


**Figure 11.1. Quasigeodesics outside horoballs.** This illustrates Theorem 11.3.1 (quasigeodesics outside horoballs), showing how a quasigeodesic stays near the union of horoballs and a hyperbolic geodesic.

The statement of the theorem is illustrated in Figure 11.1.

Before starting the proof of this theorem, we need to prove a series of lemmas.

**Definition 11.3.2 (concentric horospheres).** Let $H$ be a horoball and let $t \in \mathbf{R}$. We define $H_t$ to be the concentric horoball, whose horosphere is at a signed distance $t$ from that of $H$. If $t < 0$ then $H_t \subset H$; $H_0 = H$; and if $t > 0$ then $H_t \supset H$.

**Lemma 11.3.3 (quasigeodesics near horospheres).** *Let $k \geq 1$ and $\varepsilon \geq 0$ be given. Then there is a number $l_0 > 0$ with the following properties. Let $r > l_0$. Let $S$ be the result of removing from $\mathbf{H}^n$ a disjoint union $X$ of interiors of horoballs such that any two components of $X$ are at a hyperbolic distance at least $r$ apart. We give $S$ the path metric, not the induced metric. Let $\alpha : [a,b] \to S$ be any $(k,\varepsilon)$-quasigeodesic for $S$, and let $H$ be a component of $X$, such that*

$$0 \leq d(\alpha(a), H) = d(\alpha(b), H) = s < r - l_0.$$

*Then the image of $\alpha$ lies entirely within the horoball $H_{s+l_0}$.*

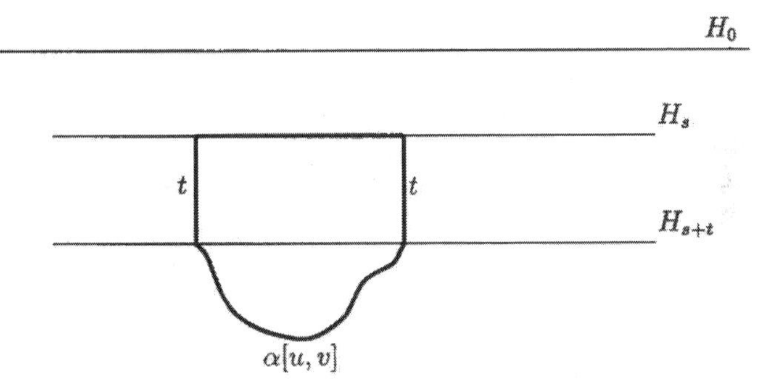


**Figure 11.2. Quasigeodesics near horospheres.** This illustrates the proof that a quasigeodesic (parametrized by pathlength) in $S$, starting and ending near a certain horoball, must stay near that horoball. We show two alternative paths from $\alpha(u)$ to $\alpha(v)$, $\alpha[u,v]$ of length $|u-v|$ and another path of length at most $2t + |u-v|e^{-t}$.

*Proof of 11.3.3:* The proof is almost identical to that of Lemma 3.4.2 (quasigeodesics near geodesics). By Lemma 3.3.5 (piecewise geodesic), there is no loss of generality in assuming that $\alpha$ is continuous, piecewise geodesic (for the metric of $S$) and parametrized by arclength. (This changes the constants for the quasigeodesic, but in a way that depends only on $k$ and on $\varepsilon$.) We fix $t = \log(2k)$. Let $(u,v)$ be a maximal open subinterval of $[a,b]$ which is disjoint from $H_{s+t}$. Orthogonal projection of $\alpha|_{[u,v]}$ onto $\partial H_s$ decreases distances by a factor of at least $e^t$. The situation is illustrated in Figure 11.2. Therefore

$$|v-u| \leq k\, d(\alpha(u), \alpha(v)) + \varepsilon \leq k(2t + |v-u|e^{-t}) + \varepsilon = 2k\log(2k) + \frac{|v-u|}{2} + \varepsilon.$$

So $|v-u| \leq 4k\log(2k) + 2\varepsilon$. We set $l_0 = t + 4k\log(2k) + 2\varepsilon$. The lemma follows. $\boxed{11.3.3}$

Now we use the following well-known lemma, illustrated in Figure 11.3.

**Figure 11.3. A piecewise geodesic gives a quasigeodesic.** This illustrates Lemma 11.3.4 (angles bounded below). If each angle is at least $\theta_1$ and each geodesic segment has a length at least $l_1$, then the result is a $(k_1, \varepsilon_1)$-quasigeodesic.

**Lemma 11.3.4 (angles bounded below).** *Let $0 < \theta_1 \leq \pi$. There are numbers $l_1 > 0$, $k_1 \geq 1$ and $\varepsilon_1 \geq 0$, depending only on $\theta_1$, with the following property. Let $\alpha$ be a piecewise geodesic in $\mathbf{H}^n$, such that each geodesic piece has length at least $l_1$ and such that successive pieces meet at an angle which is at least $\theta_1$. Let $\alpha$ be parametrized by arclength. Then $\alpha$ is a $(k_1, \varepsilon_1)$-quasigeodesic in $\mathbf{H}^n$.*

In fact we can choose $k_1$ to be any constant greater than one; the closer it is to one, the larger $l_1$ and $\varepsilon_1$ have to be. We will actually carry out the proof for $k_1 = 4$. Note that the result is false in euclidean space.

*Proof of 11.3.4:* First consider a unit speed geodesic $\beta$ in the hyperbolic plane, and let $x$ be a point which does not lie on the geodesic, as shown in Figure 11.4. Suppose that the angle between the segment $\beta(t_0)x$ and the positive tangent to $\beta$ is at least $\theta_2$, where $\pi/3 > \theta_2 > 0$. By choosing $l_2$ large enough ($\tanh(l_2/2) \geq \cos\theta_2$), we can ensure that the angle $\phi(t)$ between the segment $\beta(t)x$ and $\beta'(t)$ at $\beta(t)$ is greater than $\pi - \theta_2$ for $t > l_2$. Moreover, $\phi(t)$ is monotonic increasing as a function of $t$ for all $t$ and it varies from 0 to $\pi$ as $t$ varies from $-\infty$ to $\infty$. The derivative of $d(x, \beta(t))$ with respect to $t$ is $-\cos(\phi(t))$. So, for $t > l_2$, the derivative is greater than $\cos\theta_2 > 1/2$ and the derivative tends to 1.

Now consider the data in the statement of the lemma. We first choose $\theta_2$ so that $0 < \theta_2 < \pi/3$ and $\theta_2 < \theta_1/2$. This determines $l_2$ as above. We then choose $l_1 = 8l_2$. Let $[t_1, t]$ lie in the domain of $\alpha$. We set $x = \alpha(t_1)$ and apply the above discussion. Consider the angle between $\alpha(t)x$ and $\alpha'(t)$ at $\alpha(t)$. This is initially independent of $t$ and equal to $\pi$. By induction on $n$, the number of corners in $\alpha(t_1, t)$, the angle is at least $\theta_2$ for all $t$. We have seen that the induction starts. At a corner, the angle changes from more than $\pi - \theta_2$ to more than $\theta_1 - \theta_2 > \theta_2$, as we see in Figure 11.5, and this proves the induction step.

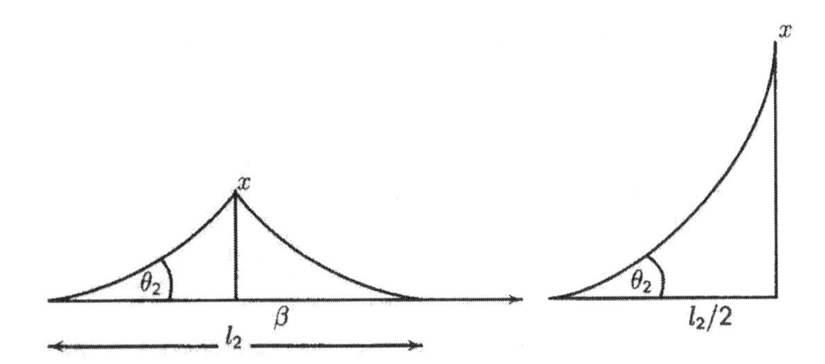


**Figure 11.4. Getting past the point $x$.** The figure on the left shows how far we need to progress along the geodesic $\beta$ in order that the angle between the segment $\beta(t)x$ and $\beta'(t)$ should become greater than $\pi - \theta_2$ after starting at an angle bounded below by $\theta_2$. The figure on the right shows the limiting case, as $x$ goes off to infinity, which gives a maximal value for $l_2/2$—the length of the side of a triangle with angles 0, $\theta_2$ and $\pi/2$. We have $l_2 = 2\text{arctanh}(\cos\theta_2)$.

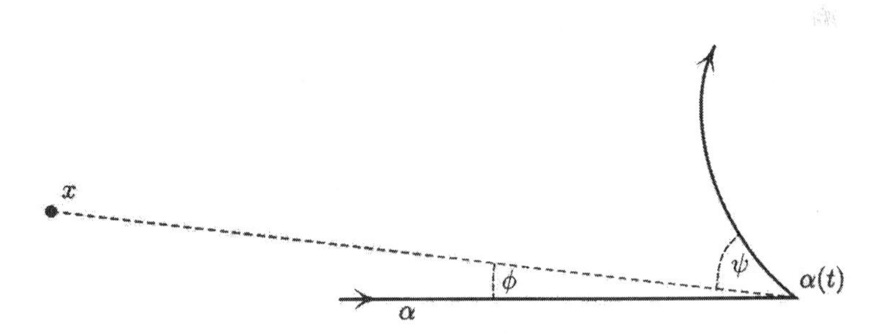


**Figure 11.5. Negotiating a corner.** At a corner, we have $\phi < \theta_2$ and the angle just before the corner is $\pi - \phi > \pi - \theta_2$. We then have $\psi \geq \theta_1 - \phi > \theta_2$.

Now $|t - t_1| \geq 8(n-1)l_2$, by our assumptions. Over a geodesic piece of $\alpha$ of length $l_1$, $d(x, \alpha(t))$ may decrease over the first stretch of the domain of length $l_2/2$ (see Figure 11.4) and it decreases by an amount which is certainly bounded above by $l_2$. Subsequently, the distance increases with a derivative at least $\cos\theta_2 \geq 1/2$. It follows that

$$\begin{aligned} d(\alpha(t_1), \alpha(t)) &\geq -nl_2 + (|t - t_1| - nl_2)\cos\theta_2 \\ &\geq |t - t_1|/2 - 2nl_2 \\ &\geq |t - t_1|/4 + 2(n-1)l_2 - 2nl_2 \\ &= |t - t_1|/4 - 2l_2 \end{aligned}$$

Since $\alpha$ is parametrized by arclength, we also have $d(\alpha(t_1), \alpha(t)) \leq |t - t_1|$. 11.3.4

Here is another important, well-known result.

**Lemma 11.3.5 (quasigeodesic over subintervals).** *Given $k \geq 1$ and $\varepsilon \geq 0$, there exist $l_3 > 0$, $k_3 \geq k$ and $\varepsilon_3 \geq \varepsilon$ with the following property. Let $\alpha$ be a path in $\mathbf{H}^n$, such that on subintervals of its domain of length at most $l_3$ it is $(k, \varepsilon)$-quasigeodesic. Then $\alpha$ is $(k_3, \varepsilon_3)$-quasigeodesic.*

*Proof of 11.3.5:* We may assume that $\alpha$ is defined on a finite closed interval. Lemma 3.4.2 (quasigeodesics near geodesics) tells us that a $(k, \varepsilon)$-quasigeodesic in $\mathbf{H}^n$ is contained in an $l_4$-neighbourhood of the geodesic connecting its endpoints, where $l_4$ depends only on $k$ and on $\varepsilon$. Any hyperbolic triangle $ABC$, such that the lengths of $AB$ and $AC$ are each at least $2l_4$, and such that $d(A, BC) \leq l_4$, has an angle at $A$ which is at least equal to $\theta > 0$, where $\theta$ depends only on $l_4$ and hence only on $k$ and on $\varepsilon$. Figure 11.6 indicates the proof.

Lemma 11.3.4 (angles bounded below) gives us $l_1$, $k_1$ and $\varepsilon_1$ when we give it $\theta_1$. We choose $l_3$, depending only on $k$ and on $\varepsilon$, larger than $l_1$ and larger than $8kl_4 + 4\varepsilon$. We can clearly restrict our attention to paths of length greater than $l_3$. We choose a monotonically increasing sequence $\{t_i\}$ of points in the domain of $\alpha$ such that $l_3/4 \leq |t_{i+1} - t_i| < l_3/2$ for each $i$. We assume that this sequence includes the endpoints of the domain of $\alpha$.

By our choice of $l_3$,

$$d(\alpha(t_{i+1}), \alpha(t_i)) \geq \frac{|t_{i+1} - t_i| - \varepsilon}{k} \geq \frac{l_3 - 4\varepsilon}{4k} \geq 2l_4.$$

We join $\alpha(t_i)$ to $\alpha(t_{i+1})$ by a geodesic segment for each $i$, as in Figure 11.7. The first paragraph of this proof then shows that the angle between these segments is at least $\theta_1$, where $\theta_1$ depends only on $k$ and on $\varepsilon$.

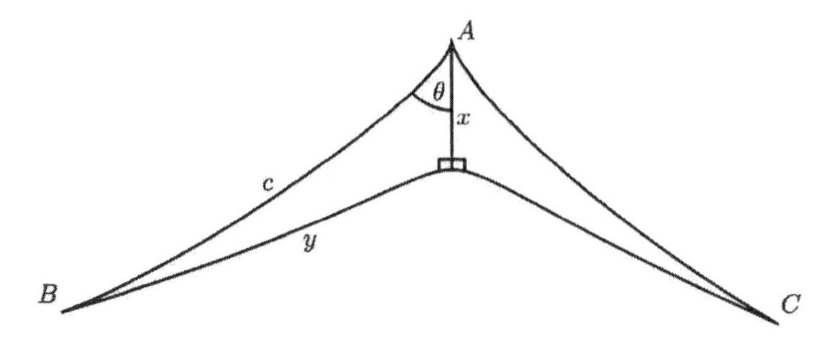


**Figure 11.6. Bounding the angle at the vertex from below.** Since $c \geq 2l_4$ and $x \leq l_4$, we have $y \geq l_4$. The angle $\theta$ increases with $y$ and decreases as $x$ increases. So the minimum value of $\theta$ is obtained with $x = y = l_4$.

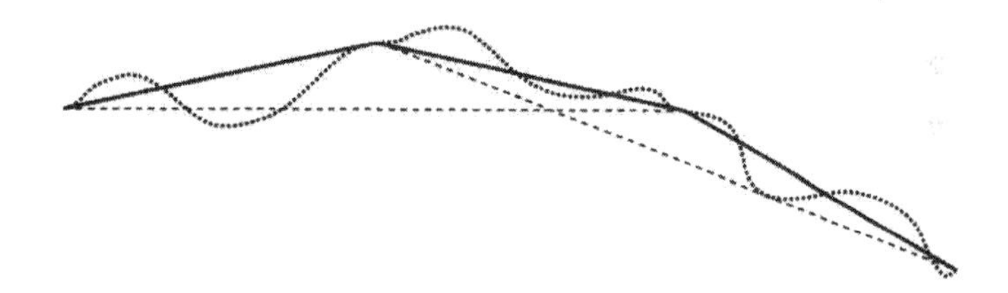

**Figure 11.7. Constructing a long quasigeodesic from short ones.** This figure illustrates Lemma 11.3.5 (quasigeodesic over subintervals). The dotted line is the quasigeodesic $\alpha$. The dashed lines are geodesic segments joining points of the form $\alpha(t_i)$ and $\alpha(t_{i+2})$. The solid lines are geodesic segments joining points of the form $\alpha(t_i)$ and $\alpha(t_{i+1})$.

Since we have chosen our constants so as to satisfy Lemma 11.3.4 (angles bounded below), the path parametrized by arclength and made by following consecutive geodesic segments $[\alpha(t_i), \alpha(t_{i+1})]$ is a quasigeodesic, and the constants for the quasigeodesic depend only on $k$ and on $\varepsilon$. Hence it is a quasigeodesic if we reparametrize the pieces linearly so that the argument is equal to $t_i$ at $\alpha(t_i)$ and once again the quasigeodesic constants depend only on $k$ and on $\varepsilon$. From this it follows easily that the whole of the original map $\alpha$ is a quasigeodesic with constants depending only on $k$ and on $\varepsilon$. $\boxed{11.3.5}$

The proof of the following lemma is immediate (see Figure 11.8).

**Lemma 11.3.6 (dipping).** *Let $H$ be a horoball with boundary $\partial H$ and let $\alpha$ be a hyperbolic geodesic from $x \in \partial H$ to $y \in \partial H$. Suppose that $\alpha$ does not intersect $H_{-t}$ where $t > 0$ (see Definition 11.3.2 for the notation). Then the*

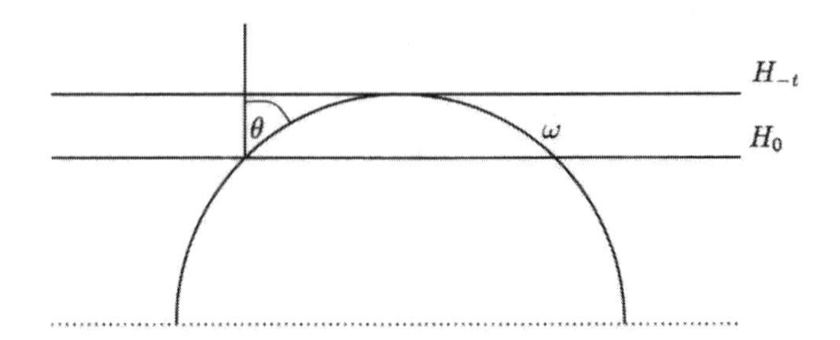


**Figure 11.8. A geodesic penetrating a horoball.** The geodesic penetrates a distance $t$ into the horoball. The angle to a perpendicular to the horosphere is $\theta$ where $\sin(\theta) = \exp(-t)$.

*angle between $\alpha$ and the perpendicular to $\partial H$ which is the geodesic from $x$ to the centre of $H$ is at least $\theta$, where $\sin(\theta) = \exp(-t)$.*

We will also need the following technical result.

**Lemma 11.3.7 (D lemma).** *Let $d > 0$ be given. Then there is a number $D > 0$ with the following property. Let $H$ be an open horoball in $\mathbf{H}^n$, and let $\beta$ and $\gamma$ be two geodesic segments in the complement of $H$, meeting the boundary horosphere of $H$ in $x$ and $y$ respectively (not necessarily at the endpoints of $\beta$ and $\gamma$), such that the hyperbolic distance satisfies $d(x, y) \leq d$. Then, for any hyperbolic geodesic segment $\rho$ from a point of $\beta$ to a point of $\gamma$, each point of $H \cap \rho$ is within a hyperbolic distance $D$ of both $x$ and $y$. If $\rho$ meets $H$, then $\rho$ is contained in the $D$-neighbourhood of $\beta \cup \gamma$.*

*Proof of 11.3.7:* The proof is illustrated in Figure 11.9. The details of the

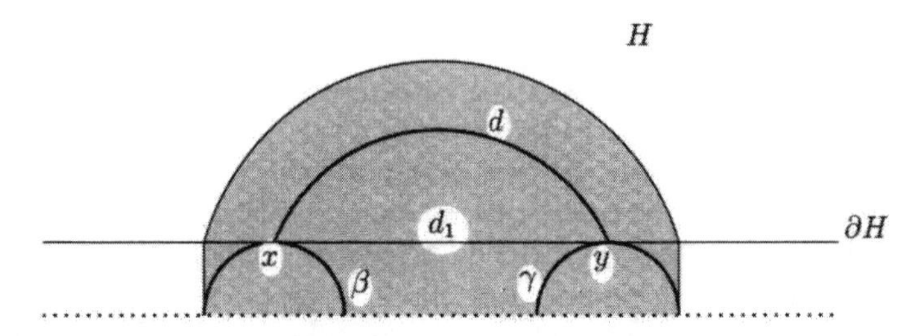


**Figure 11.9. Computing $D$.** This illustrates the proof of Lemma 11.3.7. The shaded region denotes the subspace called $A$ in the proof. $D$ is the hyperbolic diameter of the part of the shaded region in the horoball $H$.

proof are easier to explain if we work in terms of the horospherical distance $d_1 = 2\sinh(d/2)$ in $\partial H$. Orthogonal projection $\pi : \mathbf{H}^n \backslash H \to \partial H$ onto the boundary horosphere of $H$ sends the geodesic segments $\beta$ and $\gamma$ onto segments (which are geodesic with respect to the euclidean path metric on the horosphere), each of length at most 2, containing $x$ and $y$ respectively. It follows that there is a ball $B$ in the horosphere, of horospherical radius $d_1 + 2$ containing both $\pi(\beta)$ and $\pi(\gamma)$. The convex hull of $B$ is contained in the closure of $H$, and the union of the convex hull of $B$ together with the inverse image of $B$ under $\pi$ is a closed convex subset $A$ of $\mathbf{H}^n$. (To see this, consider the upper half space model.) We set $D$ equal to the hyperbolic diameter of the convex hull of $B$.

To see that $\rho$ is in a $D$-neighbourhood of $\beta \cup \gamma$, we argue as follows. The distance of a point $p \in \rho$ to $\beta$ increases monotonically as $p$ moves from $\beta \cap \rho$. So $d(p, \beta) \leq D$ while $p \in \rho$ lies between $\rho \cap \beta$ and $B$. Similarly $d(p, \gamma) \leq D$ while $p \in \rho$ lies between $\rho \cap \gamma$ and $B$. Hence $d(p, \beta \cup \gamma) \leq D$ for all $p \in \gamma$.

11.3.7

We are now in a position to start the proof of Theorem 11.3.1 (quasigeodesics outside horoballs).

*Proof of 11.3.1:* We first choose $l_0$ so that any $(k, \varepsilon)$-quasigeodesic in $\mathbf{H}^n$ stays within a distance $l_0$ of the geodesic connecting its endpoints (Lemma 3.4.2 (quasigeodesics near geodesics)) and so that Lemma 11.3.3 (quasigeodesics near horospheres) is satisfied.

Let $\alpha : [a, b] \to S$ be a continuous $(k, \varepsilon)$-quasigeodesic with respect to the metric of $S$. If $H$ is a horoball component of $X$, let $a_H$ be the smallest number and $b_H$ the largest number in the domain of $\alpha$ such that $\alpha(a_H)$ and $\alpha(b_H)$ are at a distance exactly equal to $2l_0$ from $H$. (Either both $a_H$ and $b_H$ exist or neither exists. It is possible that $b_H = a_H$.) If $d(\alpha(a), H) \leq 2l_0$, for some component $H$ of $X$, we set $a_H = a$ and, if $d(\alpha(b), H) \leq 2l_0$, we set $b_H = b$, changing the definitions of $a_H$ and $b_H$ if necessary. According to Lemma 11.3.3 (quasigeodesics near horospheres), $\alpha[a_H, b_H]$ lies entirely within the open $3l_0$-neighbourhood of $H$. We assume that $l > 3l_0$, where $l$ is as in the statement of Theorem 11.3.1 (quasigeodesics outside horoballs). Since $l_0$ depends only on $k$ and on $\varepsilon$, this is an acceptable condition.

We recall from the statement of Theorem 11.3.1 (quasigeodesics outside horoballs) that we are assuming that the distance between components of $X$ is at least $r > 3l > 9l_0$. Since the domain of $\alpha$ is a bounded interval $[a, b]$ and the intervals complementary to the $[a_H, b_H]$ have images with length at least $r - 4l_0$, we obtain a finite partition of $[a, b]$. The situation is illustrated in Figure 11.10.

**Lemma 11.3.8 (hyperbolic quasigeodesic).** *Let $(b', a'')$ be an interval complementary to the intervals of the form $[a_H, b_H]$. Then $\alpha$ restricted to this interval is a $(k, \varepsilon)$-quasigeodesic for the hyperbolic metric.*

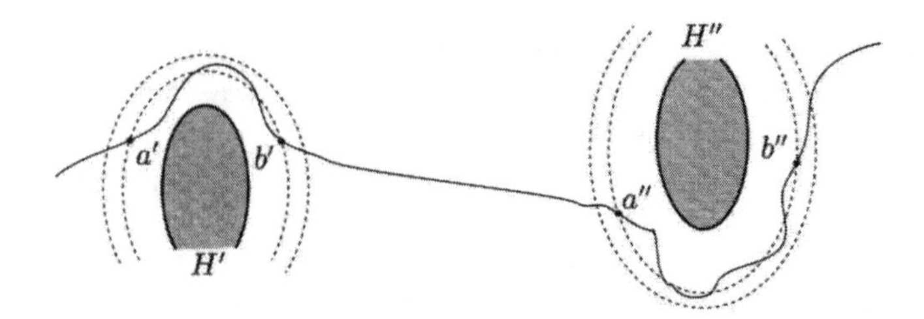


**Figure 11.10. An $S$-quasigeodesic near two horospheres.** This illustrates the first part of the proof of Theorem 11.3.1 (quasigeodesics outside horoballs). We show two horospheres, a distance at least $r > 9l_0$ apart. Each horosphere is shown with two concentric horospheres at distances $2l_0$ and $3l_0$.

The point is that one starts with a quasigeodesic for the metric on $S$, and proves that it is in fact a quasigeodesic for the hyperbolic metric. Failure to distinguish carefully between these two metrics can easily produce incorrect proofs which seem simpler than the proofs provided here.

*Proof of 11.3.8:* The proof is illustrated in Figure 11.11. Let $b' \leq t \leq a''$

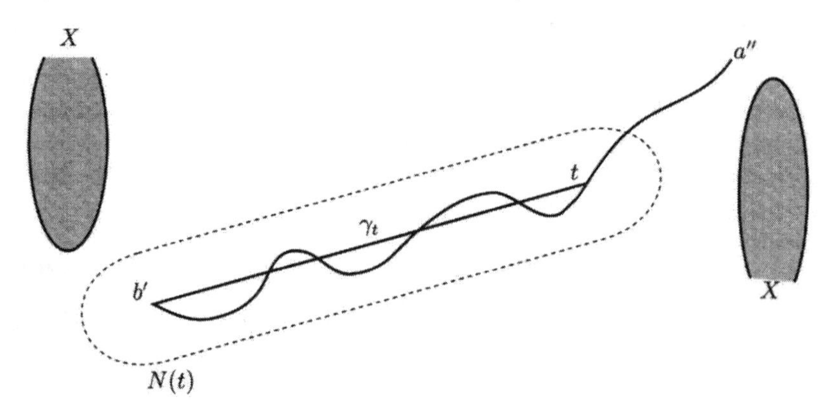


**Figure 11.11. An $S$ quasigeodesic far from any horosphere.** The geodesic $\gamma_t$ is shown joining the points $\alpha(b')$ and $\alpha(t)$. Nearby horoballs are labelled $X$. $N(t)$ is the $l_0$-neighbourhood of $\gamma_t$.

and let $N(t)$ be the closed $l_0$-neighbourhood in the hyperbolic metric of the hyperbolic geodesic $\gamma_t$ from $\alpha(b')$ to $\alpha(t)$. We choose $t$ maximal so that $\alpha[b', t] \subset N(t) \subset S$. Since $N(t)$ is hyperbolically convex, the hyperbolic geodesic between any two points of $\alpha[b', t]$ is disjoint from $X$, the union of the horoballs. Now we know that $\alpha$ is a $(k, \varepsilon)$-quasigeodesic for $S$. Since $S$-

distances are equal to hyperbolic distances between pairs of points in $N(t)$, we deduce that $\alpha|_{[b',t]}$ is a $(k,\varepsilon)$-quasigeodesic for the hyperbolic metric.

Lemma 3.4.2 (quasigeodesics near geodesics) applies to the hyperbolic quasigeodesic $\alpha|_{[b',t]}$. Therefore each point of $\alpha[b',t]$ lies in the interior of $N(t)$, and each point of $\gamma_t$ is strictly less than $l_0$ distant from $\alpha[b',t]$. Each point of $N(t)$ is at most a distance $l_0$ from the geodesic $\gamma_t$. Therefore for each $x \in N(t)$ the hyperbolic distance to $\alpha[b',t]$ is strictly less than $2l_0$. Because of the way we have chosen the interval $(b',a'')$, it follows that the compact set $N(t)$ lies in the interior of $S$. Since $t$ has been chosen maximal, we deduce that $t = a''$. 11.3.8

We continue with the proof of Theorem 11.3.1 (quasigeodesics outside horoballs). In the preceding discussion, we partitioned the domain of $\alpha$ so that intervals are alternately hyperbolic quasigeodesics joining distinct components of $X$ and quasigeodesics for $S$ lying near a fixed component of $X$. (It is possible that a quasigeodesic near a component of $X$ consists of a single point.)

We will show that there exists a $k_6 \geq 1$ and an $\varepsilon_6 \geq 0$, depending only on $k$ and on $\varepsilon$, such that if we replace each of the $S$-quasigeodesic segments just described by hyperbolic geodesic segments, then the result is a hyperbolic $(k_6,\varepsilon_6)$-quasigeodesic.The hyperbolic segment replacing an $S$-quasigeodesic which lies near a horosphere, as described in Lemma 11.3.3 (quasigeodesics near horospheres), will normally pierce the corresponding horoball. That is, it will not lie in $S$. Such a segment is called a *horospherical segment*; these alternate with *non-horospherical segments* which lie entirely in $S$. The situation is illustrated in Figure 11.12, which shows how the $S$-quasigeodesic in Figure 11.10 is replaced. For convenience we refer to a non-horospherical segment, followed by a horospherical segment, followed by a non-horospherical segment as a *3-chain*. We will show that a 3-chain is a hyperbolic quasigeodesic with constants depending only on $k$ and on $\varepsilon$.

A non-horospherical segment $\omega$ ends on the boundary of $H_{2l_0}$, where $H$ is a component of $X$. Every point of $\omega$ is within a distance $l_0$ of part of $\alpha$ which lies outside the $2l_0$-neighbourhood of $X$. Therefore $\omega$ is disjoint from the $l_0$-neighbourhood of $X$. It follows from Lemma 11.3.6 (dipping) that it meets the horosphere at an angle at least $2\theta_1 > 0$ from the perpendicular, where $\sin(2\theta_1) = \exp(-l_0)$.) Hence $\theta_1$ depends only on $k$ and on $\varepsilon$.

Next we choose $d$, such that if two points $x$ and $y$ lie on a horosphere and the hyperbolic distance between them is at least $d$, then the angle between the hyperbolic geodesic from $x$ to $y$ and the perpendicular to the horosphere is less than $\theta_1$, as shown in Figure 11.13. We will furthermore choose $d > l_1$, where $l_1$ is the constant coming from Lemma 11.3.4 (angles bounded below),

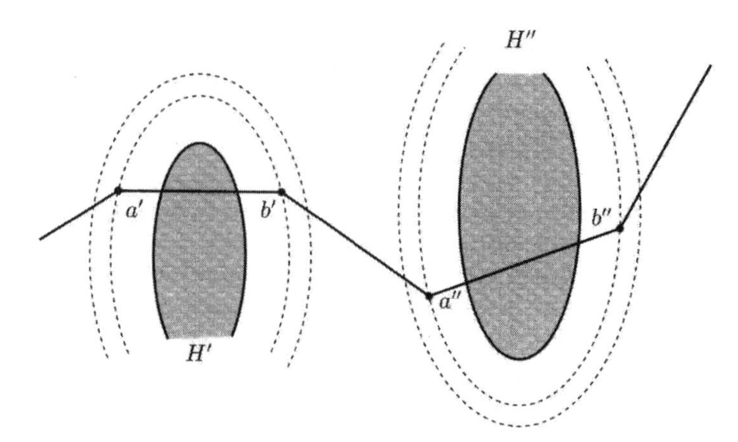


**Figure 11.12. Replacing the $S$-quasigeodesic by a piecewise hyperbolic geodesic.** We illustrate how the $S$-quasigeodesic of Figure 11.10 is replaced. The result is a quasigeodesic for the hyperbolic metric.

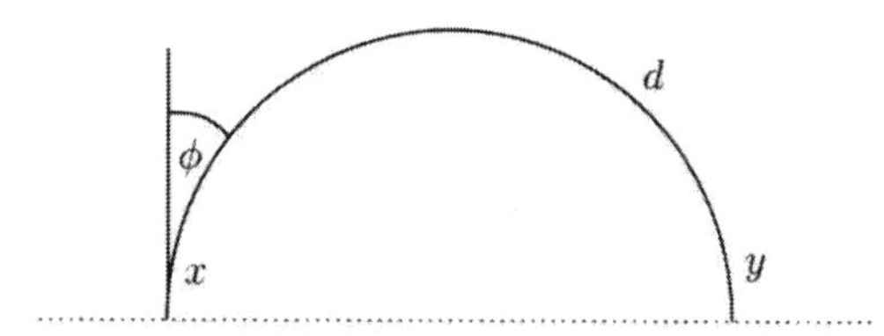


**Figure 11.13. A geodesic nearly perpendicular to a horosphere.** If the distance $d$ between $x$ and $y$ is large enough, we can ensure that $\phi < \theta_1$.

depending only on $\theta_1$. Since $\theta_1$ depends only on $k$ and on $\varepsilon$, the same is true for $d$. We will collect the set of 3-chains into two disjoint classes, depending on whether the length of the horospherical segment is greater than or less than $d$.

In the case where it is greater than $d$, the 3-chain is a quasigeodesic for some constants depending only on $k$ and on $\varepsilon$, provided $r - 4l_0 > l_1$, where $l_1$ depends only on $\theta_1$ as in Lemma 11.3.4 (angles bounded below).

We now show that the 3-chain is also a hyperbolic geodesic with constants depending only on $k$ and $\varepsilon$ if the length of the horospherical segment is less than $d$. We denote the 3-chain by $\phi$, which we regard as a path parametrized by arclength. We have $d(\phi(t_1), \phi(t_2)) \leq |t_1 - t_2|$, so we only need to consider the reverse inequality. We may assume that $t_1$ and $t_2$ are in different segments.

If $t_2$ is in the horospherical segment, let $t_3$ be the common point between the two segments. Let $D$ be as in Lemma 11.3.7. Then

$$|t_1 - t_2| \leq d(\phi(t_1), \phi(t_3)) + D \leq d(\phi(t_1), \phi(t_2)) + 2D.$$

If $t_1$ and $t_2$ are in distinct non-horospherical segments, let $t_3$ and $t_4$ be the endpoints of the horospherical segments. We may assume that $t_1 < t_3 \leq t_4 < t_2$. Let $s_1 < s_3 \leq s_4 < s_2$ be chosen in the domain of $\alpha$ such that $d(\alpha(s_1), \phi(t_1)) < l_0$, $d(\alpha(s_2), \phi(t_2)) < l_0$, $\alpha(s_3) = \phi(t_3)$ and $\alpha(s_4) = \phi(t_4)$. Now

$$t_2 - t_4 = d(\phi(t_2), \phi(t_4)) \leq d_S(\alpha(s_2), \alpha(s_4)) + l_0 \leq k(s_2 - s_4) + \varepsilon + l_0.$$

and similarly for $t_3 - t_1$. It follows that, for some constants $\varepsilon_5$ and $\varepsilon_4$, depending only on $k$ and on $\varepsilon$.

$$\begin{aligned} t_2 - t_1 &= (t_2 - t_4) + (t_4 - t_3) + (t_3 - t_1) \\ &\leq (t_2 - t_4) + D + (t_3 - t_1) \\ &\leq k(s_2 - s_4 + s_3 - s_1) + 2\varepsilon + 2l_0 + D \\ &\leq k(s_2 - s_1) + 2\varepsilon + 2l_0 + D \\ &\leq k^2 d_S(\alpha(s_2), \alpha(s_1)) + \varepsilon_5 \\ &\leq k^2 d(\phi(t_2), \phi(t_1)) + \varepsilon_4. \end{aligned}$$

The last inequality needs proof, because we are changing from $d_S$ to the smaller metric $d$. It can be deduced from the final sentence in the statement of Lemma 11.3.7. This completes the proof that for some constants $k_4$ and $\varepsilon_4$ depending only on $k$ and $\varepsilon$, each of the 3-chains is a $(k_4, \varepsilon_4)$-quasigeodesic, provided $r$, the distance between horospheres, is sufficiently large (depending only on $k$ and $\varepsilon$).

We define $l_3$ using Lemma 11.3.5 (quasigeodesic over subintervals), with $k$ replaced by $k_4$ and $\varepsilon$ replaced by $\varepsilon_4$ in the statement of that lemma. Since $l_3$ depends only on $k_4$ and on $\varepsilon_4$, and these depend only on $k$ and on $\varepsilon$, $l_3$ depends only on $k$ and on $\varepsilon$. If $r > 4l_0 + l_3$, then Lemma 11.3.5 (quasigeodesic over subintervals) shows that we can find $\varepsilon_6$ and $k_6$ as promised on page 263. Let $l > l_3 + 2l_0$ be chosen so that any $(k_6, \varepsilon_6)$-quasigeodesic lies within an $l$-neighbourhood of the hyperbolic geodesic connecting its endpoints. This value of $l$ satisfies the statement of Theorem 11.3.1 (quasigeodesics outside horoballs). 11.3.1

## 11.4. Geometrically Finite Groups of Hyperbolic Isometries

The main aim of this chapter has been to prove the following theorem, and we are able to complete the proof in this section.

**Theorem 11.4.1 (geometrically finite implies biautomatic).** *Let $G$ be a geometrically finite hyperbolic group. Then $G$ is biautomatic.*

In [Bow], Bowditch greatly clarified the notion of a geometrically finite hyperbolic group, and we will rely on his work for the equivalence of various notions of "geometrically finite" and for the consequences of this property.

Given a discrete group $G$ of isometries of hyperbolic space $\mathbf{H}^n$, the *limit set*, $L(G)$ is defined to be the set of limit points of the orbit under $G$ of any point $x$ in $\mathbf{H}^n$. $L(G)$ is independent of the choice of $x$ and it is contained in the sphere at infinity. This sphere can be identified with the unit sphere in $\mathbf{R}^n$ if we are using the Poincaré disk model or the projective model. We will usually think of hyperbolic space in terms of the Poincaré disk model, and we will sometimes talk of the *closure of hyperbolic space* as meaning the closed unit disk. (A definition entirely intrinsic to hyperbolic space can be formulated by defining a point at infinity to be a maximal set of geodesics all asymptotic to each other in the same direction.)

The *convex hull* $CH(G)$ is defined to be the (hyperbolic) convex hull of $L(G)$, from which the points of $L(G)$ itself are deleted, so that $CH(G) \subset \mathbf{H}^n$. In the projective model, the hyperbolic convex hull can be identified with the usual euclidean convex hull.

A *parabolic fixed point* for $G$ is a point $p$ on the sphere at infinity such that there exists some element $g \in G$, with $p$ the only fixed point of $g$ in the closure of the hyperbolic space.

If $G$ is geometrically finite, there are only finitely many orbits under $G$ of parabolic fixed points. Moreover, we may choose a $G$-invariant set of disjoint horoballs centred on the parabolic fixed points. We denote by $S(G)$ the complement of the interiors of these horoballs in $\mathbf{H}^n$, obtaining a $G$-invariant manifold with boundary. One way of defining a geometrically finite group is to require that a disjoint invariant set of horoballs exists and that the quotient of $CH(G) \cap S(G)$ by $G$ has a compact quotient, or, equivalently, that there is a compact fundamental domain for the action of $G$ on $CH(G) \cap S(G)$. If $M(G) = \mathbf{H}^n/G$ is a finite volume manifold or orbifold, then $S(G)$ is the universal cover of the result of cutting away the cusps of $M(G)$.

Geometrically finite groups are finitely presented. Many of the most interesting examples of finitely generated discrete groups acting on hyperbolic space are not geometrically finite. However, this is an aspect of the particular action and not necessarily a consequence of the isomorphism type of the abstract group. For all examples of finitely presented geometrically infinite groups, wherever the answer is known, it is possible to change the action just a little, without changing the isomorphism type of the group, so that it becomes geometrically finite.

We are indebted to Geoff Mess for the information in this paragraph and the next. M. Kapovich and L. Potyagailo have produced an example of a finitely generated but not finitely presentable group acting on $\mathbf{H}^4$ discretely—see [KP91]. There is also an unpublished example of the same phenomenon, due to B. H. Bowditch and G. Mess.

In dimension three, every finitely generated discrete group of isometries can be made to act in a geometrically finite way. To prove this, one first finds a core for the quotient of $\mathbf{H}^3$ by the group. This follows from Scott's Core Theorem [Sco73] if the group is torsionfree. If the group has torsion, a core orbifold still exists; the basic argument is contained in [FM91] and Geoff Mess has indicated the full argument to the authors. Once one has a compact core, Thurston's Uniformization Theorem [Thu82] can be applied.

Let $P$ be a maximal parabolic subgroup of a geometrically finite group $G$, and let $H$ be a horoball preserved by $P$. Then $P$ acts on the boundary of $H$, which we denote by $\partial H$. $\partial H$ is a euclidean space when it is endowed with its path metric, and $P$ acts by euclidean isometries. If $H$ is small enough, the quotient of $CH(G) \cap \partial H$ by $P$ is compact. In the upper half space model, with the parabolic fixed point at infinity, $\partial H$ is a horizontal plane. There is a maximal abelian normal subgroup $A$ of finite index in $P$ and an affine subspace $V$ of $CH(G) \cap \partial H$ (with respect to the euclidean structure on the horosphere), such that $A$ acts by translations on $V$, provided $H$ has been chosen small enough. In general, $V$ is not unique—there is a compact set of alternative choices, each of which is parallel to $V$. (Caution: $A$ does not in general act by translations on $\partial H$—there can be a rotation component in the direction orthogonal to $V$.)

We fix a finite set of generators for $A$ which is setwise invariant under conjugation by elements of $P$. Using $A$, we can construct generators for $P$ and a biautomatic structure for $P$ as in Theorem 4.2.1 (automatic structure with invariance).We recall that the accepted strings for this automatic structure are geodesics in the Cayley graph of $P$.

We take a basepoint in $V$ and, map the Cayley graph of $P$ into $CH(G) \cap \partial H$ equivariantly as in Theorem 3.3.6 (pseudoisometry of Cayley graph), such that each edge is mapped to a straight euclidean segment. By Theorem 3.3.4 (accepted implies quasigeodesic) or directly, we can find constants $k_1$ and $\varepsilon_1$ such that the accepted strings represent paths in $CH(G) \cap \partial H$ which are $(k_1, \varepsilon_1)$-quasigeodesics (for the euclidean metric).

Now what happens if we shrink $H$ to the concentric horosphere $H_{-k\log 2}$, where $k > 0$ is an integer? We have the same picture, except that the scale is reduced by a factor $2^k$. In order to be able to discuss the situation invariantly, we compensate by changing the lengths of the edges in the Cayley graph to $2^{-k}$. The result is that accepted strings in the automatic structure of $P$ now represent paths in $CH(G) \cap \partial H$ which are $(k_1, 2^{-k}\varepsilon_1)$-quasigeodesics. In particular, they are still $(k_1, \varepsilon_1)$-geodesics.

We now consider all parabolic subgroups simultaneously as follows. We first choose a disjoint union of horoballs, each centred on a parabolic fixed point, which is setwise invariant under $G$. We choose a finite set of representatives of $G$-orbits of these horoballs. For each horoball $H$ of this finite set, we have a parabolic subgroup $P$ and a basepoint in a subspace $V \subset CH(G) \cap \partial H$ as above.

We choose generators for each parabolic subgroup $P_i$ in this finite set of representatives, and make $k_1$ and $\varepsilon_1$ larger so that they are quasigeodesic constants simultaneously in each of the horospheres, in the manner already described.

Let $H_1, \dots, H_p$ be our chosen set of horoballs (representing distinct orbits under $G$). We denote by $x_i$ the basepoint chosen as explained above in $V_i \subset CH(G) \cap \partial H_i$ $(1 \leq i \leq p)$. We also choose a basepoint $x_0$ in the intersection of the convex hull $CH(G)$ with the interior of $S(G)$. Let $\gamma_i$ be a shortest hyperbolic geodesic segment from $x_0$ to $x_i$. We denote by $A_i$ the subgroup of the parabolic subgroup $P_i$ acting by translations on $V_i$.

By making the horoballs smaller, we may assume that each $\gamma_i$ lies entirely in $S(G)$ and only meets the boundary of $S(G)$ at its endpoint $x_i$. We choose a set of generators for $G$. (It is immaterial whether this set contains the parabolic generators we have chosen or not.) By making the horoballs smaller, we can ensure that hyperbolic geodesics from $x_0$ to $gx_0$ are entirely contained in the interior of $S(G)$ for each generator $g$ of $G$.

Let $N$ be a positive integer. We have a finite abstract graph $Q(N)$ associated to the above constructions. Each edge of $Q(N)$ will be assigned a length which is a rational number. $Q(N)$ has vertices $y_0^0, \dots, y_p^0$ associated with the basepoints $x_0, \dots, x_p$. $Q(N)$ also has vertices $y_i^j$ for $1 \leq i \leq p$ and $1 \leq j \leq N$. Corresponding to each generator of $G$, we have a loop in $Q(N)$, from $x_0$ to itself, and we assign length one to this loop. Corresponding to each $\gamma_i$, we have an edge in $Q(N)$ joining $y_0^0$ to $y_i^0$, and we assign length one to this edge. For each $i$ $(1 \leq i \leq p)$ and $j$ $(0 \leq j < N)$ $Q(N)$ has an edge of length one joining $y_i^j$ to $y_i^{j+1}$. For each $i$ $(1 \leq i \leq p)$ and $j$ $(0 \leq j \leq N)$, $Q(N)$ has loops of length $2^{-j}$ from $y_i^j$ to itself, one loop for each parabolic generator of $P_i$. Figure 11.14 shows $Q(N)$ for the punctured torus.

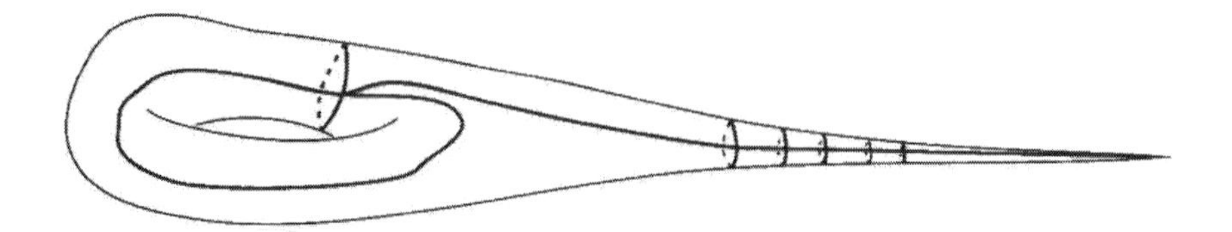

**Figure 11.14. The graph associated to a punctured torus.** The picture shows a punctured torus, with the graph $Q(4)$ embedded in it.

There is a unique maximal tree in $Q(N)$, which can be used to determine the fundamental group of $Q(N)$. There is an obvious homomorphism of the fundamental group $\pi_1(Q(N))$ onto $G$, which maps each loop corresponding to a generator of $G$ to that generator and each loop from $y_i^j$ to itself to the corresponding generator of $P_i \subset G$ (so that $N$ distinct generators in $\pi_1(Q(N))$ are all sent to the same generator of $P_i$).

The covering $\tilde{Q}(N)$ of $Q(N)$ corresponding to the kernel of this homomorphism is an analogue of the Cayley graph of a group—see Definition 11.1.2 (Cayley graph of a groupoid). In fact, it is the Cayley graph of a groupoid with $pN + p + 1$ objects. $G$ acts as the group of covering translations of $\tilde{Q}(N)$. There is a $G$-equivariant map $f : \tilde{Q}(N) \to CH(G) \subset \mathbf{H}^n$, defined as follows.

We write $x_i^0 = x_i \in \mathbf{H}^n$, for $0 \leq i \leq p$. For each $i$, $(1 \leq i \leq p)$, we draw the geodesic ray $\rho_i$ in $CH(G)$ from $x_i^0$ to the centre of its horosphere, that is, to the fixed point of the parabolic subgroup $P_i$. Let $x_i^j$ be the point on this ray, a distance $j \log 2$ from $x_i^0$. We lift the unique maximal tree to a tree $T$ in $\tilde{Q}(N)$ and continue to denote the vertices by $y_i^j$. $T$ is mapped in the obvious way into the union of the $\gamma_i$ and the $\rho_i$, with $y_i^j$ being sent to $x_i^j$. The loops corresponding to generators of $G$ are lifted to hyperbolic geodesic segments in $CH(G)$. The loops corresponding to generators of $P_i$ are lifted to horospherical geodesics.

We define $S(N)$ to be the closed $N \log(2)$-neighbourhood of $S(G)$.

**Proposition 11.4.2 (geodesics in $\tilde{Q}(N)$).** *There are numbers $k \geq 1$ and $\varepsilon \geq 0$ depending only on $G$, such that, for $N$ sufficiently large, the following result is true. Let $\alpha$ be a geodesic parametrized by arclength in $\tilde{Q}(N)$ between two vertices of $\tilde{Q}(N)$. Then $f \circ \alpha$ is a $(k, \varepsilon)$-quasigeodesic for $S(N)$.*

*Proof of 11.4.2:* For each edge $e$ of $\tilde{Q}(N)$, let $r(e)$ be the hyperbolic length of $f(e)$ divided by the length of $e$ in $\tilde{Q}(N)$. Our construction ensures that there are only a finite number of distinct ratios as $e$ varies over all possible edges. Let $k$ be chosen so that $r(e) < k$ for each edge $e$. Then $d_{S(N)}(f\alpha(t_1), f\alpha(t_2)) \leq k|t_1 - t_2|$.

We now prove the other part of the quasigeodesic condition. Let $\phi : [0, t_1] \to S(N)$ be a geodesic for the path metric on $S(N)$, parametrized by arclength, from $x_a^b$ to $gx_i^j$ where $g \in G$. We will show that there is a path in $\tilde{Q}(N)$ from $y_a^b$ to $gy_i^j$ of length bounded by $kt_1 + \varepsilon$, where $k$ and $\varepsilon$ are chosen independently of $N$. This will show that any geodesic path in $\tilde{Q}(N)$ between these endpoints has length at most $kd_S(x_a^b, gx_i^j) + \varepsilon$.

In order to have a uniform treatment when $t_1$ is not an integer, we extend the definition of $\phi$ to $[0, M]$, where $M$ is the smallest integer such that $M \geq t_1$, by making $\phi$ constant on $[t_1, M]$. For each integer $q$ $(0 \leq q \leq M)$, we will choose a vertex $v_q$ of $\tilde{Q}(N)$. We insist that $v_q$ be in the $G$-orbit of $y_0^0$ or in the $A_i$-orbit of $y_i^j$, and that, subject to this, $fv_q$ is as near as possible to $\phi(q)$. We call such a vertex of $\tilde{Q}(N)$ an *estimable vertex*. Our first task is to estimate $d_{S(N)}(fv_q, \phi(q))$.

Let $\delta > 1$ be chosen so that any point in $CH(G) \cap S(0)$ can be joined to some point in the orbit of $x_0$ under $G$ by a path in $CH(G) \cap S(0)$ of length less than $\delta$. We also choose $\delta$ so that any point in $CH(G) \cap \partial H_i$ can be joined to some point in the orbit of $x_i^0$ under $A_i$ by a path in $CH(G) \cap \partial H_i$ of length less than $\delta$.

Let $\beta$ be a hyperbolic geodesic joining the endpoints of $\phi$. Then $\beta \subset CH(G)$. We apply Theorem 11.3.1 (quasigeodesics outside horoballs) with $k = 1$ and $\varepsilon = 0$, to obtain a number $l$, independent of $N$, such that $\phi$ is contained in the union of an $l$-neighbourhood of $\beta$ and an $l$-neighbourhood $U_l$ of $\partial S(N)$, provided $N$ is large enough. We set $\beta_l = \beta \setminus U_l$. Since $\phi$ is an $S(N)$-geodesic, it meets any component of $\partial S(N)$ at a tangent. Let $d$ be the maximal length of any segment $\phi(t', t'')$ which is disjoint from both the $l$-neighbourhood of $\beta_l$ and from $\partial S(N)$. Then $\tanh(d/2) \leq \sqrt{1 - \exp(2l)}$. There is a path in $S(N)$ of length $l$ from either $\phi(t')$ or $\phi(t'')$ to a point of $\beta$. So every point of $\phi$ either lies on $\partial S(N)$ or is at an $S(N)$-distance at most $d + l$ from $\beta_l$.

Let $x \notin \partial S(N)$ be a point of $\phi$. Since $\beta \subset CH(G)$, we know that there is an estimable vertex $v$ of $\tilde{Q}(N)$, such that $d_{S(N)}(x, fv) \leq d + l + \delta + \log(2)$. We set $l_1 = d + l + \delta + \log(2)$; $l_1$ is independent of $N$.

Now let $x$ be the common endpoint of a segment of $\phi$ disjoint from $\partial S(N)$ and a segment of $\phi$ in $\partial S(N)$. Let $\gamma$ be an $S(N)$-geodesic from $x$ to a nearest $fv$ for some estimable vertex $v \in \tilde{Q}(N)$. If $N$ is large enough, $\gamma$ will lie in $\partial S(N)$, where the geometry is euclidean. After possibly increasing $l_1$ by an amount which is independent of $N$, we deduce that, for every point $y$ of $\phi$ on $\partial S(N)$, there is an estimable vertex $v = gy_i^N$ with $g \in A_i$, such that $d_{S(N)}(v, y) \leq l_1$. This implies that for each integer $q$ satisfying $0 \leq$

$q < t_1 + 1$, $d_{S(N)}(fv_q, \phi(q)) \leq l_1$. Hence $d_{S(N)}(fv_q, fv_{q+1}) \leq l_2 = 2l_1 + 1$; $l_2$ is independent of $N$. We choose $k$ so that it satisfies all the inequalities previously imposed, and, in addition, if $u$ and $v$ are two vertices of $\tilde{Q}(2)$, and $d_{S(N)}(fu, fv) \leq l_2$, then there is a path in $\tilde{Q}(2)$ from $u$ to $v$ of length $k$. This is possible since $G$ acts properly discontinuously. Note that we can choose $k$ independently of $N$. There is a path of length $k$ in $\tilde{Q}(N)$ joining $v_q$ to $v_{q+1}$, provided $fv_q$ and $fv_{q+1}$ are in $S(2)$.

Otherwise $v_q = gy_i^j$ and $v_{q+1} = hy_i^k$, where $g, h \in A_i$. We want to bound above their distance apart in $\tilde{Q}(N)$. For this purpose, there is no loss of generality in taking $g$ to be the identity and assuming $j \leq k$. Let $u = fv_q$ and $w = fv_{q+1}$. Then $d_{S(N)}(u, v) \leq l_2$. Consider the path in $S(N)$ from $u$ to $w$ which consists of a geodesic segment $\sigma_1$ from $u$ along the ray $\rho_i$ (see page 269 for the definition of this ray), followed by a horospherical segment $\sigma_2$ which is geodesic with respect to the euclidean structure on the horosphere. Then the length of $\sigma_1$ is bounded above by $l_2$ and the length of $\sigma_2$ is bounded above by $2\sinh(l_2/2)$. Therefore there is a path in $\tilde{Q}(N)$ joining $v_q$ to $v_{q+1}$ of length

$$\frac{l_2}{\log 2} + k_1 2\sinh(l_2/2) + \varepsilon_1,$$

where $k_1$ and $\varepsilon_1$ are defined on page 267. We take $k$ larger than this value and then $v_q$ is joined to $v_{q+1}$ by a path in $\tilde{Q}(N)$ of length at most $k$.

So we have found $k > 1$, independent of $N$, such that (in the notation introduced on page 270) there is a path in $\tilde{Q}(N)$ from $x_a^b$ to $gx_i^j$ whose length is bounded by $kM < k(t_1 + 1)$. Taking $\varepsilon = k$, the proof of the proposition is complete. 11.4.2

We can now give a more precise statement of Theorem 11.4.1 (geometrically finite implies biautomatic)

**Theorem 11.4.3 (geometrically finite implies biautomatic: details).** *Let $G$ be a geometrically finite automatic group. Let $L(N)$ be the language consisting of shortest paths in $Q(N)$ which follow the automatic structure coming from Theorem 4.2.1 (automatic structure with invariance). This structure on horospherical paths is described in our context on page 267. Then for sufficiently large $N > 0$, $L(N)$ defines a biautomatic structure on $G$.*

Actually this defines a biautomatic structure on a groupoid, and $G$ is the vertex group of the groupoid (see Section 11.1 (Groupoids))

*Proof of 11.4.3:* In Proposition 11.4.2 (geodesics in $\tilde{Q}(N)$) we showed that $k$ and $\varepsilon$ could be chosen independent of $N$, so that the elements of $L(N)$

map to $(k, \varepsilon)$-quasigeodesics in $S(N)$, provided $N$ is large enough. Let $\alpha$ and $\beta$ be two paths in $\tilde{Q}(N)$, starting within distance one of each other, and represented by elements of $L(N)$. Suppose $\alpha$ and $\beta$ end no more than a distance one apart.

By Theorem 11.3.1 (quasigeodesics outside horoballs), if $N$ is chosen large enough, then the hausdorff distance between $\alpha \cap \tilde{Q}(N-2l)$ and $\beta \cap \tilde{Q}(N-2l)$ is bounded by a constant $l$ which is independent of $N$.

Since a path in $L(N)$ is required to be geodesic in $\tilde{Q}(N)$, its behaviour near a horoball has to satisfy very restrictive rules. It can only enter a fixed cusp region once. In this cusp region, it must first travel orthogonally to the horospheres towards the centre of the horoball, then horospherically, following the given automatic structure, and then orthogonally to the horospheres, away from the centre of the horoball. It is easy to see that any other behaviour would mean that we could replace the path by a shorter path.

It follows that the hausdorff distance between $\alpha$ and $\beta$ inside the cusps, and while they are travelling vertically, is also bounded by $l$. Now we use the fact that the automatic structure is biautomatic for the parabolic subgroups, to ensure that the hausdorff distance between the paths on a horosphere is bounded in terms of $l$, $k$ and the lipschitz constants of the automatic structure on the parabolic subgroups.

We apply Theorem 11.2.3 (geodesic automaton 4) to complete the proof. [11.4.3]

**Corollary 11.4.4.** *The group of a hyperbolic link complement is automatic. The automatic structure extends any given automatic structure on the boundary tori.*

*Proof of 11.4.4:* Any automatic structure on an abelian group is biautomatic. Hyperbolic link complements are geometrically finite. [11.4.4]

# Chapter 12

# Three-Dimensional Manifolds

In this chapter we investigate which compact three-dimensional manifolds have fundamental groups which are automatic. We are not able to solve this problem in general—we need to restrict to the class of three-dimensional manifolds satisfying Thurston's Conjecture (see [Sco83]), but this may turn out to be no restriction at all. Note that Thurston's Conjecture has been proved in a number of cases (see [Thu82]).

One of the main points of this chapter is to show that most Seifert fibre spaces have automatic fundamental groups—this is a problem which we *can* solve in full generality; the solution was provided by Thurston. Indeed, we are able to solve this problem for any compact three-manifold, with or without boundary, which is modelled on some geometry. We also show that the pieces in the standard decomposition theorems for three-manifolds have automatic fundamental groups if and only if the whole manifold has an automatic fundamental group. The upshot, for compact three-manifolds satisfying Thurston's Conjecture, is that there is an automatic structure unless one of the prime factors is a closed manifold modelled on nilgeometry or a closed manifold modelled on solvgeometry, in which case there is definitely no automatic structure. These results are joint work of Thurston and Epstein.

## 12.1. Taking the Problem to Pieces

In this section we simplify the problem of finding our whether the fundamental group of a three-manifold is automatic, by simplifying the three-manifold. At first we will make no assumption, except that the three-manifold is compact, but later, when we start proving non-trivial results, we will assume that it satisfies Thurston's Conjecture.

By Theorem 4.1.4 (finite index), we may take double covers if necessary, and so we may assume that our three-manifold $M$ is oriented. According to a theorem of H. Kneser [Hem76], $M$ can be written as the connected sum and boundary sum of compact irreducible and boundary irreducible three-manifolds. When the three-manifold is cut up using the two-spheres and disks of Kneser's Theorem, van Kampen's Theorem tells us that the fundamental group of the original manifold is the free product of a finitely generated free group with the fundamental groups of each of the components. By Theorem 12.1.8 (factor is automatic) and Theorem 12.1.4 (free product with amalgamation), proved below, we know that a free product of a finite number of groups is automatic if and only each factor is automatic. Since finitely generated free groups are automatic, the fundamental group of each of the pieces is automatic if and only if the original fundamental group is automatic. It follows that there is no loss of generality in assuming that $M$ is irreducible and boundary irreducible.

The Torus Theorem [Joh79, JS79] says that there is a collection of disjoint tori cutting $M$ into pieces $M_i$ with certain properties (see [Sco83]). We will now restrict ourselves to those manifolds $M$ which satisfy the Thurston Geometrization Conjecture. This means that the interior of each of the $M_i$ is a complete riemannian manifold modelled on one of eight three-dimensional geometries. We have to investigate the various possibilities.

The three-dimensional geometries are: $\mathbf{R}^3$, three-dimensional euclidean geometry; $S^3$, three-dimensional spherical geometry; $\mathbf{H}^3$, three-dimensional hyperbolic geometry; $S^2 \times \mathbf{R}$, the product of two-dimensional spherical geometry with the reals; $\mathbf{H}^2 \times \mathbf{R}$, the product of two-dimensional hyperbolic geometry with the reals; $N$, three-dimensional nilgeometry; $S$, three-dimensional solvegeometry; and $F$, the universal cover of the special linear group $SL(2, \mathbf{R})$, which models Seifert fibre spaces.

The restriction to irreducible three-manifolds excludes the geometry $S^2 \times \mathbf{R}$, because the only such compact oriented three-manifold is $S^2 \times S^1$, which is reducible. In any case, its fundamental group is infinite cyclic, and so automatic.

We have reduced to a situation where there are no spherical boundary components and each boundary component is incompressible. In particular, if there is a boundary the fundamental group must be infinite. A complete manifold modelled on $S^3$ has finite fundamental group, and so cannot have a boundary. Finite groups are automatic, so we may from now on exclude the geometry $S^3$.

A compact three-manifold $M$ with boundary whose interior is modelled on nilgeometry or on solvegeometry can be modelled instead on euclidean geometry [Sco83]. By Theorem 8.2.8 (nilpotent implies not automatic), a closed manifold modelled on $N$ cannot have an automatic fundamental group.

By Theorem 8.1.3 (soluble implies not automatic), a closed three-dimensional manifold modelled on $S$ cannot have an automatic fundamental group.

We have now reduced to the situation where our three-manifold $M$ is compact, oriented, irreducible and boundary irreducible. We have a family of incompressible tori embedded in $M$, and when we cut along these tori we obtain three-manifolds $M_i$ (with boundary), whose interiors are modelled on $\mathbf{H}^3$ or on $\mathbf{H}^2 \times \mathbf{R}$ or on $F$, the universal cover of $SL(2, \mathbf{R})$, or on $\mathbf{R}^3$. We may assume that if $M_i = S^1 \times S^1 \times I$, then there is only one $M_i$.

If there is only one $M_i$ and the geometry is modelled on $\mathbf{R}^3$, then the fundamental group is automatic by Corollary 4.2.4 (euclidean implies biautomatic). So we assume that this is not the case.

We next get rid of embedded Klein bottles. An embedded Klein bottle has a neighbourhood which is a twisted interval bundle, with boundary a torus. The number of disjoint embedded Klein bottles is bounded by the rank of $H_1(M; \mathbf{Z}_2)$. (One way to see this is to use intersection numbers modulo 2 of a one-cycle with the Klein bottles.) If necessary, we take a double cover, unwrapping simultaneously each twisted interval bundle to a product bundle over the torus. So we may now assume that there are no embedded Klein bottles. One consequence is that we may assume that none of the $M_i$ is modelled on $\mathbf{R}^3$.

To sum up, we have reduced to the following situation. Our three-manifold $M$ is compact, oriented, irreducible and boundary irreducible. We have a family of incompressible tori embedded in $M$, and when we cut along these tori we obtain three-manifolds $M_i$ (with boundary), whose interiors are modelled on $\mathbf{H}^3$ or on $\mathbf{H}^2 \times \mathbf{R}$ or on $F$, the universal cover of $SL(2, \mathbf{R})$. When the three-manifold is modelled on $\mathbf{H}^3$, the manifold is geometrically finite.

The remainder of this section is devoted to the proof of Theorems 12.1.8 (factor is automatic) and 12.1.4 (free product with amalgamation), which we used at the beginning of this section, and of related results. Although these theorems are more general than what is needed here, we include them for completeness.

**Lemma 12.1.1 (regular transversal).** *Let $G$ be a group with a finite subgroup $K$. Let $A$ be an ordered set of semigroup generators for $G$, such that $A$ contains all elements of $K$. Let $L'$ be a language over $A$ which contains exactly one representative for each element of $G$. Suppose that we have a combing of $G$, in which the path of the combing assigned to $g \in G$ defines the representative $w \in L'$. Then there is a set $L$ of representatives of non-trivial left cosets $gK$, such that $LK \cup K$ also defines a combing of $G$. If $(A, L')$ defines an automatic structure on $G$ with unique representatives of elements of $G$, then so does $(A, LK \cup K)$.*

*Proof of 12.1.1:* According to Lemma 3.6.5 (parametrize by pathlength), we may assume that the paths of the combing are edge paths, parametrized by pathlength. There is no loss of generality in assuming that $L'$ contains the nullstring (whether we are dealing with the combing or the automatic structure). For each coset $gK$, we choose the string $w \in L'$ which represents $gK$ and is smallest for the ShortLex-ordering. Let $L$ be the set of these strings $w$, as the coset $gK$ varies over non-trivial elements of $G/K$. To see that $LK \cup K$ corresponds to a combing, suppose that $w_1, w_2 \in L \cup \{\varepsilon\}$, $k_1, k_2 \in K$ and that $d_\Gamma(\overline{w_1 k_1}, \overline{w_2 k_2}) = 1$, where $d_\Gamma$ is the distance function in the Cayley graph $\Gamma$ of $G$. It follows that $d_\Gamma(\overline{w_1}, \overline{w_2}) \leq 3$. Since $L'$ is a combing, the corresponding paths $\widehat{w_1}$ and $\widehat{w_2}$ are uniformly near each other. It follows that the paths $\widehat{w_1 k_1}$ and $\widehat{w_2 k_2}$ are uniformly near each other.

To see that, if $(A, L')$ defines an automatic structure, then $(A, LK \cup L)$ also does, we need only show that $LK \cup K$ is a regular language. For each $k \in K$, let $L_k$ be the set of pairs $(w_1, w_2)$ of strings in $L'$, such that $\overline{w_1 k} = \overline{w_2}$ in $G$. By the definition of an automatic structure, $L_k$ is a regular language. We have

**12.1.2.** $$L = \bigcap_k \{w \in L' \ : \ (\forall w')((w, w') \in L_k \implies (w, w') \in \mathsf{ShortLex}\}.$$

It follows from Corollary 1.4.7 (predicates closed), that $L$ is a regular language. We deduce that $LK \cup K$ is a regular language. $\boxed{12.1.1}$

Now we discuss a similar result for the case of biautomatic and bicombable groups. $G$ acts on the left on the set $G/K$ of left cosets $gK$. For each element $g \in G$, let $S_g \subset K$ be the stabilizer of the coset $gK$ in $K$. Since $K$ is finite, there are a finite number of possibilities $S$ for $S_g$. For each of these we choose a transversal $T_S$ for the set $K/S$ of left cosets $kS$. Let $C = \cup_S T_S \subset K$ be the union of all these transversals, as $S$ varies.

Let $L''$ be a language over $A$ which contains exactly one representative for each element of $G$. Suppose that we have a bicombing of $G$, in which the path of the combing assigned to $g \in G$ defines the representative $w \in L''$. For each double coset $KgK$, let $b \in L''$ be the element in the double coset which is smallest in the ShortLex ordering. Let $B$ be the set of such strings $b$. The elements $\overline{b} \in G$ form a complete set of double coset representatives. As in Lemma 12.1.1 (regular transversal), it is easy to see that if $L''$ is regular (with $L''$ coming from a biautomatic structure), then $B$ is regular. Now let $L$ be the set of strings of the form $cbk$, where $c \in C$, $b \in B$, $k \in K$, and, if $S$ is the stabilizer of $\overline{b}K$, then $c \in T_S \subset C$.

**Lemma 12.1.3 (double cosets).** *If $L''$ defines a bicombing, then $L$ defines a bicombing. If $L''$ defines a biautomatic structure, then so does $L$.*

*Proof of 12.1.3:* To show that $L \to G$ is injective, suppose $\overline{c_1 b_1 k_1} = \overline{c_2 b_2 k_2}$. Then $b_1 = b_2$, since we have chosen a unique string in each double coset. Let $S \subset K$ be the stabilizer of $\overline{b_1}K$. Then $c_1$ and $c_2$ are elements of $T_S$. Since $\overline{c_1 b_1}K = \overline{c_2 b_2}K$, we deduce that $c_1 = c_2$. Therefore $k_1 = k_2$.

To show that $L \to G$ is surjective, take $g \in G$. Then $g$ lies in a unique double coset $KgK$. Let $b \in B$ be the unique representative of the double coset. Then $g = \overline{k_1 b k_2}$, for some $k_1, k_2 \in K$. Let $S \subset K$ be the stabilizer of $\bar{b}K$. Then $k_1 = cs$, where $c \in C$, $c \in T_S$ and $s \in S$. Let $s\bar{b} = \bar{b}k_3$, with $k_3 \in K$. Then $g = \overline{cbk_3k_2}$.

We now show that $L$ is a regular language in the biautomatic case. For each $k \in K$, let

$$L_k = \{w \in L'' \mid k\overline{w}K = \overline{w}K.\}$$

We claim that $L_k$ is a regular language. To see this, we note that, for any $k_1 \in K$, the language

$$\{(w_1, w_2) \mid w_1 \in L'',\ w_2 \in L'',\ k\overline{w_1} = \overline{w_2}k_1\}$$

is regular, because $(A, L'')$ is biautomatic. The union of these, with $k$ fixed and $k_1$ varying, is also a regular language. Since the condition $w_1 = w_2$ is regular, our claim is proved.

For each subgroup $S$ of $K$, we define the regular language $L_S = \bigcap_{k \in S} L_k$. We have already seen that there is a finite state automaton which can check whether a string $cbk$ has the form $c \in C$, $b \in B$ and $k \in K$. We combine this with a finite state automaton which reads $b$ and checks that $b \in L_S$ for some $S$ such that $c \in T_S$. 12.1.3

**Theorem 12.1.4 (free product with amalgamation).** *Let $G_1$ and $G_2$ both be combable, automatic, bicombable or biautomatic groups. Let $K = G_1 \cap G_2$ be a finite subgroup of each of them. Then the free product with amalgamation, $G_1 *_K G_2$, is combable, automatic, bicombable or biautomatic respectively.*

*Proof of 12.1.4:* First we consider the combable case. For $i = 1, 2$ let $A_i$ be a set of semigroup generators of $G_i$ containing each element of $K$. In Lemma 12.1.1 (regular transversal), we produced languages $L_i$, such that $L_i K \cup K$ gives a combing of $G_i$.

We now define $L$, a language over $A_1 \cup A_2$ (where the $A_i$ are assumed to have intersection exactly $K$) by

**12.1.5.** $$L = (L_2 \cup \{\varepsilon\})(L_1 L_2)^*(L_1 \cup \{\varepsilon\})K.$$

Note that the identity element is included as one of the elements in $K$, and that any string in $L$ *must* end with an element of $K$. The map from $L$ to $G$, the free product with amalgamation, is bijective.

We now have to show that the map, specified by $L$, from $G$ to the space of paths in the Cayley graph of $G$ is lipschitz, where the metric on the space of paths is given by Definition 3.6.1. Let $w \in L$ and $g \in A_1 \cup A_2$. We have to find the unique representative in $L$ of $\overline{wg}$. This means writing down a product of elements of $L_1$, $L_2$ and $K$ as in Equation 12.1.5. It is easy to see that $w$ and the representative in $L$ of $\overline{wg}$ are identical, except possibly for the last two factors of $w$. It follows that the two paths are uniformly near each other, with the distance being given by the fact that $G_1$ and $G_2$ are combable.

If $G_1$ and $G_2$ are automatic, then $L_1$ and $L_2$ are regular languages by Lemma 12.1.1 (regular transversal). It follows that $L$ is also regular.

Now we deal with the bicombable case. Once again, let $L_1$ and $L_2$ be the languages over $A_1$ and $A_2$ for the groups $G_1$ and $G_2$. For $i = 1, 2$, let $C_i$ and $B_i$ be derived from $L_i$ according to the recipe just before the statement of Lemma 12.1.3. We define $L$ to be the set of all strings of the form

**12.1.6.** $$c_1 b_1 \ldots c_n b_n k,$$

where one of the following conditions are satisfied.

(1) For all $i$ $(1 \le i \le n)$, if $i$ is odd, then $c_i \in C_1$ and $b_i \in B_1$ and if $i$ is even, then $c_i \in C_2$ and $b_i \in B_2$.

(2) For all $i$ $(1 \le i \le n)$, if $i$ is even, then $c_i \in C_1$ and $b_i \in B_1$ and if $i$ is odd, then $c_i \in C_2$ and $b_i \in B_2$.

(If both conditions are satisfied, we must have $n = 0$.) In addition, we require that $k \in K$, and that if $S \subset K$ is the stabilizer of $\overline{b_i}K$ for some $i$, then $S \in \mathcal{S}(c_i)$.

To check that $L$ gives a bicombing, we first need to show that $L \to G$ is bijective. This is easy. Given $w \in L$ and $x \in A_1 \cup A_2$, we next need to find the representatives of $\overline{xw}$ and of $\overline{wx}$ in $L$. The case of $\overline{wx}$ is very similar to the case already discussed for combable groups, so we discuss only the case of $\overline{xw}$.

Let $v \in L$ represent $\overline{xw}$. Then $\widehat{v}$ is near $\widehat{xw}$. This can either be seen either algebraically or geometrically. The geometric proof goes as follows. Let $Y$ be formed by taking a wedge of circles, one for each element of $K$, and adding 2-cells so that the fundamental group of $Y$ becomes equal to $K$. Let $y_0$ be the unique vertex of $Y$. We then add further 1-cells corresponding to the elements of $A_1 \backslash K$ and then further 2-cells to get $X_1$ with fundamental group $G_1$. We similarly add further 1-cells to another copy of $Y$ and further 2-cells to get $X_2$ with fundamental group $G_2$. We glue $X_1$ and $X_2$ together along $Y$ to give $X$ with fundamental group $G = G_1 *_K G_2$. We work in the universal cover $\tilde{X}$ of $X$, whose one-skeleton is the Cayley graph of $G$. We get a slightly clearer picture (see Figure 12.1) for our purposes by thickening up $Y$ to $Y \times [1,2]$, and constructing $X$ by gluing $Y \times \{1\}$ to $X_1$ and $Y \times \{2\}$ to $X_2$. This thickens up the basepoint of $X$ (and therefore of $\tilde{X}$) to the interval $y_0 \times [1,2]$. Formally speaking, our new basepoint will be $y_0 \times \{3/2\}$. From now on, when we talk of $Y$, we will mean $Y \times \{3/2\}$. Each component of the inverse image of $Y$ in $\tilde{X}$ is a finite complex which universally covers $Y$.

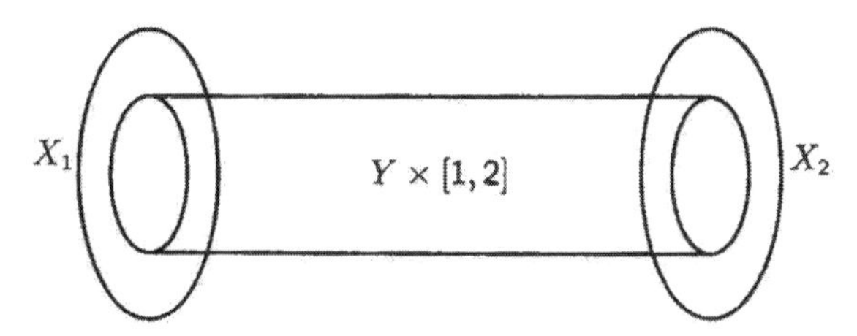


**Figure 12.1. Free product with amalgamation.** This illustrates the proof of Theorem 12.1.4. By van Kampen's Theorem, this space, denoted $X$, has fundamental group $\pi_1(X_1) *_{\pi_1(A)} \pi_1(X_2)$. Thickening up $Y$ to $Y \times [1,2]$ removes a number of technical difficulties in proofs about free products with amalgamation.

Consider the paths $\widehat{xw}$ and $\widehat{v}$ in $\tilde{X}$. (When making these into edge paths, we only cross the inverse image of $Y \times \{3/2\}$ when it is essential to do so.) Now note the sequences of components of the inverse image of $Y = Y \times \{3/2\}$ which these paths cross when moving from the inverse image of $X_1$ to the inverse image of $X_2$ or vice versa. We ignore the starting component, which the paths do not cross. The next one or two components met by $\widehat{xw}$ may not be encountered by $\widehat{v}$, but, after this point, the sequence of components is sequences are identical. Let $Y'$ and $Y''$ be two components of the inverse image of $Y$ in the same component of the inverse image of $X_1$, say, and suppose that $b'$ and $b''$ are corresponding minimal segments of the two paths, each joining $Y'$ to $Y''$. Our careful choice of the language $L$ ensures that $b'$ and $b''$ carry the same label in $B_1$, because they represent the same double coset, though they probably have different starting points in $Y'$ and different ending points in $Y''$. As paths they are uniformly near each other because of the bicombing hypothesis.

Each of the two paths alternates a segment of length one in the inverse image of $Y$, corresponding to a factor $c_i$ in Equation 12.1.6 and a segment corresponding to a factor $b_i$ in either the inverse image of $X_1$ or the inverse image of $X_2$. By the previous paragraph, we see that we have managed to control the lengths of the paths so that they keep in step. The only time

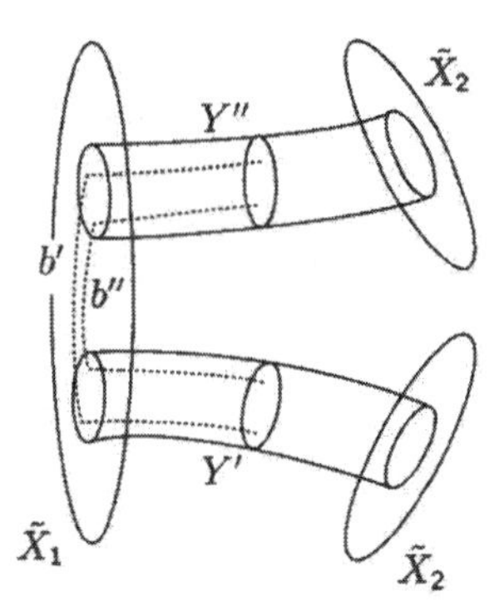


**Figure 12.2. The universal cover $\tilde{X}$ is tree-like.** This illustrates a portion of the universal cover $\tilde{X}$ of the space $X$ in Figure 12.1. The inverse image of $X_1$ in $\tilde{X}$ consists of a number of components, each homeomorphic to the universal cover $\tilde{X}_1$. One of these components is shown. Similarly for $X_2$; two components of the inverse image of $X_2$ are shown. The two spaces $Y'$ and $Y''$ are each homeomorphic to the universal cover of $Y$. Note the tree-like nature of $\tilde{X}$—if each component of the inverse image of $X_1$ is pinched to a point, and if each component of the inverse image of $X_2$ is pinched to a point, and if each component of the inverse image of $Y \times (1,2)$ is pinched to a copy of $(1,2)$, then we do in fact get a tree. The paths $b'$ and $b''$ from the proof of Theorem 12.1.4 are indicated by dotted lines.

they can get out of step is at the beginning. But the bicombing hypothesis makes sure that the paths are also near here and that they are only a few steps out of step at most.

In order to deal with the biautomatic case, we need only show that $L$ is a regular language. This follows easily from Lemma 12.1.3 (double cosets).

12.1.4

**Corollary 12.1.7 (free groups).** *Finitely generated free groups are biautomatic.* 12.1.7

The next theorem is due to M. Shapiro and others [BGSS]. In fact, [BGSS] contains many new results concerning free products with amalgamation, and how the property of a group being word hyperbolic or automatic or combable in various variants of the definitions relates to the same property for the factors.

**Theorem 12.1.8 (factor is automatic).** *Let $G = G_1 *_K G_2$, be a free product with amalgamation of two groups $G_1$ and $G_2$ over a finite group $K$, and suppose $G$ is automatic. Then $G_1$ and $G_2$ are automatic.*

*Proof of 12.1.8:* Let $A$ be the union of an ordered set $A_1$ of generators for $G_1$ and an ordered set $A_2$ of generators for $G_2$. Let $(A, L)$ be an automatic structure for $G$, and suppose that $L$ has unique representatives for elements of $G$. Let $W$ be a minimal deterministic finite state automaton over $A$ accepting $L$. For each pair of states $(s,t)$ of $W$, where $t$ is not a failure state, define $W_{s,t}$, to be the set of all paths of arrows from $s$ to $t$ representing elements of $K$. The uniqueness condition implies that, for fixed $s$ and $t$, no two of these paths represent the same element of $K$. In particular, the number of such paths is finite as $s$ and $t$ vary.

We now adjoin new generators for $F$, denoted by $x_{w,s,t}$, one for each string $w \in W_{s,t}$. We denote the new, enlarged alphabet by $B$. The generator $x_{w,s,t}$ maps to $\overline{w} \in K \subset G$. We turn $W$ into a generalized finite state automaton $W'$ over $B$ as follows. For each word $w \in W_{s,t}$, we add an arrow labelled $x_{w,s,t}e^{|w|-1}$ from $s$ to $t$, where $e$ represents the identity element. $W'$ accepts a language $L'$ over $B$.

Let $L_s$ be the set of all paths from the start state of $W$ to $s$ and let $L^t$ be the set of all paths from $t$ to some accept state of $W$. For each $w \in W_{s,t}$ let $L_{w,s,t} = L_s w L^t$. Then $L_{w,s,t}$ is regular. Let $L''$ be the result of removing from $L'$ the union of the $L_{w,s,t}$. Clearly $L''$ is a regular language. It is easy to see that $(B, L'')$ is an automatic structure on $G$, since we have taken care to ensure that corresponding words in $L'$ and $L''$ have the same length. It is not the case that $L''$ necessarily has unique representatives.

Now let $L_1$ be the set of elements of $L''$ representing elements of $G_1$. It is easy to see, for example by tracing out the path in $\tilde{X}$—see Figure 12.2—that this is the same as the set of elements of $L''$ which do not contain any generator in $A_2$. Since $(B, L'')$ is an automatic structure for $G$, we deduce that $(B \setminus A_2, L_1)$ is an automatic structure for $G_1$. $\boxed{12.1.8}$

Let $G$ be a group and let $K_1$ and $K_2$ be finite subgroups. Let $\phi : K_1 \to K_2$ be an isomorphism. Then we have the HNN extension $G_\phi$.

**Theorem 12.1.9 (HNN).** *If $G$ is combable, automatic, bicombable or biautomatic, then $G_\phi$ is the same.*

*Proof of 12.1.9:* For $i = 1, 2$, let $A_i$ be an ordered set of semigroup generators of $G$, such that each element of $K_i$ lies in $A$. Let $L_i$ be a regular language over $A_i$, such that $L_i K_i$ is disjoint from $K_i$ in $G$ and $L_i K_i \cup K_i$ gives an automation for $G$ with uniqueness.

The group $G_\phi$ is formed from $G$ by adjoining an element $z$, not already in $G$, such that $zk_1 = k_2 z$, if $k_2 = \phi(k_1)$. Let $A$ be the disjoint union of $A_1$, $A_2$, together with two new letters, $z$ and $z^{-1}$. We define a language $L$ over $A$ as the set of strings which can be written $u_1 \cdots u_r$, satisfying the following conditions. Each $u_i$ lies in $L_1$ or in $L_2$, or is equal to an element of $K_1$ or $K_2$, or is equal to $z$ or to $z^{-1}$. If $i < r$, then $u_i$ does not lie in $K_1$ or in $K_2$. If $u_i$ is in $L_1$, $u_{i+1}$ may equal $z^{-1}$, or it may be an element of $K_1$, and there is no other choice. If $u_i$ is in $L_2$, $u_{i+1}$ may equal $z$, or it may be an element of $K_2$ and there is no other choice. If $u_i$ is equal to $z$, $u_{i+1}$ is not equal to $z^{-1}$ and vice versa. $L$ is well-known to map bijectively onto the HNN extension.

If $L_1$ and $L_2$ are regular languages, we want to show that $L$ is a regular language. The easiest way to do this is to construct a generalized non-deterministic automaton which accepts exactly this set of strings.

To prove the bicombable or the biautomatic property, we follow the lines of the proof of Theorem 12.1.4 (free product with amalgamation). 12.1.9

## 12.2. The Basic Seifert Fibre Space

In this section we treat the case of a three-manifold modelled on $\mathbf{H}^2 \times \mathbf{R}$ or on the universal cover of $SL(2, \mathbf{R})$. This means that the manifold is a Seifert fibre space, or, in more modern terminology, a fibre bundle over a two-dimensional orbifold. Actually, in this section we will complete the analysis only for very special cases of a Seifert fibre space, and we defer to a later section the more general case. The result that such a manifold has an automatic fundamental group is due to Thurston.

We denote this fibre bundle by $\pi : M \to P$. We are assuming, from Section 12.1 (Taking the Problem to Pieces), that $M$ is orientable, and there are no embedded Klein bottles. It follows that the two-dimensional orbifold $P$ contains no mirrors.

One possibility is that $P$ is a two-sphere with three cone points, of orders $p_1$, $p_2$ and $p_3$, with $1/p_1 + 1/p_2 + 1/p_3 < 1$. In this case $M$ has no boundary. We can take a finite cover of $P$ which has no cone points (Selberg's Theorem). By Theorem 4.1.4 (finite index), we only need to prove that the total space of the pullback fibre bundle has an automatic fundamental group. This reduces the situation to that which we are about to discuss, when we will show that there is an automatic structure. So we will assume from now on that $P$ is a two-dimensional hyperbolic orbifold which is not a two-sphere with three cone points. (Actually, the method we are about to discuss can be made to work directly for the case where $P$ is a two-sphere with three cone points, with a little more work.)

We can find disjoint embedded simple closed geodesics cutting $P$ into pieces, each of which is a 3-holed two-sphere, or a cylinder with one cone point of order $p_1 > 1$, or a disk with two cone points, one of order $p_2 > 1$

and the other of order $p_1 \geq p_2$, such that $1/p_1 + 1/p_2 < 1$. The last condition means that we cannot have $p_1 = 2 = p_2$, but any other possibility is allowed.

Let $Q$ be one of these irreducible two-dimensional hyperbolic orbifolds with geodesic boundary obtained by cutting $P$. We have the induced fibre bundle $\pi : R \to Q$ which is a Seifert fibration, as opposed to an ordinary fibre bundle, if and only if $Q$ has cone points. Our task in this section is to show that $R$ has an automatic fundamental group. Later we will change the automatic structure to allow such pieces to be glued together in such a way that the automatic structures match up on each torus which two such pieces have in common.

**Theorem 12.2.1 (basic Seifert automatic).** *Let $R$ be a Seifert fibre space over an irreducible hyperbolic. Then the fundamental group $\pi_1 R$ is automatic.*

*Proof of 12.2.1:* Let $C$ be a component of the boundary of $R$. Then $C$ is a torus. The *vertical foliation* of $C$ is the foliation given by the fibres of $\pi$. In due course we will want to glue some other three-manifold to $R$ along $C$. As a result, we may be given a foliation of $C$ which is equal for example to the foliation which is vertical for the *other side* of $C$ from $R$. This may or may not match up with the vertical foliation from $R$. For the moment, we will not bother about all the circumstances from which the second foliation might arise, but we simply assume that a second foliation of $C$ by circles is given, and that it is transverse to the vertical foliation induced by the Seifert fibre space structure on $R$. We will call the second foliation the *slanting foliation*. As a torus, $C$ has an affine structure. Both vertical and slanting foliations will be taken to be linear with respect to the affine structure.

We do not assume that typical leaves intersect once only—see Figure 12.3. If we regard $C$ as the quotient of $\mathbf{R}^2$ by $\mathbf{Z}^2$ acting by translation, then the vertical foliation might be given by curves satisfying $x = c$, for various values of $c$, and the slanting foliation might be given by curves satisfying $3y - 2x = c$, for various values of $c$. Then leaves intersect three times.

The orbifold $Q$ is a hyperbolic two-dimensional orbifold with geodesic boundary which is a sphere with three holes or a cylinder with one cone point or a disk with two cone points. The three possibilities are illustrated in Figure 12.4. Consider the set $X$ of points of $Q$ which are not manifold points. $X$ consists of boundary points and cone points and is the union of three components, which we call $\sigma$, $\gamma_1$ and $\gamma_2$. The component $\sigma$ will always be chosen to be a geodesic. If $Q$ has two or three boundary components, $\gamma_2$ is chosen to be a geodesic. If $\gamma_i$ is a cone point, let $\beta_i$ be a small circle in $Q$ with $\gamma_i$ at its centre, and, if $\gamma_i$ is a geodesic, let $\beta_i = \gamma_i$. For $i = 1, 2$, let $\alpha_i$ be the shortest geodesic from $\beta_i$ to $\sigma$. Then $\alpha_i$ meets $\sigma$ and $\beta_i$ orthogonally. The situation is illustrated in Figure 12.5.

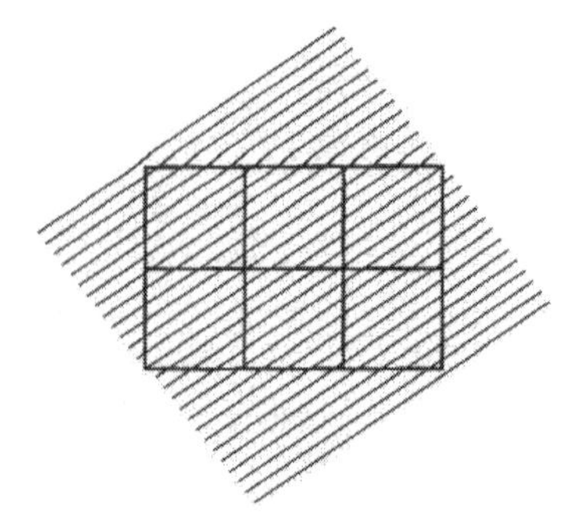

**Figure 12.3. The slanting foliation.** The squares in the picture are copies of the fundamental domain for the covering of the torus by the plane. The horizontal and vertical foliations are not drawn. In the situation drawn, a typical slanting leaf meets a typical vertical leaf in three points and a typical slanting leaf meets a typical horizontal leaf in two points.

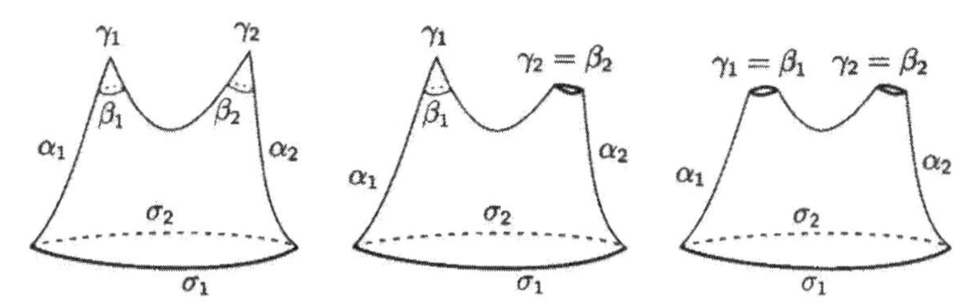


**Figure 12.4. Possibilities for $Q$.** We illustrate the three possibilities for $Q$. If $\beta_i = \gamma_i$, then this component is a geodesic. Otherwise, $\gamma_i$ is a cone point and $\beta_i$ is a circle of small radius surrounding $\gamma_i$. $\sigma = \sigma_1 \cup \sigma_2$ is a geodesic boundary component of $Q$. For $i = 1, 2$, $\alpha_i$ is the shortest geodesic segment joining $\beta_i$ to $\sigma$.

We cut $Q$ along the curves $\beta_i$, and we let $S$ denote the component of $Q$ containing $\sigma$. So $S$ is a three-holed sphere with three boundary circles, $\sigma$, $\beta_1$ and $\beta_2$. A fundamental domain $D$ for $S$ is obtained by cutting along $\alpha_1$ and $\alpha_2$. This operation cuts $\sigma$ into two edges, which we call $\sigma_1$ and $\sigma_2$.

A cell decomposition of $S$ with a single 2-cell $D$ is implicit in the above description. We illustrate $D$ in Figure 12.6. The boundary of $D$ is divided into eight edges. We get one edge from each $\sigma_i$ and from each $\beta_i$ and two edges from each $\alpha_i$. $S$ has four 0-cells, six 1-cells and one 2-cell.

Over $S$, $\pi : R \to Q$ is a fibre bundle in the classical sense, with circle fibres. Since $S$ can be deformation retracted to the wedge of two circles and the fibres are oriented circles, the inverse image in $R$ of $S$ is a trivial fibre

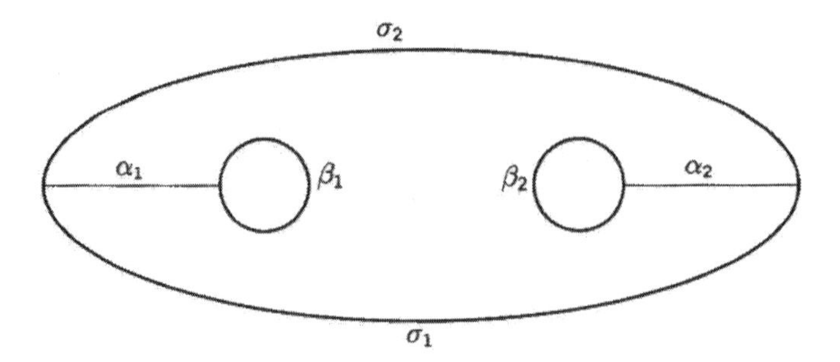


**Figure 12.5. Cutting up $Q$.** We have a geodesic boundary component $\sigma = \sigma_1 \cup \sigma_2$ of $Q$. The curves $\beta_1$ and $\beta_2$ are either geodesic boundaries of $Q$ or small circles surrounding cone points. For $i = 1, 2$, $\alpha_i$ is the shortest geodesics from $\beta_i$ to $\sigma$.

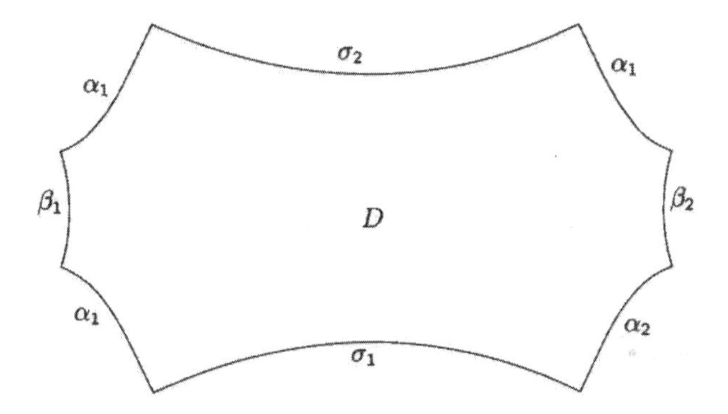


**Figure 12.6. A picture of $D$.** We illustrate $D$, with its boundary edges labelled by the edges of $S$ from which they come.

bundle over $S$. However, it is not good tactics to use this trivialization. To see why, consider a closed surface of genus two, with a fibre bundle over it which is a circle bundle with non-trivial euler class. When we cut the surface up into pairs of pants, we get a torus over each of the boundary circles for the pairs of pants. These tori have the same vertical foliation. However, the trivializations from the two sides do not match up, providing some complications when explaining what is happening.

Instead, we proceed as follows. Over each of the two endpoints of $\alpha_1$ in $S$ and the point $\alpha_2 \cap \beta_2$ (see Figure 12.5) we choose a point in $R$. Then we use the locally linear structure in the corresponding boundary torus of $\pi^{-1}S$ to draw a locally linear circle in the boundary torus from the chosen point to itself. The circle is chosen to map homeomorphically under $\pi$ to the corresponding circle in the boundary of $S$. The point plus the linear structure in the torus plus the homotopy class of the circle determine the circle uniquely, but there are an infinite number of distinct homotopy classes to choose from. The choice of homotopy class of the circle in the torus is done in a random way—two such choices differ by some multiple of the fibre. The circle above $\sigma$ intersects the vertical fibre above $\alpha_2 \cap \sigma$ in exactly one point. So now we have fixed four points in $R$, each lying above one of the four vertices of $S$, and we have fixed a lift to $R$ of each of the three boundary components of $S$.

Each of the three boundary tori of $\pi^{-1}S$ is now foliated by taking the set of all circles parallel to the lifted circle we have defined and we get a foliation called the *horizontal foliation*. We next extend the horizontal foliation to the inverse image in $R$ of the entire one-skeleton of $S$. Let $e$ equal $\alpha_1$ or $\alpha_2$. Then $\pi^{-1}e$ is an annulus in $R$. We choose some lift of $e$ to $R$ so that each endpoint goes to one of the four vertices in $R$, and then foliate the annulus with intervals which are parallel to this lift in the annulus.

Suppose that $\gamma_i$ is a cone point of order $p_i > 1$ (where $i = 1, 2$). Let $C_i = \pi^{-1}\beta_i$. Then $C_i$ is a torus with a vertical foliation by circles and $C_i$ is the boundary of a solid torus in $R \setminus \pi^{-1}(Q)$. The simple closed curves on $C_i$ which bound a disk in the solid torus define a linear foliation by circles. Each of these circles has intersection number $p_i$ with each vertical circle. We apply the term "slanting foliation" to this foliation, extending the usage already introduced on page 283. So each of the three boundary components of $\pi^{-1}S$ has both a slanting and a vertical foliation by circles and these circles are (locally) straight with respect to an affine structure on the boundary component. Each boundary torus also has a horizontal linear foliation by circles, but for our purposes this is less significant than the slanting foliation. The slanting and vertical foliations are likely to be intrinsic to the situation, whereas there is normally no canonical choice for the horizontal foliation. (We have already seen that the slanting foliation is intrinsic on the boundary of a solid torus neighbourhood of a singular fibre of the Seifert fibration.) A horizontal leaf has intersection number one with a vertical leaf, and a slanting leaf has a nonzero intersection number with a vertical leaf.

We have already fixed an orientation for the fibres over $Q$. Since the boundary of $S$ is oriented, we get an induced orientation for each boundary component of $\pi^{-1}S$, by thinking of the boundary component as the product of the boundary component of $S$ with $S^1$. We assume that all the fibres over $S$ have length 1. The slanting and horizontal foliations are oriented so that $\pi$ preserves the orientations between the leaves in the domain and the boundary circles of $S$ in the range.

Lifting to the horizontal foliation gives us a section of $\pi : R \to Q$ over the one-skeleton. We take as the one-skeleton of $R$ the union of the image of this section, together with the fibres over the four vertices of $S$. We call this one-skeleton $X_H$. Let $\tilde{R}$ be the universal cover of $R$. The inverse image of the one-skeleton of $S$ has an equivariant horizontal foliation in $\tilde{R}$, induced from that of $R$. If we lift to the induced horizontal foliation in $\tilde{R}$ a clockwise circuit of the boundary of $D$, we do not necessarily obtain a closed curve in $\tilde{R}$. In order to obtain a closed curve, we need to compose the lift with an integral number of circuits of the fibre. Let $b_1$ be the number of such circuits. (Then $b_1$ gives the euler class of the bundle, which can also be looked at as the two dimensional cohomology class of the central extension we are studying.)

Now map $[0, 1]$ onto a boundary component of $S$ in such a way that the endpoints of the interval are both mapped to the basepoint on the boundary, but the map of the interval is otherwise injective. In addition the map must preserve orientation. We lift this map to lie inside the slanting foliation, with 0 mapping to a vertex (one of the four discussed above) in the boundary component of $\pi^{-1}S$. Projecting onto the vertical circle through the basepoint by sending each horizontal circle to a point, we obtain a map $[0, 1] \to S^1 = \mathbf{R}/\mathbf{Z}$. We lift this map to a map $[0, 1] \to \mathbf{R}$ which sends 0 to 0 and 1 to some number $r$, and $r$ can be easily shown to be rational. This means that we have assigned to each oriented boundary component of $S$ a rational number.This is illustrated in Figure 12.3, where $r = 2/3$ (except that orientations are not given, so $r$ is only determined up to sign in the figure).

If we take universal covers, we have a fibre bundle $\tilde{R} \to \tilde{Q}$, with fibre the real numbers. Our objective is to define a section $\tilde{Q} \to \tilde{R}$ which is lipschitz. In fact we want the section to be determined, in a certain sense, by a finite state automaton. The section must agree with the lift to the slanting foliations on the tori over the boundary components of $S$. This task will be accomplished by thinking of the data contained in the numbers $r$ just defined as being the values of a partially defined 1-cochain.

Our discussion can be viewed in the context of several different areas of mathematics. Understanding of these points of view would give the reader a deeper appreciation of what we are doing, but our treatment is self-contained for the benefit of those who do not have the background. In group theory, central extensions have associated cohomology classes. A Seifert fibration gives a central extension. The cochains we consider are an aspect of this theory. There is also a differential geometry point of view. The horizontal foliations we defined above can be regarded as the horizontal lifts associated to a connection on a circle bundle over $S$. Our slanting foliations are associated to a connection which is flat over a neighbourhood of the non-manifold points of $Q$ (that is the boundary and the cone points). We have to construct a flat connection over $\tilde{Q}$ which restricts to the given (slanting) flat connection. The condition that the section $\tilde{Q} \to \tilde{R}$ be lipschitz is the vital restriction on the connection.

Let $\tilde{Q}$ be the universal cover of $Q$ and let $\tilde{S}$ be the inverse image of $S$ in $\tilde{Q}$. The complement of $\tilde{S}$ in $\tilde{Q}$ is the disjoint union of disks, each mapping as a cone point $\gamma_i$ of order $p_i$. The boundary of such a disk is a $p_i$-fold covering of the corresponding boundary component of $S$. The cell structure of $S$ induces a cell structure on $\tilde{S}$ in which the 1-cells are oriented and labelled by the symbols $\sigma_1$, $\sigma_2$, $\beta_1$, $\beta_2$, $\alpha_1$ and $\alpha_2$.

We have the groupoid $\pi_1 Q$ with generators

$$\sigma_1,\ \sigma_1^{-1},\ \sigma_2,\ \sigma_2^{-1}, \beta_1,\ \beta_1^{-1},\ \beta_2,\ \beta_2^{-1},\ \alpha_1,\ \alpha_1^{-1},\ \alpha_2,\ \alpha_2^{-1}.$$

We order the generators as they are listed. This groupoid has an automatic structure with uniqueness given by assigning length one to each generator and then taking ShortLex paths starting at the basepoint. To see that this is an automatic structure, we argue as follows. From Proposition 11.2.4 (Cayley graph of group and groupoid) we see that the Cayley graph of the groupoid is pseudoisometric to the Cayley graph of a vertex group. By Theorem 3.3.6 (pseudoisometry of Cayley graph), the Cayley graph of the vertex group is pseudoisometric to a subset of the hyperbolic plane with geodesic boundary, namely the universal cover of $Q$. Such a subset is necessarily convex. Hence a groupoid geodesic gives a quasigeodesic for the hyperbolic metric. We can then apply Lemma 3.4.2 (quasigeodesics near geodesics) to see that a groupoid geodesic is near to hyperbolic geodesic connecting its endpoints.

Suppose we have a finite state automaton $W$ which is a word acceptor for this structure. We fix a basepoint which is a 0-cell of $\tilde{S}$. Each 0-cell $v$ of $\tilde{S}$ corresponds to a ShortLex string $w_v$ over the alphabet of generators which can be viewed an edge path starting from the basepoint and ending at $v$. In this way we assign to $v$ a state $s(v)$ of the finite state automaton.

**Theorem 12.2.2 (bounded cochain).** (1) *Let $d^1$ be a cellular 1-cochain on $\partial S$, and let $c^2$ be a cellular 2-cochain on $S$. Let $p : \tilde{S} \to S$ be the covering map described above. Then there is a bounded cellular 1-cochain $c^1$ on $\tilde{S}$, such that $\delta c^1 = p^* c^2$ and such that the restriction of $c^1$ to $\partial\tilde{S}$ is equal to $p^* d^1$. The bound for $c^1$ depends only on the bounds for $d^1$ and $c^2$ and the geometry of $\tilde{S}$.*

(2) *If the given cochains are rational with common denominator $q$, then so are the values of $c^1$.*

(3) *We can construct a finite state automaton for the automatic structure, described on page 288, such that the state $s(v)$ assigned to a vertex $v$ of*

*$\tilde{S}$ carries with it the values of the constructed rational valued 1-cochain on each edge with endpoint $v$.*

Note that we are not asserting that $c^1$ can be chosen equivariant for the action of $\pi_1 Q$, although $d^1$ and $c^2$ are equivariant. Also note that we are constructing a cochain with good properties for $\tilde{S}$, not for $\tilde{Q}$. In fact the value of the cochain on the boundary of any disk in $\tilde{Q} \setminus \tilde{S}$ will be a non-zero integer in the applications.

A version of Theorem 12.2.2 is in fact true for any orbifold $S$ with negative euler characteristic, but we will stick to the simplest possible situation.

*Proof of 12.2.2:* We fix a basepoint for $Q$ and we choose a basepoint in $\tilde{Q}$ above the basepoint of $Q$. The developing map for $\tilde{Q}$ embeds it in the hyperbolic plane with geodesic boundary. If $\gamma_1$ is a cone point, let $\{U_i\}$ be the set of disks in $\tilde{Q}$ which are components of the inverse image of the singular disk bounded by $\beta_1$. If $\gamma_2$ is a cone point, let $\{V_j\}$ be the set of disks in $\tilde{Q}$ which are components of the inverse image of the disk in $Q$ bounded by $\beta_2$. We now cut $\tilde{Q}$ up into pieces which we will denote by $S_i$, where $i$ varies over a countable indexing set. $S_i$ is illustrated in Figure 12.7 and in Figure 12.8. The different $S_i$ meet at most along their boundaries. The details of the construction will be slightly different depending on how many cone points there are in $Q$ and the values of the cone angles as determined by $p_1$ and $p_2$.

We have to distinguish between different uses of the word "boundary". One use is the boundary of a manifold and the second is the pointset boundary. We will use the term "boundary" in the first case and "frontier" in the second case.

First suppose that $Q$ has no cone points. Then a lift of $D$ to $\tilde{Q}$ has a boundary with exactly four edges which are not in the frontier, namely $\sigma_1$, $\sigma_2$, $\beta_1$ and $\beta_2$, and these are separated from each other in the boundary of $D$—see Figure 12.6. It follows that the frontier of $D$ has four components. In this case, we define the $S_i$ as $i$ varies over a countable indexing set, to be the various lifts of $D$.

If $Q$ has cone points then $\gamma_1$ is a cone point of order $p_1$. To define $S_i$, we fix one of the disks $U_i$ and define $S_i'$ to be the union of $U_i$ and each lift of $D$ whose boundary meets $U_i$. If $\gamma_2$ is a geodesic, we define $S_i = S_i'$. If $\gamma_2$ is a cone point, we define $S_i$ to be the union of $S_i'$ and each disk $V_j$ whose boundary meets the boundary of $S_i'$ and whose interior lies in a component of $\tilde{Q} \setminus S_i'$ containing the basepoint of $\tilde{Q}$. (There is exactly one such component unless the basepoint lies in $S_i'$ in which case there are no such components.) The situation where both $\gamma_1$ and $\gamma_2$ are cone points is shown in Figure 12.7.

We claim that the frontier of $S_i$ has at least three components. To see this, note that the boundary of $S_i$ contains $p_1$ components labelled $\sigma$ and

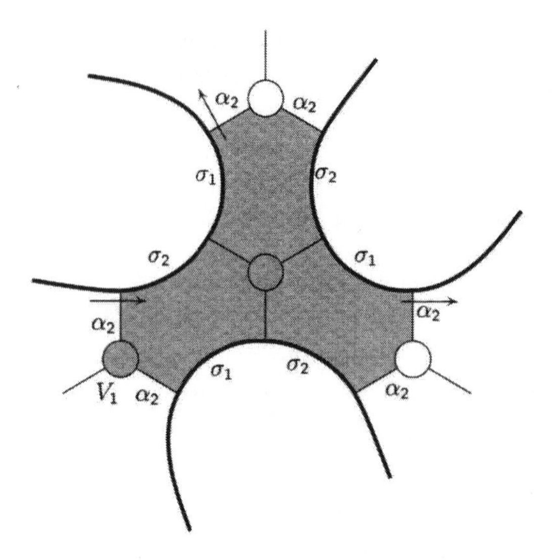


**Figure 12.7.** $S_i$ **with a cone point.** We show both $\gamma_1$ and $\gamma_2$ as cone points of order three. $S_i$ is shaded and includes the circle $V_j$ at the bottom left. The circle in the centre is $U_i$. The arrows show the first incoming edge and the first and second outgoing edges.

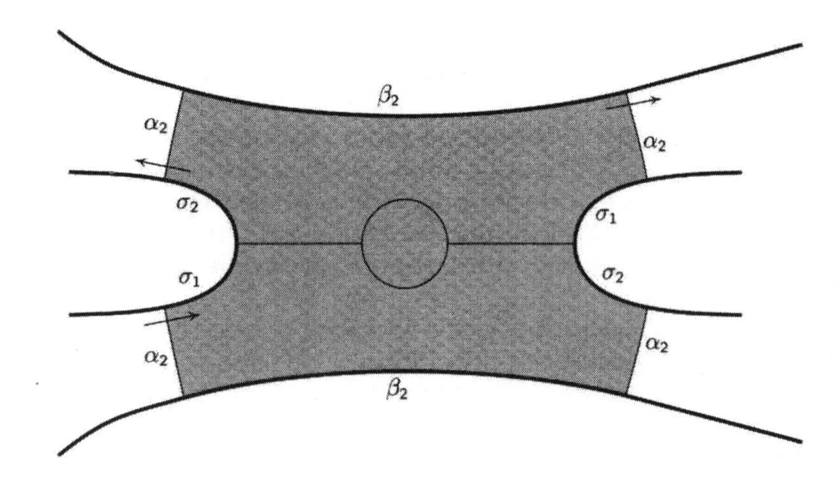


**Figure 12.8. A cone point of index two.** Here $\gamma_1$ is a cone point of order two, and therefore $\gamma_2 = \beta_2$ is a geodesic. The central circle is $U_i$. $S_i$ is the shaded region. The arrows indicate the incoming edge and the first and second outgoing edges.

these edges are separated by the frontier. So our claim is true if $p_1 \geq 3$. If $p_1 = 2$, it must be the case that $\gamma_2$ is a geodesic, since $p_2 \leq p_1$ and the case $p_1 = p_2 = 2$ is excluded. Then the frontier of $S_i$ has four components, each labelled $\alpha_2$, as in Figure 12.8. If $\gamma_1$ and $\gamma_2$ are both geodesics, then the frontier of $S_i$ has four components, two labelled $\alpha_1$ and two labelled $\alpha_2$.

For each vertex $v$ of $\tilde{Q}$ we have the unique string $w_v$, defined on page 288, which leads to it from the basepoint. Let $v \in S_i$. We can associate to $v$ a bounded history which tells us which vertex of $S_i$ was first entered by $w_v$. We call such a vertex an *entry vertex* for $S_i$. The construction of $w_v$ is such that all subsequent vertices *en route* to $v$ also lie in $S_i$ .

In view of the fact that $\tilde{Q}$ is simply connected, each frontier component $F$ of $S_i$ separates $\tilde{Q}$. $F$ is an arc which is a finite union of edges, and assigning an orientation to $F$ is the same as choosing a direction, or drawing an arrowhead, on $F$. We choose the direction so that it goes from left to right, if we position ourselves on the basepoint side of $F$. ($F$ may contain copies of lifts of $\beta_2$, which has already been assigned an orientation—we do not care whether the orientations on $F$ and on $\beta_2$ agree.) The orientation enables us to say which is the *least* edge in $F$. If $F$ happens to contain the basepoint of $\tilde{Q}$, we do not bother to orient it. (This can happen at most three times, because at most three of the $S_i$ meet at a point.)

For each $S_i$, except for at most three, which contain the basepoint, we pick the frontier component nearest to the basepoint of $\tilde{Q}$, and call it the *incoming component* of $S_i$. We define the *incoming edge* of $S_i$ to be the least edge in the incoming component.

We call the other frontier components of $S_i$ *outgoing components.* The outgoing frontier components of $S_i$ are ordered starting with the incoming frontier component and proceeding round the boundary of $S_i$ in a clockwise direction. We define the *first outgoing edge* to be the least edge in the first outgoing component and the *second outgoing edge* to be the least edge in the second outgoing component. These edges are illustrated in Figure 12.7 and Figure 12.8.

There is one case where this does not work, namely when $S_i$ contains the basepoint. In that case there is no incoming frontier component and no incoming edge, and we order the outgoing frontier components in some arbitrary way.

We will consider the $S_i$'s in turn, according to the graph distance from the basepoint. Suppose $S_i$ is the next region to be considered. Let $e$ be an edge of $S_i$ which is a first or second outgoing edge of some other $S_j$. Then the least vertex of $e$ lies in the boundary of $\tilde{Q}$. Let $F_i$ be the frontier component of $S_i$ in which $e$ lies, and let $F_j$ be the outgoing frontier component of $S_j$ containing $e$. Note that in general $F_i$ and $F_j$ do not coincide, as we can see by looking at Figure 12.7. Clearly $F_j$ is an outgoing frontier component of $S_j$ and $F_i$ is the incoming frontier component of $S_i$. Hence $e$ is the incoming edge of $S_i$.

All of these constructions can be determined by using only a finite amount of state information. By this we mean that one may have to split the states of the finite state automaton $W$, defined on page 288, using a finite number of bits of extra information, and then we will be able to tell from the state $s(v)$ associated to a vertex $v$, what its incoming vertex is, and which are the incoming edge and the first and second outgoing edges of each $S_i$ containing $v$.

The proof of the theorem is now straightforward. The hypotheses of the theorem give us a partially defined 1-cochain, defined only on the edges labelled $\sigma_1$, $\sigma_2$, $\beta_1$ and $\beta_2$. Let $B_0$ be an integer which is greater than the sum of the moduli of the values of the partially defined 1-cochain on the edges of any $S_i$. Let $B_0$ also be greater than the absolute value of $c^2$ on any $S_i$. We assign to any boundary edge of $S_i$, which is not a first or second outgoing edge of some $S_i$, the value zero. We will define the value of $c^1$ on the first and second outgoing edges of the $S_i$, by induction on the graph distance of $S_i$ from the basepoint.

The inductive hypothesis is that the value of the 1-cochain, on each first or second outgoing edge whose value has been assigned, has a modulus which is no larger than $2B_0$, and that $(\delta c^1 - c^2)(S_i) = 0$ whenever it is defined.

All boundary edges of $\tilde{S} \cap S_i$, except for the first and second outgoing edges, have by now had a value assigned. The value of the cochain on the incoming edge is bounded by $2B_0$ and the sum of the absolute values of the cochain on the other edges, where its value is already known, is bounded by $B_0$. We denote by $s_1$ and $s_2$ the still unknown values of $c^1$ on the first and second outgoing edges. Removing these unknowns, we are left only with known values: we set

$$s = (\delta c^1 - c^2)(S_i) - s_1 - s_2.$$

Then $s$ is determined by the value of $c^1$ on the incoming edge, the value of $c^1$ on the edges where it was initially defined, and the value of $c^2(S_i)$. So $-4B_0 \leq s \leq 4B_0$, by the definition of $B_0$ and the induction hypothesis.

We can now define $s_1$ and $s_2$ as follows.

$$\begin{aligned} s_1 &= 2B_0 && \text{if } s \leq -2B_0, \\ s_1 &= -s && \text{if } -2B_0 < s < 2B_0, \\ s_1 &= -2B_0 && \text{if } s \geq 2B_0, \\ s_2 &= -s - s_1. \end{aligned}$$

Then $|s_1| \leq 2B_0$ and $|s_2| \leq 2B_0$, establishing the inductive hypothesis.

We next extend the partially defined 1-cochain to all the edges of $S_i$. The edges on which the cochain is not yet defined are those in the interior of $\tilde{S} \cap S_i$. The procedure is to first determine the order in which these values will be decided (using a finite state automaton to decide the order). We assign the value zero if possible, while insisting that the chain boundary of any lift of $D$ is assigned the value $c^2(D)$, which means that some non-zero values may be forced on certain interior edges. Since the values previously assigned to the edges are bounded by $2B_0$, every edge has a value which is bounded in modulus by some integer $B_1$.

The statements about rational values are clearly proved by the same induction. The statement about the finite state automaton follows from the fact that there are only a finite number of different situations to consider, and all relevant factors can be recorded using a finite number of bits. This completes the proof of the theorem. 12.2.2

We now use Theorem 12.2.2 (bounded cochain) to construct the automatic structure on $\pi_1 R$. Let $\tilde{R}$ be the universal cover of $R$. Then we have a fibre bundle $\tilde{\pi} : \tilde{R} \to \tilde{Q}$, whose fibre is $\mathbf{R}$. We have already constructed the slanting foliation on the boundary of $\tilde{\pi}^{-1}(\tilde{S})$. We define the 2-cochain $c^2$ by setting its value on $D$ to be $b_1$ (this was defined on page 287). Theorem 12.2.2 now gives us the bounded cochain $c^1$, whose values are given by a finite state automaton.

We extend the slanting foliation to the inverse image of the one-skeleton of $\tilde{S}$, by using $c^1$ to assign slopes relative to the horizontal foliation. The slanting foliation is not equivariant.

We now proceed as follows. We will construct an automatic structure on $\pi_1 R$ using a lipschitz section $s : \tilde{Q} \to \tilde{R}$ of $\tilde{\pi} : \tilde{R} \to \tilde{Q}$, which we call the *standard slanting section*, and which we now define. The standard slanting section maps each boundary component of $\tilde{S}$ into a leaf of the slanting foliation in $\tilde{R}$.

Recall that we have defined a one-skeleton $X_H$ for $R$ (see page 287). We have an induced one-skeleton $\tilde{X}_H$ in $\tilde{R}$. We fix a basepoint $r_0 \in \tilde{R}$ which is a vertex of $\tilde{X}_H$ lying over the basepoint of $\tilde{S}$.

One can map the universal cover of the one-skeleton of $\tilde{S}$ into a leaf of the slanting foliation so that the basepoint is sent to $r_0$. Using Theorem 12.2.2 (bounded cochain) and comparing with the horizontal foliation, it is easy to see that this actually gives us a section $s$ on the one-skeleton of $\tilde{S}$. We now extend $s$ to the 2-cells of $\tilde{Q}$. This is possible since the boundary of each cell has been lifted to a circle, and the inverse image in $\tilde{R}$ of the 2-cell in $\tilde{Q}$ is simply connected. (We need to refer to the construction of the slanting foliation near cone points of $Q$ as explained on page 286.)

We now use this information to construct an automatic structure on $\pi_1 R$. We will use a finite number of *marked points* in $R$ and regard $\pi_1 R$ as a groupoid whose objects are the marked points. First we define the marked points. The bounded cochain $c^1$, provided by Theorem 12.2.2 and used to construct the standard slanting section, has values which are rational with a common denominator $q > 0$. Let $B_1$ be an integer which is an upper bound for the values of $c^1$. We choose as marked points the image in $R$ of each 0-cell of $S$ under the horizontal section, and in addition every point obtained by travelling from such a point a vertical distance which is an integral multiple of $1/q$. The inverse image in $\tilde{R}$ of the marked points in $R$ give an infinite collection of points in $\tilde{R}$ which we call "vertices". Given such a point $v$, it projects to a point $v'$ in $\tilde{S}$.

Next we choose generators for the groupoid. We will use the following alphabet, with letters ordered lexicographically in the order chosen

**12.2.3.** $\sigma_1,\ \sigma_1^{-1},\ \sigma_2,\ \sigma_2^{-1},\ \beta_1,\ \beta_1^{-1},\ \beta_2,\ \beta_2^{-1},\ \alpha_1,\ \alpha_1^{-1},\ \alpha_2,\ \alpha_2^{-1}, z, z^{-1}.$

Each of the symbols except for $z$ and $z^{-1}$ has also been used to represent a directed edge in $S$. In this context each such letter is used to label any horizontal lift (*not* a slanting lift) of such an edge. We label by $z$ any vertical edge which travels a distance $1/q$ in the positive vertical direction and by $z^{-1}$ the inverse of any such edge. We give each of these generators weight one. Note that if any of these edges starts at a marked point, it will also end at a marked point. In the definition of an automatic groupoid (see Section 11.1 (Groupoids)), we insisted that there should be only one morphism with a given name. This is not the case in the present situation. We can rectify this formally by including with the label the name of the marked points of $R$ at which an edge starts and ends. If we know at which marked point a path starts then the sequence of labels listed in 12.2.3 gives us sufficient information to follow it without ambiguity. Therefore it is sufficient to describe an automatic structure on the groupoid $\pi_1 R$ using the symbols from 12.2.3.

It is convenient to adjoin a number of other letters to the alphabet, each of weight one. For each horizontal generator $x$, we adjoin $1 + 2B_1 q$ letters labelled $x_i$, where $0 \le |i| \le B_1 q$ and $x_i = xz^i$.

Let $e$ be a directed edge starting at $v$, with label $x_i$ where $x$ is a horizontal generator. In order for this to be a legal portion of a path in the automatic structure, we insist that $e$ has the correct slope. Explicitly, let $j/q$ be the value of $c^1$ on the projection of $e$ to $\tilde{S}$. Then we insist that $i = j$.

Note that each vertex of $\tilde{R}$ gives rise to a well-defined slanting section of $\tilde{S}$ going through this vertex. This is obtained from the standard slanting section described above by vertical translation by some multiple of $1/q$. These slanting sections have a cell structure coming from $\tilde{Q}$. By taking the product of the cell structure on the slanting section with intervals of length $1/q$, we obtain a cell structure $C(\tilde{R})$ for $\tilde{R}$ (which is not invariant under $\pi_1 R$). Another way of stating the necessary condition of the preceding paragraph is that each legal path must lie in the 1-skeleton of $C(\tilde{R})$.

We can now describe the automatic structure on the groupoid $\pi_1 R$. We start with the automatic structure for the groupoid $\pi_1 Q$ described on page 288. Given a string in the horizontal generators, we check that it is acceptable by running the string simultaneously through the automaton for $Q$ and the automaton provided in the statement of Theorem 12.2.2 (bounded cochain). We replace each occurrence of a horizontal generator $x$ by the unique legal $x_i$. We define the accepted strings for the groupoid $\pi_1 R$ to be any accepting slanting sequence as just described, followed by $z^m$, where $m$ is any integer.

We need to prove this set of strings defines an automatic structure. We have already shown that it is a regular language. It only remains to show that there is a number $k$ with the following property. Let $w_1$ and $w_2$ be accepted strings, starting at the same vertex of $\tilde{R}$, and such that $\overline{w_1 y} = \overline{w_2}$ where $y$ is a generator. Then $w_1$ and $w_2$, regarded as paths, are a uniform distance at most $k$ apart.

Let $w_j = u_j v_j$, where $u_j$ is slanting and $v_j$ is vertical. Then $u_1$ and $u_2$ are the images under the standard slanting section of accepted strings in $\tilde{Q}$. Therefore there is a bound on their uniform distance apart. In view of the way the automatic structure on $Q$ has been chosen, using shortest paths, $u_1$ and $u_2$ differ in length by at most 1. (We are talking here of the length in terms of the number of generators involved, rather than in terms of the geometric length of a rectifiable path. The weighting of each of the $x_i$ is one.) Since $z$ commutes with the other generators, we can deduce that $d(\widehat{w_1}(i), \widehat{w_2}(i))$ is also uniformly bounded, independently of $w_1$, $w_2$ and $i$.

**Lemma 12.2.4 (geodesic in Seifert complex).** *The accepted paths are geodesic in the 1-skeleton of $C(\tilde{R})$ (but not necessarily in the Cayley graph of $\pi_1 R$).*

*Proof of 12.2.4:* To see this, let $w$ be a geodesic in the 1-skeleton of $C(\tilde{R})$, and let $uv$ be the path in the automatic structure described above with the same endpoints, with $u$ slanting and $v$ vertical. We will show that $w$ and $uv$ have the same length. Since $z$ commutes with other generators, we may assume that $w = u_1 v_1$, where $u_1$ is slanting and $v_1$ is vertical. Projecting to $\tilde{Q}$, we see that $u_1$ and $u$ define paths in $\tilde{Q}$ with the same endpoints. Therefore $v = v_1$. Since $w$ is geodesic, $u_1$ is a geodesic in $\tilde{Q}$, and so $u_1$ and $u$ have the same length. 12.2.4

It follows from Theorem 11.2.2 (geodesic automaton 3), with the role of $\Theta$ there being played here by the one-skeleton of $C(\tilde{R})$ that we have defined an automatic structure. 12.2.1

The construction just be described can also be used to prove the following result about a Seifert fibration over a general two-dimensional compact hyperbolic orbifold, rather than over the very special orbifolds we have considered above.

**Theorem 12.2.5 (Seifert implies automatic).** *Let $\pi : R \to Q$ be a compact Seifert fibre space over a hyperbolic orbifold $Q$ with geodesic boundary. Let $\tilde{\pi} : \tilde{R} \to \tilde{Q}$ be the induced bundle with fibre $\mathbf{R}$, where $\tilde{R}$ and $\tilde{Q}$ are the universal covers of $R$ and $Q$ respectively. Let $S$ be the complement of the union of small disjoint disks in $Q$ whose centres are the singular points of $Q$, and let $\tilde{S}$ be the inverse image of $S$ in $\tilde{Q}$. Then there is a cell structure on $S$, inducing a cell structure on $\tilde{S}$, and a bounded 1-cochain $c^1$ on $\tilde{S}$ with rational values and denominator $q$ on $\tilde{S}$, inducing the appropriate slanting foliations above the boundaries of the disks of $Q \setminus S$ and the boundary of $R$. The values of $c^1$ are given by a finite state automaton. It can be used to define a slanting section of $\tilde{S}$ in $\tilde{R}$, with bounded slope. There is an automatic structure on the groupoid $\pi_1 R$ where an accepted string is of the form $uv$, where $u$ is the image of a geodesic in $\tilde{Q}$ under the slanting section and $v$ is vertical.*

We will only give a very sketchy treatment, because the theorem that these groups are automatic will in any case be a consequence of Theorem 12.4.7 (3-manifolds and automatic structures). In particular we will omit for now the proof that $c^1$ has values which are given by a finite state automaton, as this will be discussed later in detail.

*Proof of 12.2.5:* We cut $Q$ up into pieces each of which is a hyperbolic two-dimensional orbifold with geodesic boundary and is a sphere with three holes or a cylinder with one cone point or a disk with two cone points. We make sure that the cell structure matches up along the cuts. $\tilde{S}$ is the union of a number of copies of spaces like the version of $\tilde{S}$ discussed above. The bounded 1-cochain is constructed separately on each of these copies. The rest of the construction is the same as that already discussed.

We have not yet discussed the awkward case, when $Q$ is a two-sphere with three cone points. We can find a finite cover of $Q$ which is an orientable two-manifold. We cut this cover up into pairs of pants, obtaining various simple closed geodesics in the cover. Projecting these geodesics to $Q$, we obtain a certain one-dimensional complex. We subdivide $Q$ in such a way that this one-dimensional complex is contained in the one-skeleton of $Q$. We are now in a position to follow the same procedure as before. 12.2.5

## 12.3. Fitting Pieces Together along the Boundaries

The results in this section are due to Epstein and Thurston.

In Section 12.2, we discussed the following situation. $Q$ is a hyperbolic orbifold, which is a pair of pants, or a cylinder with one branch point, or a disk with two branch points. The boundary components of $Q$ are geodesic. $R$ is a fibre bundle over $Q$ with fibre a circle. $\pi : R \to Q$ is a Seifert fibre space. We produced an automatic structure on the fundamental groupoid $\pi_1 R$. We will use other notation from that section. In particular, $S$ is the result of removing from $Q$ neighbourhoods of the cone points. The reader is referred to Figure 12.4 for other notation.

The next task is to use three-manifolds like $R$ as building blocks for more general three-manifolds, and to deduce the automatic structure of the whole from the automatic structure on the pieces. The boundary of $R$ consists of one or two or three tori. The building blocks are put together by gluing along the tori. If the automatic structures on the various blocks are to be glued together, they need to coincide on the boundaries, and this is not the case for the automatic structure defined in Section 12.2.

When two blocks are glued along a torus, then each side will in general give rise to its own vertical foliation. In Section 12.2, the vertical foliation for $R$ gave rise to a generator $z$ for the groupoid $\pi_1 R$, where $z^q$ represents one circuit of a fibre of $R$. The positive integer $q$ cannot in general be deduced from the structure of a single boundary component; one needs to look at the structure over the various boundary components of $S$, and take the common denominator.

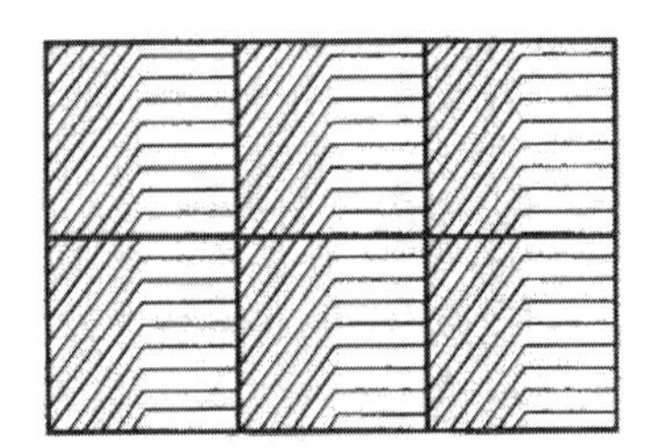

**Figure 12.9. The slanting foliation after subdividing the base.** We show how the slanting foliation changes as a result of subdividing the boundary geodesic of $Q$. This picture should be compared with Figure 12.3, from which it is obtained by an isotopy.

The slanting foliation over the geodesic boundary component $\sigma$ of $R$ may come from the vertical foliation on the "other side" of $\pi^{-1}(\sigma)$ from

$R$. Analysis of the block on the other side may therefore lead us to want to subdivide $\sigma_1$ and $\sigma_2$ further into edges of equal length. So we redo the analysis of Section 12.2 (The Basic Seifert Fibre Space), where each of the geodesic boundary components of $R$ is subdivided into $q$ edges of equal length. We take $q$ to be even, because the geodesic boundary $\sigma$ is already subdivided into $\sigma_1$ and $\sigma_2$.

Suppose we need to consider the properties of a geodesic in the graph metric of a geodesic joining two points in the same component of the boundary of $\tilde{Q}$, the universal cover of $Q$. We want to make it clear that the most efficient way is to stay in the boundary. Suppose $x$ is the hyperbolic length of a boundary component of $Q$, and that $e$ is an interior edge of $Q$, with hyperbolic distance $y$ between its endpoints. Then we divide $e$ up into more than $q([y/x]+1)$ equal pieces and assign each piece length $1/q$ in the graph metric. This is done simultaneously with respect to each geodesic boundary component of $Q$. This means that the edge $e$ of hyperbolic length $y$ is assigned graph length greater than $[y/x]+1$.

**Lemma 12.3.1 (boundary efficiency).** *There is a constant $k > 1$, with the following property. Let $A$ and $D$ be two points on the same boundary component $L$ of $\tilde{Q}$ and let $\alpha$ be any edge path joining them, such that $\alpha$ meets $L$ only at the endpoints of $\alpha$. Then the graph length of $\alpha$ is at least $k$ times the graph length of the edge path in $L$ from $A$ to $D$.*

*Proof of 12.3.1:* We compare the hyperbolic length of each edge of $\tilde{Q}$, as now subdivided, to the hyperbolic length of its orthogonal projection to $L$. If the edge does not meet $L$, then the distance $d$ between $L$ and the edge is positive and orthogonal projection to $L$ reduces the length of the edge by a factor of at least $\cosh d$. The hyperbolic length of each edge with exactly one endpoint on $L$ is also reduced by some definite factor, and there are only a finite number of possibilities, because of invariance under action of the fundamental group.

This proves that the hyperbolic length of $\alpha$ is some definite factor $k > 1$ longer than the hyperbolic distance $d(A, D)$, which is measured along $L$. The graph distance along $L$ from $A$ to $D$ is $d(A, D)/x$. Let $y_1, \ldots, y_m$ be the hyperbolic lengths of the successive edges of $\alpha$, in the original subdivision. Then $\sum y_i \geq kd(A, D)$. We therefore have

$$\sum([y_i/x]+1) \geq \sum(y_i/x) \geq kd(A, D)/x.$$

It follows that the graph length of $\alpha$ is at least $k$ times the graph distance along $L$ from $A$ to $D$. $\boxed{12.3.1}$

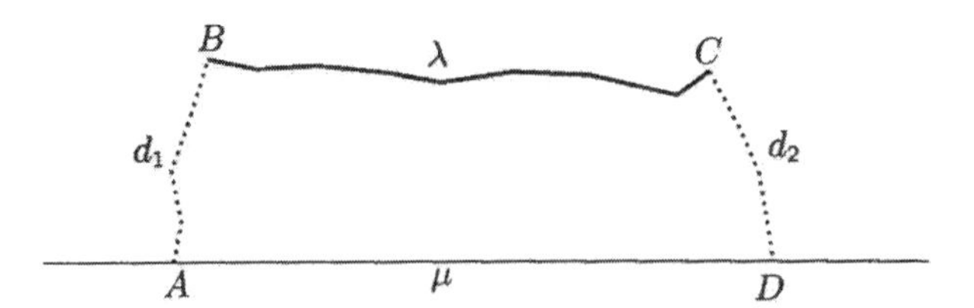


**Figure 12.10. Graph geodesic not in the boundary.** This figure illustrates Lemma 12.3.2 (boundary disjoint). The straight line $AD$ is part of the geodesic boundary $L$. The graph distance from $B$ to $C$ is $\lambda$ and the graph geodesic from $B$ to $C$ is disjoint from $L$. The graph distance from $A$ to $D$ is $\mu$. The graph distance from $B$ to $L$ is $d_1$ and the graph distance from $C$ to $L$ is $d_2$.

**Lemma 12.3.2 (boundary disjoint).** *Let $L$ be a boundary component of $\tilde{Q}$. Let $\gamma$ be a graph geodesic which is disjoint from $L$, except possibly at its endpoints $B$ and $C$ (see Figure 12.10). Let $A \in L$ be the nearest point to $B$ in the graph metric at a graph distance $d_1$, and let $D \in L$ be the nearest point to $C$ in the graph metric at a graph distance $d_2$. Let the graph distance from $A$ to $D$ be $\mu$ and let the graph length of $\gamma$ be $\lambda$. Let $k$ be as in Lemma 12.3.1 (boundary efficiency). Then*

$$\mu \leq \frac{2(d_1+d_2)}{k-1} \qquad \text{and} \qquad \lambda \leq \frac{(k+1)(d_1+d_2)}{k-1}.$$

*Proof of 12.3.2:* The proof is obvious by looking at Figure 12.10 and applying Lemma 12.3.1 (boundary efficiency). We immediately deduce the inequalities

$$k\mu - d_1 - d_2 \leq \lambda \leq d_1 + \mu + d_2.$$

The inequalities in the statement follow. $\boxed{12.3.2}$

**Lemma 12.3.3 (angular efficiency).** *There is an angle $\theta_0 > 0$ with the following property. Let $w$ be a graph geodesic in $\tilde{Q}$, whose only intersection with the boundary $L$ is at its initial point. Let $\beta$ be a hyperbolic geodesic joining the endpoints of $w$. Then the angle between $\beta$ and $L$ is at least $\theta_0$.*

The statement and proof of the lemma are illustrated in Figure 12.11.

*Proof of 12.3.3:* We know by Proposition 11.2.4 (Cayley graph of group and groupoid) and Theorem 3.3.6 (pseudoisometry of Cayley graph) that there are constants $k_0 > 1$ and $\varepsilon > 0$, such that $w$ is a $(k_0, \varepsilon)$-quasigeodesic. Hence there is a number $d$, depending only on $k$ and $\varepsilon$, such that $\beta$ and $w$ are a

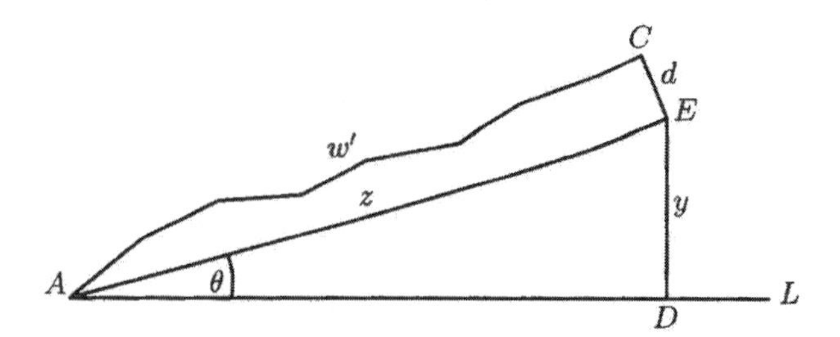


**Figure 12.11. Lower bound to the angle.** This picture illustrates the proof of Lemma 12.3.3 (angular efficiency). The segment $AD$ lies in the geodesic boundary component $L$ of $\tilde{Q}$. The segment $AE$ is an initial segment of the geodesic $\beta$. The angle between $AD$ and $AE$ is $\theta$. We show an initial segment $w'$ of $w$, ending at a point $C$. The distance from $C$ to $E$ is at most $d$. The segment $ED$ is perpendicular to $L$ and has length $y$.

hausdorff distance at most $d$ apart. We choose

$$\lambda = \frac{(k+1)(k_0(d+1)+\varepsilon)}{k-1}$$

and choose an initial segment $w'$ of $w$ of graph length greater than $\lambda$. (If this is not possible, $w$ must be short and there are only a finite number of possibilities to consider; so we obtain a lower bound on the angle.) Let $C$ be the endpoint of $w'$, let $E$ be the nearest point to $C$ on $\beta$, and let $D$ be the nearest point on $L$ to $E$. Let the hyperbolic distance $d(E, D) = y$. Then the graph distance from $C$ to $L$ is at most $k_0(d+y)+\varepsilon$. By the second inequality of Lemma 12.3.2 (boundary disjoint), it follows that $y \geq 1$.

Let $z$ be the hyperbolic distance from the initial point $A$ of $w$ to $E$ and let $0 < \theta \leq \pi/2$ be the angle between $\beta$ and $L$. Then $z \leq k_0\lambda + \varepsilon + d$. By hyperbolic trigonometry,

$$\sin\theta = \frac{\sinh y}{\sinh z} \geq \frac{\sinh 1}{\sinh(k_0\lambda + \varepsilon + d)} .$$

This proves the result, because the right-hand side is now independent of choices. 12.3.3

Recall the definition of $C(\tilde{R})$ from page 295. We subdivide the cell complex $C(\tilde{R})$ by insisting that each fibre over each of the new vertices of $\tilde{Q}$ should be a subcomplex, divided into $q$ edges. Also the inverse image of each new edge of $\tilde{Q}$ is a cylinder cut up into $q$ rectangles.

We assign the same length $1/q$ to each vertical edge labelled $z$. The analysis of Section 12.2 goes through unchanged, except that the number of generators of the fundamental groupoid $\pi_1 R$ is increased. When we subdivide an edge $e$, we concentrate the value of the cochain $c^1$ of Theorem 12.2.2 (bounded cochain) on a single edge of the subdivision of $e$, setting the value equal to zero on all the other edges into which $e$ is subdivided. The effect on the slanting foliation is to make it coincide with the horizontal foliation over all edges except the first—the effect is shown in Figure 12.9.

The next alteration to facilitate gluing two pieces along a boundary component, is to remove the distinction between the vertical and slanting foliations on each boundary component of $R$. The point is that the role of slanting and vertical is likely to be interchanged if one considers the two building blocks on each side of a boundary torus, and so symmetry in their treatment is very desirable.

Previously, we introduced generators $x_i$ for each horizontal generator $x$, where $0 \leq |i| \leq B_1 q$ and $x_i = xz^i$. Now we introduce further generators, $x_i^+ = x_i z$ and $x_i^- = x_i z^{-1}$, for the groupoid $\pi_1 R$. We adjoin to $C(\tilde{R})$ edges with these labels and with appropriate endpoints. These additional edges are assigned length $2/q$.

Here are the rules for the language of strings in the generators, regarded as paths in $\tilde{R}$ starting at the basepoint:

(1) An edge of the path labelled $x_i$ or $x_i^+$ or $x_i^-$ projects to an edge of $\tilde{Q}$ which is assigned the value $i/q$ by the cochain $c^1$ of Theorem 12.2.2 (bounded cochain), in the same manner as was described on page 294.

(2) If an $x_i^+$ occurs, its successor, if any, must be of the form $y_j^+$ for some horizontal generator $y$ and some $j$, or of the form $y_j$ or equal to $z$.

(3) If an $x_i^-$ occurs, its successor, if any, must be of the form $y_j^-$ for some horizontal generator $y$ and some $j$, or of the form $y_j$ or equal to $z^{-1}$.

(4) If a $z$ occurs, its successor, if any, is equal to $z$.

(5) If a $z^{-1}$ occurs, its successor, if any, is equal to $z^{-1}$.

(6) If an $x_i$ occurs, its successor, if any, is of the form $y_j$ for some horizontal generator $y$ and some $j$.

(7) The projection of the path to $\tilde{Q}$ is accepted by the automaton $W$ described on page 288 (ignoring the projection of any $z$ or $z^{-1}$).

This is analogous to the automatic structure on an abelian group, which is invariant under a finite group of automorphisms, introduced in Theorem 4.2.1 (automatic structure with invariance). In Section 12.2 (The Basic Seifert Fibre Space) we constructed a cell complex $C(\tilde{R})$ by taking edges of the form $x_i$ over an edge $e$ labelled $x$ in $\tilde{Q}$, where the cochain of Theorem 12.2.2 (bounded cochain) assigns the value $i/q$ to $e$. Let $C'$ denote the abstract graph obtained by adding to the 1-skeleton of $C(\tilde{R})$ edges of the form $x_i^+$ and $x_i^-$ and of length $2/q$.

**Theorem 12.3.4 (improved structure).** *The above structure for accepted strings is an automatic structure on the fundamental groupoid $\pi_1 R$. The accepted strings are geodesic in $C'$.*

*Proof of 12.3.4:* Let $w$ be any geodesic string in $C'$ starting at the basepoint of $\tilde{R}$. We will show that there is an accepted path from the basepoint to the endpoint of $\widehat{w}$, of the same length as $w$. First note that $w$ can be assumed not to contain simultaneously symbols of the form $x_i^+$ and $y_j^-$, using the fact that $z$ is central. Next, if $w$ contains $z$, it can be assumed not to contain any $x_i$ nor any $x_i^-$. Similarly, if $w$ contains $z^{-1}$, it can be assumed not to contain any $x_i$ nor any $x_i^+$. Without loss of generality, we may assume that in $w$ all symbols of the form $x_i^+$ or $x_i^-$ occur at the beginning. Next we may without loss of generality assume that the projection of $w$ to $\tilde{Q}$ is accepted by the automaton $W$ of page 288. This step entails using the fact that the slanting section lifts the 1-skeleton of $\tilde{Q}$ into $C(\tilde{R})$, so that any two geodesic paths in $\tilde{Q}$ with the same endpoints lift to paths in $C(\tilde{R})$ with the same endpoints. It follows that we may assume without loss of generality that $w$ is accepted.

Clearly the set of accepted strings is a regular language. It is easy to see that accepted paths, ending within distance one of each other, lie in a uniform hausdorff neighbourhood of each other. This means that the hypotheses of Theorem 11.2.2 (geodesic automaton 3) are fulfilled, which proves the result. 12.3.4

## 12.4. The Automatic Structure on a Three-Manifold

The results in this section are due to Epstein and Thurston.

In Section 12.1 (Taking the Problem to Pieces), we considered the problem of finding an automatic structure of a general compact three-manifold satisfying Thurston's Conjecture. Recall that we have reduced the situation to the case where our three-manifold $M$ is obtained from a finite number of compact three-manifolds with boundary. Each of these compact three-manifolds has either a Seifert fibre space structure as in Section 12.2 (The Basic Seifert Fibre Space), or the structure of a geometrically finite hyperbolic three-manifold with the cusps cut off to make the manifold compact. Each such piece is called a *structure component*. The structure components are glued together along boundary tori to obtain $M$.

Let $C$ be a torus locally separating two of our structure components. ($C$ could also be a torus which results from gluing together two boundary components of a single structure component.) Then $C$ is modelled on a real affine space of dimension 2 (that is, a vector space in which we have forgotten the position of the zero). If $C$ is the boundary of a cusp, then the affine structure comes from the euclidean structure on a horosphere. If $C$ is a boundary component of a Seifert fibre space like $R$, the fibre bundle $\pi : R \to Q$ induces an affine structure on $C$. Since any diffeomorphism of a torus can be isotoped to an affine diffeomorphism, the two affine structures on $C$ can be assumed to agree with each other, giving a well-defined affine structure on $C$. We call $C$ a *structure torus*.

If $C$ is a boundary component of $R_1$ and of $R_2$, where both $R_1$ and $R_2$ are Seifert fibre spaces, then $R_i$ induces a vertical foliation $V_i$ on $C$, for $i = 1, 2$. $V_i$ is linear with respect to the affine structure on $C$. If $V_1 \neq V_2$, then, when considering the structure on $C$ from the point of view of $R_1$, we take the vertical foliation on $C$ to be $V_1$ and the slanting foliation to be $V_2$. From the point of view of $R_2$, we take the vertical foliation to be $V_2$ and the slanting foliation to be $V_1$. If $V_1 = V_2$, we choose any transverse linear foliation by circles each meeting each vertical circle exactly once, and call this the slanting foliation for each side.

If $C$ is a structure torus which is a component of the boundary of a geometrically finite volume structure component, then we choose an identification $C = S^1 \times S^1$, and use this identification to help define the automatic structure. If $C$ is also a boundary component of a Seifert fibre space $R$, then one of the factors is chosen to match the vertical foliation from $R$.

Each geometrically finite hyperbolic structure component leads to an integer $N$, as described in Section 11.4 (Geometrically Finite Groups of Hyperbolic Isometries). The edges of the graph carrying the fundamental groupoid were assigned lengths which were multiples of $2^{-N}$. We multiply the length of each edge in the graph by $2^N$, so that the length of each edge is a positive integer, and the length of an edge corresponding to a generator in the boundary torus has length one.

Each Seifert fibre space leads to a positive integer, which is the common denominator $q$ for the $c^1$ cochain of Theorem 12.2.2 (bounded cochain) as discussed in Section 12.3 (Fitting Pieces Together along the Boundaries). We choose $q$ to be twice the product of all the common denominators which occur for the Seifert structure components. Each boundary torus is then cut up into a mesh of slanting and vertical edges, each of length $1/q$ (but recall that "slanting" from one point of view becomes "vertical" in the other, and *vice versa*).

Now we take the universal cover $\tilde{M}$ of our three-manifold $M$. Each component of the inverse image of a structure torus is called a *structure plane*. The structure planes separate $\tilde{M}$ into components, each of which is the universal cover of some structure component, either a geometrically finite hyperbolic three-manifold from which the cusps have been removed, or a Seifert component. The components in the universal cover will also be called structure components. Each structure plane has two transverse linear foliations. Let $\{R_i\}_{i\in I}$ be the structure components which are universal covers of the Seifert structure components, and let $\{H_j\}_{j\in J}$ be the structure components which are universal covers of the geometrically finite three-dimensional hyperbolic structure components.

We will now describe how to choose a basepoint in each structure plane and a graph in each $R_i$ and each $H_j$, which is suitable for computing the automatic structure in that structure component, according to the method given in Section 12.3 (Fitting Pieces Together along the Boundaries) or Section 11.4 (Geometrically Finite Groups of Hyperbolic Isometries). We call the graph the *component structure graph*. In particular, the component structure graph carries the fundamental groupoid of the structure component, and contains the basepoints in the structure planes on the boundary of the structure component. Each of the edges in the component structure graph is cut up edges of length $1/q$ or $2/q$. (Generators of the form $x_i^+$ and $x_i^-$ have length $2/q$ as on page 301.)

We start with a random basepoint in one of the structure planes. This will be the main basepoint for the entire universal cover. The basepoints in the structure planes and the component structure graphs in the various structure components are chosen by induction, using as a basis for the induction the number of structure planes one has to cross to get to the main basepoint.

Once a basepoint $*_i$ has been chosen in a boundary component of $R_i$, we choose a standard slanting section in $R_i$ containing $*_i$, using the method described on page 293. This gives us a component structure graph for each $R_i$. Let $L$ be the boundary component of the standard slanting section which contains $*_i$. In the standard slanting section, we select, for each boundary plane $P$ of $R_i$, the least path, according to the ShortLex-ordering, lying in the standard slanting section, and joining $L$ to $P$. The other end of the path gives the basepoint in $P$. Notice that we have also chosen a basepoint in each boundary component of the slanting section. This gives a basepoint in each structure plane bounding $R_i$, and therefore provides part of the inductive process.

Once a basepoint $*_j$ has been chosen in one boundary component of $H_j$, we choose a component structure graph containing the basepoint and carrying the fundamental groupoid of $H_j$, as described in Section 11.4 (Geometrically Finite Groups of Hyperbolic Isometries). For each structure plane $P$ bounding $H_j$, we take the ShortLex-least accepted path (with respect to the automatic structure on $H_j$), in the component structure graph of $H_j$

from the component of the boundary containing $*_j$ to $P$, and we define the endpoint in $P$ to be the basepoint in $P$. This completes our description of the inductive procedure defining the basepoints in the structure planes and the component structure graphs in each structure component of $\tilde{M}$.

In each of the above two cases, such a ShortLex-path in a structure component, used to define a basepoint, will be called a *basepoint designator*.

**Definition 12.4.1 (structure graph).** The union of all the component structure graphs will be called the *structure graph* of $\tilde{M}$.

To define the automatic structure, we take geodesics in the structure graph of $\tilde{M}$ which start at the basepoint of $\tilde{M}$ and which follow the automatic structure in each structure component. In $H_j$, it is clear what this means: the automatic structure is in fact biautomatic (see Theorem 11.4.1 (geometrically finite implies biautomatic)), so it does not matter at which point in the boundary of $H_j$ the geodesic begins. In $R_i$, we need to be more careful. We have fixed a slanting section, which is computed by starting at the basepoint of $R_i$. "Following the automatic structure of $R_i$" means that our geodesic has to satisfy the conditions listed on page 301, *with respect to the one fixed slanting section*, no matter where the geodesic starts. The standard slanting section in $R_i$ is defined using the basepoint $*_i$, but our geodesic may enter $R_i$ at some other vertex. For this reason, it is not immediately clear that the condition that a path follows the automatic structure in each structure component is a condition which can be recognized by a finite state automaton—this is a condition we will have to prove. There is no problem, if our path enters a structure component at a basepoint of a structure plane, but if it enters at some other point, there is no immediate way of finding the values of the $c^1$ cochain of Theorem 12.2.2 (bounded cochain), which is necessary in order to be able to follow the automatic structure in a Seifert structure component.

Notice that any portion of the path which lies in one of the structure planes participates in the automatic structure of two different structure components. That is to say, the intersection of the geodesic with a *closed* structure component is required to follow the automatic structure of that structure component. There is normally an overlap section of the geodesic, as it moves from one structure component to another, when it satisfies conditions from two different sides at the same time.

**Lemma 12.4.2 (nearby basepoints).** *There is a positive integer $k_2$ with the following properties:*

(1) *Let $Q_i$ be the base of the simply connected Seifert fibre space $R_i$. Let $L'$ and $L''$ be two boundary components of $Q_i$. Let $P'$ be the structure*

*plane corresponding to $L'$. Suppose that either $P'$ separates $R_i$ from the main basepoint or that $P'$ contains the main basepoint. Let $\alpha$ be a graph geodesic in $Q_i$ joining $L'$ to $L''$ and suppose that the only point of $\alpha$ in $L'$ is the initial point, and the only point of $\alpha$ in $L''$ is the endpoint. Let $\beta$ be the basepoint designator in $Q_i$ (see page 305 for the definition). Then $\alpha$ and $\beta$ are a hausdorff distance at most $k_2$ apart. In addition the initial points of $\alpha$ and $\beta$ are a distance at most $k_2$ apart, and similarly for the endpoint of $\alpha$ and the endpoint of $\beta$. (Note that the endpoint of $\beta$ is the basepoint in $L''$. See Figure 12.12.)*

(2) *Let $H_j$ be the universal cover of a geometrically finite hyperbolic structure component, from which the cusps have been removed. Let $P'$ and $P''$ be two boundary components of $H_j$. Suppose that either $P'$ separates $H_j$ from the main basepoint or that $P'$ contains the main basepoint. Let $\alpha$ be a graph geodesic in $H_j$ joining $P'$ to $P''$ and suppose that the only point of $\alpha$ in $P'$ is the initial point, and the only point of $\alpha$ in $P''$ is the endpoint. Let $\beta$ be the basepoint designator in $H_j$ (see page 305 for the definition). Then $\alpha$ and $\beta$ are a hausdorff distance at most $k_2$ apart. In addition the initial points of $\alpha$ and $\beta$ are a distance at most $k_2$ apart, and similarly for the endpoint of $\alpha$ and the endpoint of $\beta$. (Note that the endpoint of $\beta$ is the basepoint in $P''$. )*

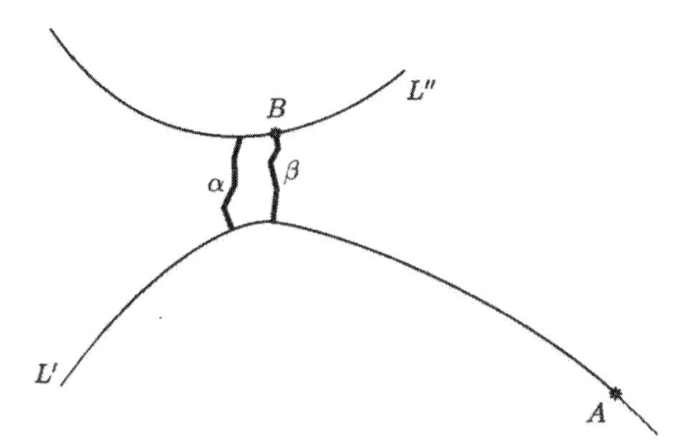


**Figure 12.12. Graph geodesics in $\tilde{Q}$.** This illustrates the statement of Lemma 12.4.2 (nearby basepoints) in the case of a Seifert component. Two components $L'$ and $L''$ of the boundary of $\tilde{Q}$ are shown. Each is a hyperbolic geodesic. $A$ is the basepoint in $L'$. We show $\beta$ as the basepoint designator. The point $B$ where $\beta$ meets $L''$ is defined to be the basepoint for $L''$.

*Proof of 12.4.2:* The graph geodesics $\alpha$ and $\beta$ give rise to quasigeodesics in $Q_i$. By Lemma 3.4.2 (quasigeodesics near geodesics) together with Proposition 11.2.4 (Cayley graph of group and groupoid), $\alpha$ is near a hyperbolic geodesic with the same endpoints, and similarly for $\beta$. By Lemma 12.3.3 (angular efficiency), the hyperbolic geodesics have angle to $L''$ which is bounded below. Because these geodesics have their other endpoints on $L'$, we can easily deduce that there is an upper bound to how far their endpoints on $L''$ are apart. Similarly, the distance apart on $L'$ is bounded. It then follows that there is a bound on the hausdorff distance between $\alpha$ and $\beta$.

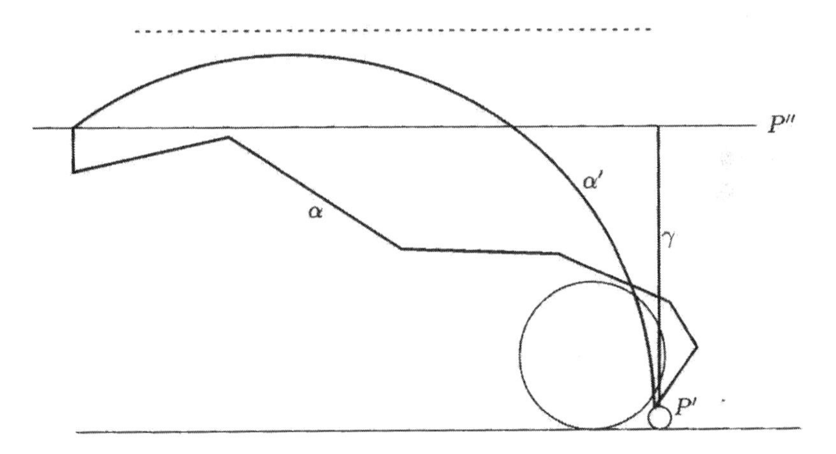


**Figure 12.13. Where $\alpha$ ends.** This picture illustrates the proof of Lemma 12.4.2 (nearby basepoints) in the case of a geometrically finite hyperbolic component $H_j$, using the upper half space model. $P'$ and $P''$ are horospheres. The horoball bounded by $P'$ is shaded and another horoball which is not in the universal cover of $H_j$ is also shaded. The geodesic $\gamma$ is the shortest hyperbolic geodesic joining $P'$ to $P''$ and $\alpha'$ is also a hyperbolic geodesic. The top dotted line shows the theoretical maximal limit of penetration of a geodesic like $\alpha'$ into the horoball bounded by $P''$—the limit results because $\alpha'$ is near the graph geodesic $\alpha$, which is disjoint from the horoball bounded by $P''$.

The proof for $H_j$ is similar. Instead of Lemma 3.4.2 (quasigeodesics near geodesics) we apply Theorem 11.3.1 (quasigeodesics outside horoballs), which enables us to approximate graph geodesics by hyperbolic geodesics. Let $\alpha'$ be the hyperbolic geodesic with the same endpoints as $\alpha$ and let $\gamma$ be the shortest hyperbolic geodesic joining $P'$ to $P''$. Because the distance between $\alpha$ and $\alpha'$ is bounded, $\alpha'$ must meet the horosphere $P''$ inside a certain fixed disk centred at the endpoint of $\gamma$ on $P''$ (see Figure 12.13). Since $\beta$ is one possible choice for $\alpha$, the result follows. 12.4.2

We refer to a generator of the form $x_a^+$ or $x_a^-$, discussed in Theorem 12.3.4 (improved structure) as a *composite generator*. There are also composite generators in the geometrically finite case. In that case, the universal cover of a boundary torus has a linear mesh, tesselating the plane with parallelograms. The sides of the parallelograms lie either in the slanting foliation or in the vertical foliation. Not every vertex in the plane covers the basepoint in the torus. The image of the mesh in the torus is a finite decomposition of the torus into parallelograms, with a finite number of vertices. Each edge of a parallelogram is assigned length $1/q$. A composite generator is obtained by concatenating two sides of some parallelogram with a vertex in common. A composite generator is assigned length $2/q$.

**Lemma 12.4.3 (geodesic following automatic structure).** *Given any vertex $v$ in the structure graph of $\tilde{M}$, there is a graph geodesic from the basepoint to $v$, which follows the automatic structure in each structure component, namely the structure of Theorem 12.3.4 (improved structure) for Seifert components, and the structure of Theorem 11.4.3 (geometrically finite implies biautomatic: details). We can also require that the predecessor in the graph geodesic of a composite generator is either a composite generator or a generator in the interior of a geometrically finite hyperbolic structure component. (This is an additional restriction only when we cross a structure plane, with a single point of intersection with the structure plane.) Where two Seifert components meet, if the vertical foliations agree (in which case we have arranged for the slanting foliations also to agree), the signs of composite generators on each side of the structure plane match up, as in the rules on page 301.*

*Proof of 12.4.3:* Let $w$ be a geodesic string from the basepoint to $v$. We can write

$$w = u_0 v_1 u_1 \dots v_r u_r,$$

where each $u_i$ lies in some structure plane, and each $v_i$ meets such a plane only at its endpoints. It is possible for $u_i$ to be the nullstring. We attach to $w$ the $(2r+1)$-tuple $(\lambda_1, \dots, \lambda_{2r+1})$, where $\lambda_{2i+1}$ is the length of $u_i$ and $\lambda_{2i}$ is the length of $v_i$. Amongst all geodesics from the basepoint to $v$, we choose one which is largest in the lexicographic ordering.

By induction on $i$, we may assume that $u_{j-1}v_j u_j$ follows the automatic structure for $j < i$. Let $u = u_{i-1}v_i u_i$. As in Theorem 12.3.4 (improved structure) or Theorem 11.4.1 (geometrically finite implies biautomatic), we may replace $u$ by a geodesic which follows the automatic structure. If $u$ lies in a structure component $H_j$, this replacement will not change $u_{i-1}$ and

$v_i$, though $u_i$ may change to another path in the same structure plane of the same length. The reason that $u_i$ may have to change is because the automatic structure in the structure plane containing $u_i$ has to be followed. This constraint is already satisfied by induction for $u_{i-1}$, and the only other constraint is that $u_{i-1}v_iu_i$ be a geodesic, which is also satisfied. Such changes do not affect $(\lambda_1, \ldots, \lambda_{2r+1})$.

If $u$ lies in a structure component $R_i$, then consider the possibility that $u_{i-1}$ changes. This can only happen if the central element $z$ or $z^{-1}$ is moved backwards to change a letter $x_a$ to $x_a^+$ or $x_a^-$, as described in the proof of Theorem 12.3.4 (improved structure). But this increases $\lambda_{2i-1}$, without changing $(\lambda_1, \ldots, \lambda_{2i-2})$, which is impossible, since $(\lambda_1, \ldots, \lambda_{2r+1})$ has been chosen as large as possible in the lexicographical ordering. It is also possible that $v_i$ and $u_i$ might change, but their lengths cannot change.

Proceeding in this way, we eventually arrange for each $u_{i-1}v_iu_i$ to follow the automatic structure of the structure component in which it lies. 12.4.3

**Proposition 12.4.4 (3m automaton).** *Let*

$$w = u_0v_1u_1 \ldots v_ru_r,$$

*be a path in $\tilde{M}$, starting at the main basepoint, such that each $u_{i-1}v_iu_i$ follows the automatic structure of the structure component in which it lies. We also suppose that the predecessor of any composite generator is either a composite generator of the same sign, where this is an appropriate restriction, or a generator in the interior of a geometrically finite hyperbolic component. These conditions can be checked by passing $w$ through a finite state automaton.*

Note that we do not assume that $w$ is a geodesic globally, though it is a graph geodesic within each structure component.

*Proof of 12.4.4:* Let $k_2$ be the constant found in Lemma 12.4.2 (nearby basepoints).

There are a finite number of structure components in our three-manifold $M$, each of which is compact. In particular, there are a finite number of Seifert structure components. Let $\pi : R \to Q$ be a typical projection for one of these Seifert fibrations. Since $Q$ is compact, there are a finite number of choices of basepoint for $Q$, if the basepoint is to be a vertex in a geodesic boundary component of $Q$. Let us fix a basepoint in a boundary component $B$ of $Q$, and let us orient $B$. We label each directed edge of $B$ with the same letter $x$, and the reverse direction is labelled $x^{-1}$.

In Section 12.2 (The Basic Seifert Fibre Space), we explained how one can associate, to each hyperbolic orbifold like $Q$ together with a basepoint, a finite state automaton which accepts geodesic paths in the universal cover

$\tilde{Q}$ starting at the basepoint. This can be enhanced so as to accept geodesic paths starting from any point. The finite state automaton can be assumed to carry enough information, so that we know, up to covering translations, from which vertex of $\tilde{Q}$ the geodesic starts. (We need to adjoin to each label for an edge of $Q$ information as to what its endpoints are.)

In order to complete the proof of Proposition 12.4.4 (3m automaton), we prove the following lemma.

**Lemma 12.4.5 (concatenating geodesics).** *Given $Q$ with the cell structure defined in Section 12.3 (Fitting Pieces Together along the Boundaries), there exist $k_1 \geq 1$ and $\varepsilon_1 \geq 0$, with the following property. Let $u = x^n$ $(n \geq 0)$ be a geodesic edge path along the boundary component $L$ of $\tilde{Q}$, where $x$ is the label of every edge in a certain boundary component of $Q$. Let $w$ be a geodesic edge path in the 1-skeleton of $\tilde{Q}$ which starts where $u$ ends. Suppose further that $w$ does not start with $x^{-1}$. Then the concatenation $uw$, parametrized by graph length, is a $(k_1, \varepsilon_1)$-quasigeodesic for the hyperbolic structure on $Q$.*

*Proof of 12.4.5:* There is no loss of generality in assuming that $w$ does not start with an $x$. Whatever value of $\varepsilon_1$ we choose, we may confine our attention to the case where $u$ and $w$ each have length at least $\varepsilon_1$. So, we may assume that $u$ and $w$ are fairly long. Let $\beta$ be a hyperbolic geodesic segment joining the starting point $w_0$ of the path $w$ to the endpoint $w_1$ of $w$. By Lemma 12.3.1 (boundary efficiency), a graph geodesic joining two points of $L$ must lie in $L$. Since $w$ does not start with an $x$ or an $x^{-1}$, it follows that $w$ is disjoint from $L$, except at $w_0$. We can therefore apply Lemma 12.3.3 (angular efficiency) to find that the angle between $\beta$ and $L$ is bounded below by some constant $\theta_0$.

Lemma 11.3.4 (angles bounded below) now implies that $u$ concatenated with $\beta$ is a quasigeodesic. A little arithmetic now implies the stated result.

12.4.5

We know from Lemma 3.4.2 (quasigeodesics near geodesics) that there is a number $L$, such that any $(k_1, \varepsilon_1)$-quasigeodesic in the hyperbolic plane lies within an $L$-neighbourhood of the geodesic connecting its endpoints. We do the construction for each of the Seifert structure components, and choose $L$ large enough so that any graph geodesic in the one-skeleton of $\tilde{Q}$ lies within a distance $L$ of the corresponding hyperbolic geodesic. We also choose $L$ larger than the diameter of the sets called $S_i$ in Section 12.2 (The Basic Seifert Fibre Space).

We have already discussed the automaton for an orbifold $Q$, giving the values of the cochain $c^1$ of Theorem 12.2.2 (bounded cochain). We augment the automaton so that it contains a picture of an $L$-neighbourhood of whatever vertex one happens to be at, together with information saying what the geodesics from the basepoint through the neighbourhood look like. We do this for each of the basic Seifert components.

By restricting the input alphabet of this automaton to be $x$, where $x$ has the same meaning as in Lemma 12.4.5, we convert our automaton to a one-variable automaton. We also get a one-variable automaton by restricting the input to be $x^{-1}$. These are finite state automata over alphabets in one variable, accepting infinite languages. Therefore they have the form indicated in Figure 1.11 (SimplyStarred) on page 20, that is to say, there is a finite initial segment, followed by a cycle.

We will construct an automaton which keeps track of a certain integer $I$, which either increases monotonically or decreases monotonically as additional input is accepted. The automaton cannot remember $I$ itself—finite state automata have finite memories. Instead it remembers $I$ when $|I| \leq c_0$ for a certain number $c_0$, and, when $|I|$ is large, it remembers $I$ modulo some fixed integer $J$. The fixed integer $J$ is the least common multiple of all the cyclic lengths of all the possible one-variable finite state automata arising from the finite set of compact Seifert structure components and the finite set of choices of basepoint in the quotient orbifolds and the choice of the direction in which we circle a boundary component. We choose $c_0$ equal to the maximum of the lengths of the initial segments of each of these one-variable automata. We call the combined automaton $F$. $F$ also remembers whether $I$ is increasing or decreasing.

When $w$ passes through a geometrically finite hyperbolic structure component, we keep track of all accepted geodesics within a fixed uniform distance of $w$ and starting in the same boundary component as $w$ using the method explained in the proof of Theorem 11.2.2 (geodesic automaton 3). This fixed distance is equal to the product of the constant $k_2$ from Lemma 12.4.2 (nearby basepoints) and the lipschitz constant of the biautomatic structure. When $w$ passes through a Seifert structure component $R_i$, we keep track of all geodesics in $Q_i$ within a fixed uniform distance of the projection of $w \cap R_i$ into $Q_i$ and starting in the same boundary component of $Q_i$ as $w$. In each case we take particular note of which of the geodesics are ShortLex-geodesics and the ordering of these in the ShortLex-ordering.

Now let us discuss the operation of the automaton accepting $w$, as in the statement of Proposition 12.4.4 (3m automaton). The only point that really needs elaboration is the behaviour as we pass through a structure plane $P$. There is no problem if the component on the far side of $P$ is on a geometrically finite hyperbolic component, since then the structure is biautomatic, and it is a matter of indifference where we start. So we may restrict to the case where the component on the far side of $P$ is a Seifert component $R_2 \to Q_2$. We recall that $Q_2$ is embedded in $R_2$ as the standard slanting section.

Suppose we are travelling in a Seifert component $R_1$ and we come to $P$. We assume inductively that we know the state of $F$, which measures the height above the standard slanting section in $R_1$. We also know whether the fibre coordinate is positive or negative and whether it is increasing (the current letter in $w$ is an $x_i^+$ or $z$), decreasing (the current letter is an $x_i^-$ or $z^{-1}$) or constant (the current letter is an $x_i$).

Let $\pi : R_1 \to Q_1$ be the Seifert projection. Suppose we reach $P$ for the first time at a vertex $p$. Let $\pi p = q \in Q_1$. By Lemma 12.4.2 (nearby basepoints), the graph distance of $q$ from the basepoint in $Q_1$ is less than $k_2$. We are keeping track of nearby geodesics, and taking note of which of them is a ShortLex-geodesic in $Q_1$. This means we can determine which of these paths is the basepoint designator. It follows that we can determine the basepoint.

If the vertical foliations of $R_1$ and $R_2$ are equal on $P$, then, by construction, so are their slanting foliations. Therefore the standard slanting sections in $R_1$ and $R_2$ will have a boundary component in common, which lies in $P$. In this case, as we move from $R_1$ to $R_2$, the state of $F$ is handed over intact. (Note that the orientation in the fibres is consistent between $R_1$ and $R_2$, since we have ensured that our three-manifold contains no embedded Klein bottles.)

If the vertical foliations of $R_1$ and $R_2$ do not agree, then vertical and slanting foliations interchange as we move from $R_1$ to $R_2$. Let $s_2 \in \mathbf{R}$ and $v_2 \in \mathbf{R}$ be slanting and vertical coordinates for $P$ with respect to $R_2$, with the basepoint in $P$ having coordinates $(0,0)$. The value of $s_2$ cannot be determined using finite state information. However, sufficient information is contained in the state of $F$ so that we can determine the state of the machine looking at geodesics in the standard slanting section $Q_2$ in $R_2$. (As we have already pointed out, $F$ is able to remember this information because of the eventually periodic behaviour of one-variable finite state automata.) In particular, we can know, in an $L$-neighbourhood of where we are, what the geodesics from the basepoint in $Q_2$ are, the values of the bounded cochain, and other similar information. We have shown that $|v_2| \leq k_2$. Since we are keeping track of all the accepted paths in a disk of radius $k_2$ around the point we are visiting, we can determine $v_2$ by the information carried in our finite state machine. This completes the discussion when we have a Seifert component $R_1$ on one side of $P$ and the structure component on the other side of $P$ is a Seifert component.

Next suppose that we are passing through a geometrically finite hyperbolic component, and the component on the other side of $P$ is the Seifert component $R_2$. By Lemma 12.4.2 (nearby basepoints), we can determine where the basepoint in $P$ is by following the progress of $w$ in a finite state machine, and this basepoint is within a distance $k_2$ of the first point of contact of $w$ with $P$. So this enables us to reset the finite state automaton $F$ and also to determine which state we are in for the finite state automaton which determines geodesics and the bounded cochain in $Q_2$.

This is sufficient to give the proper control over $w$ as it passes through $R_2$. $\boxed{12.4.4}$

**Theorem 12.4.6.** *Suppose we have a finite disjoint set of compact three-dimensional manifolds possibly with boundary, each of which is a Seifert fibre space or has an interior with a geometrically finite complete hyperbolic metric. Let $M$ be a connected compact three-manifold formed from these three-manifolds by identifying certain pairs of torus boundary components (so one of the original pieces may have two torus boundary components which are glued to each other). Then $M$ has an automatic fundamental group.*

*Proof of 12.4.6:* The language we will use consists of all geodesics which also lie in the language of Proposition 12.4.4 (3m automaton). By Theorem 11.2.2 (geodesic automaton 3), we need only show that two such geodesics $\alpha$ and $\beta$, ending within a distance one of each other, are a bounded hausdorff distance apart.

We start by amalgamating Seifert components where vertical foliations from the two sides of a structure plane coincide. In a Seifert component $R$ we have the vertical coordinate $v$, which is the signed distance from the standard slanting section. The two paths $\alpha$ and $\beta$ each have vertical coordinates, and so we get functions $v(\alpha)(t)$ and $v(\beta)(t)$ where $t$ is the number of generators inside $R$. Each of these functions has a graph which is piecewise linear, with at most one corner, and the possible slopes of the graph are $+1$, $-1$, $0$ and $\infty$. These correspond to generators of the form $x_j^+$, $x_j^-$, $x_j$ and $z$ respectively. Moreover, slope 1 or $-1$ must be the slope that occurs first (see page 301 to remind yourself of why this should be). It follows that the difference $v(\alpha)(t) - v(\beta)(t)$ is monotonic increasing or monotonic decreasing. If the paths do not end in $R$, one can see from Lemma 12.4.2 (nearby basepoints) that the final value inside $R$ of this function is bounded in absolute value by some constant $k_2$. By induction this is true for the initial value inside $R$ as well and so

$$|v(\alpha)(t) - v(\beta)(t)| \leq k_2$$

throughout. (The previous inequality needs some interpretation because the two paths may not have the same length in $R$. In that case and *only* for the purpose of giving a meaning to this inequality, we extend the domain of the shorter path by making it constant for a while at its final point in $R$.)

We denote by $s$ the slanting coordinate in the initial structure plane $P_1$ and the final plane $P_2$. We have just seen that $|s(\alpha)(t) - s(\beta)(t)|$ is bounded by some constant on $P_1$ and $P_2$. Therefore the projections of $\alpha$ and $\beta$ into the quotient two-orbifold $Q$ are near each other throughout $R$.

This proves the required result, except over the final segments of $\alpha$ and $\beta$. Since they end a distance one apart, they end in the same closed structure component. Moreover, they start in the last component a bounded distance apart. In the case of a geometrically finite component, we can use the fact that the automatic structure is biautomatic (Theorem 11.4.1 (geometrically finite implies biautomatic)). In the case of a Seifert component, we can use the fact that the automatic structure on a hyperbolic surface given by all geodesics is biautomatic—it is straightforward to fill in the details of the argument which take into account the vertical direction. 12.4.6

Collecting together the information about three-manifolds, we have the following result.

**Theorem 12.4.7 (3-manifolds and automatic structures).** *Suppose we have a finite disjoint set of compact three-manifolds, $M_1, \ldots, M_k$, such that the interior of each is modelled on one of the eight three-dimensional geometries (see [Sco83]). Let $M$ be a connected compact three-manifold, formed from the $M_i$ by the operations of connected sum, disk sum and identifying boundary tori in pairs. Then $M$ has an automatic structure if and only if none of the $M_i$ is closed and modelled on nilgeometry or solvgeometry.*

# Bibliography


[ABC$^+$00] J. M. Alonso, T. Brady, D. Cooper, V. Ferlini, M. Lustig, M. Mihalik, M. Shapiro, and H. Short. Notes on word hyperbolic groups. In *Group theory from a geometric viewpoint.* World Scientific, Singapore, to appear. Proceedings of the ICTP conference in summer 1990. Cited on page 80.

[Alo89] J. M. Alonso. Combings of groups. To appear in MSRI Proceedings of the workshop on algorithms, word problems and classification in combinatorial group theory, 1989. Cited on page 220.

[Ani86] D. J. Anick. On the homology of associative algebras. *TAMS*, 296:641–659, 1986. Cited on page 220.

[Art47] E. Artin. Theory of braids. *Annals of Mathematics*, 47:101–126, 1947. Cited on pages 181, 204.

[BGSS] G. Baumslag, S. M. Gersten, M. Shapiro, and H. Short. Automatic groups and amalgams. Unpublished. Cited on pages 135, 280.

[Bie12] Bieberbach. Über die bewegungsgruppen der euklidischen räume II. *Mathematische Annalen*, 72:400–412, 1912. Cited on page 88.

[Bir75] J. S. Birman. *Braids, Links and Mapping Class Groups.* Princeton Univ. Press, 1975. Cited on pages 181, 204.

[BL82] B. Buchberger and R. Loos. Algebraic simplification. In B. Buchberger, G. E. Collins, and R. Loos, editors, *Computer Algebra, Symbolic and Algebraic Computation, Second Edition*, pages 11–43. Springer-Verlag, New York, 1982. Cited on page 113.

[Bow] B. H. Bowditch. Geometric finiteness for hyperbolic groups. preprint, originally Warwick Ph.D. thesis 1988. Cited on page 266.

[Bow00] B. H. Bowditch. Notes on Gromov's hyperbolicity criterion. In *Group theory from a geometric viewpoint.* World Scientific, Singapore, to appear. Proceedings of the ICTP conference in summer 1990; IHES preprint, June 1990. Cited on pages 79–80.

[Bro89] K. S. Brown. The geometry of rewriting systems: a proof of the Anick-Groves-Squier theorem. Preprint, 1989. Cited on page 220.

[BS79] R. Bowen and C. M. Series. Markov maps associated with Fuchsian groups. *Publ. IHES*, 50:153–170, 1979. Cited on page v.

[Can84] J. Cannon. The combinatorial structure of cocompact discrete hyperbolic groups. *Geom. Dedicata*, 16:123–148, 1984. Cited on pages v, 28, 63, 70.

[CDP90] M. Coornaert, T. Delzant, and A. Papadopoulos. *Notes sur les groupes hyperboliques de Gromov.* Springer, Berlin-Heidelberg-New York, 1990. Lecture Notes 1441. Cited on page 79.

[CEG87] R. D. Canary, D. B. A. Epstein, and P. Green. Notes on notes of Thurston. In *Analytical and geometric aspects of hyperbolic space, LMS Lecture Notes Series 111*, pages 3–92. Cambridge University Press, Cambridge, 1987. Cited on pages 65, 75.

[Cha86] Leonard Charlap. *Bieberbach groups and flat manifolds.* Springer-Verlag, New York, 1986. Cited on page 88.

[Deh12] Max Dehn. Transformationen der Kurven auf zweiseitigen Flächen. *Math. Annalen*, 72:413–421, 1912. Cited on page 63.

[Deh87] Max Dehn. *Papers on group theory and topology.* Springer, 1987. Translated and introduced by J. Stillwell. Cited on pages iii, 63.

[EHR91] D. B. A. Epstein, D. F. Holt, and S. E. Rees. The use of Knuth-Bendix methods to solve the word problem in automatic groups. *Journal of Symbolic Computation*, 1991. Cited on pages 113, 133.

[ET88] D. B. A. Epstein and W. P. Thurston. Combable groups. *Rend. dell'Univ. Cagliari*, 58 (Supplement):423–429, 1988. Proceedings of conference in Cala Gonone, Sept. 1988. Cited on page 211.

[Fed69] H. Federer. *Geometric Measure Theory.* Springer-Verlag, Berlin-Heidelberg-New York, 1969. Cited on page 222.

[FF60] H. Federer and W. Fleming. Normal and integral currents. *Annals of Mathematics*, 72:458–520, 1960. Cited on pages 211, 223.

[FM91] M. Feighn and G. Mess. Conjugacy classes of finite subgroups of kleinian groups. *Amer. J. of Math.*, 113:179–188, 1991. Cited on page 267.

[Gar69] F. A. Garside. The braid group and other groups. *Oxford Quart. J. of Math*, 20:235–254, 1969. Cited on pages 181, 186, 204.

[GdlH89] E. Ghys and P. de la Harpe, editors. *Sur les groupes hyperboliques d'après Mikhael Gromov.* Available from editors, 1989. Cited on page 79, 80.

[Ger91] S. M. Gersten. Dehn functions and $\ell_1$-norms of finite presentations. In C. F. Miller III and G. Baumslag, editors, *Proceedings of the workshop on algorithmic problems*. Springer-Verlag, 1991. MSRI series. Cited on pages 154, 162.

[Gil79] R. H. Gilman. Presentations of groups and monoids. *Journal of Algebra*, 57:544–554, 1979. Cited on page vi.

[Gil84a] R. H. Gilman. Computations with rational subsets of confluent groups. In *Eurosam '84 Proceedings*, pages 207–212. Springer, 1984. Lecture Notes in Computer Science 174. Cited on page vi.

[Gil84b] R. H. Gilman. Enumerating infinitely many cosets. In M. D. Atkinson, editor, *Computational Group Theory (Durham, 1982)*, pages 51–55. Academic Press, 1984. Cited on pages vi, 114.

[Gil87] R. H. Gilman. Groups with a rational cross-section. In S. M. Gersten and J. R. Stallings, editors, *Combinatorial Group Theory and Topology*. Princeton, 1987. Cited on pages vi, 114.

[Gri80] R. I. Grigorchuk. On the Burnside problem for periodic groups. *Functional Anal. Appl.*, 14:41–43, 1980. Cited on page 61.

[Gro87] M. Gromov. Hyperbolic groups. In S. M. Gersten, editor, *Essays in group theory, MSRI Publ. 8*, pages 75–263. Springer, 1987. Cited on pages 79, 113, 174.

[GS00a] S. M. Gersten and H. Short. Rational subgroups of biautomatic groups. *Annals of Mathematics*, to appear. Cited on pages 161, 173, 177.

[GS00b] S. M. Gersten and H. Short. Small cancellation theory and automatic groups, ii. *Inventiones Mathematicae*, to appear. Cited on page 59.

[Hem76] J. Hempel. *3-manifolds*. Princeton, 1976. Annals of mathematics studies, 86. Cited on page 274.

[Hem79] G. Hemion. On the classification of homeomorphisms of 2-manifolds and the classification of 3-manifolds. *Acta Math.*, 142:123–155, 1979. Cited on page 181.

[HR00] D. F. Holt and Sarah Rees. Testing for isomorphism between finitely presented groups. In *Proceedings of Conference on Groups and Geometries, Durham, July 1990*. Cambridge University Press, Cambridge, to appear. Cited on page 27.

[HU69] J. E. Hopcroft and J. D. Ullman. *Formal languages and their relation to automata*. Addison-Wesley, 1969. Cited on page 32.

[Joh79] K. Johannson. *Homotopy equivalence of 3-manifolds with boundaries.* Springer-Verlag, Berlin, New York, 1979. Lecture notes in mathematics, 761. Cited on page 274.

[JS79] W. Jaco and P. Shalen. *Seifert fibre spaces in 3-manifolds.* AMS, 1979. Memoirs of the AMS. Cited on page 274.

[KB70] D. E. Knuth and P. B. Bendix. Simple word problems in universal algebra. In J. Leech, editor, *Computational problems in abstract algebras*, pages 263–297. Pergamon Press, 1970. Cited on page 116.

[Koe27] P. Koebe. Allgemeine Theorie der Riemannischen Mannigfaltigkeiten. *Acta Math.*, 50:157, 1927. Cited on page vii.

[Koe29] P. Koebe. Riemannische Mannigfaltigkeiten und nichteuklidische Raumformen. iv. *Sitzungberichte der Preussichen Akad. der Wiss.*, pages 414–457, 1929. Cited on page vii.

[KP91] M. Kapovich and L. Potyagailo. On the absence of Ahlfors' Finiteness Theorem for Kleinian groups in dimension-3. *Topology and its applications*, 40:83–91, 1991. Cited on page 267.

[LW69] A. T. Lundell and S. Weingram. *The topology of CW complexes.* Van Nostrand, New York, 1969. Cited on page 211.

[Mac68] I. D. Macdonald. *The theory of groups.* Oxford University Press, Oxford, 1968. reprinted 1988 by Krieger. Cited on page 167.

[Mar70] G. Margulis. The isometry of closed manifolds of constant negative curvature with the same fundamental group. *Dokl. Akad. Nauk SSSR*, 192:736–737, 1970. Cited on page 71.

[Mil65] J. W. Milnor. *Topology from the differentiable viewpoint.* University of Virginia, Charlottesville, 1965. Cited on page 228.

[Mor87] Marston Morse. *Collected papers.* World Scientific, Singapore, 1987. Edited by Raoul Bott. Cited on page vii.

[Mos] L. Mosher. Conjugacy invariants of mapping class groups of surfaces. in preparation. Cited on page 181.

[Mos68] George D. Mostow. Quasi-conformal mappings in $n$-space and the rigidity of hyperbolic space forms. *Publications Mathématiques Institut des Hautes Études Scientifiques*, 34:53–104, 1968. Cited on page 71.

[Mos87] L. Mosher. Classification of pseudo-Anosovs. In *Low-dimensional Topology and Kleinian Groups, LMS Lecture Notes Series 112*, pages 13–75. Cambridge University Press, Cambridge, 1987. Cited on page 181.

[Mun66] J. R. Munkres. *Elementary differential topology.* Princeton University Press, 1966. Annals of mathematics studies, no. 54a. Cited on page 211.

[Mun84] J. R. Munkres. *Elements of Algebraic Topology.* Benjamin-Cummings, Menlo Park, California, 1984. Cited on page 223.

[PR00] M. Paterson and A. Razborov. The set of minimal braids is co-np-complete. *J. of Algorithms*, to appear. Cited on page 209.

[Sco73] G. P. Scott. Compact submanifolds of 3-manifolds. *J. L. M. S.*, 7:246–250, 1973. Cited on page 267.

[Sco83] G. P. Scott. The geometries of 3-manifolds. *Bulletin of the London Mathematical Society*, 15:401–487, 1983. Cited on pages 273–274, 314.

[Sho90] H. B. Short. Groups and combings. ENS Lyon preprint, 1990. Cited on page 80.

[Sim84] L. Simon. *Lectures on Geometric Measure Theory.* Centre for Mathematical Analysis, Australian National University, 1984. Cited on pages 221, 223.

[Squ87] C. G. Squier. Word problems and a homological finiteness condition for monoids. *J. Pure and Applied Algebra*, 49:201–217, 1987. Cited on page 220.

[Thu82] W. P. Thurston. Three dimensional manifolds, Kleinian groups and hyperbolic geometry. *BAMS*, 6:357–381, 1982. Cited on pages 267, 273.

[Thu88] W. P. Thurston. On the geometry and dynamics of diffeomorphisms of surfaces. *BAMS*, 19:417–431, 1988. Cited on page 273.

[Wol68] Joseph A. Wolf. Growth of finitely generated solvable groups and curvature of Riemannian manifolds. *Journal of Differential Geometry*, 2:421–426, 1968. Cited on page 169.

# Index


$w(t)$: 4
$K^*$, $*$, $\vee$: 6
$\neg L, L \vee L', L \wedge L'$: 22
\$, $\$^i$, $L^\$$: 24
$f_*(L_B)$: 25
$\overline{w}$: 28
$x^{-1}$: 29
$\langle A/R \rangle$: 31
$H \backslash G$: 33
$(Hg_1, x, Hg_2)$: 34
$|g|$: 37
$[w]$: 102
$\triangle$: 184
$\neg$: 185

abelian group: 87 ff., **168**; *see also* euclidean group
– – is ShortLex-automatic: 58, 61, 80, **96**
– – has symmetric automation: 58, **89**
accept state: 7, 10
accepted string: 5, 8, 12, 137
– – is quasigeodesic: **73**, 83, 267
Adobe *Illustrator*: vii
Alonso, J. M.: [ABC+00, Alo89]
alphabet: **4**, 7, 10
alternating group: 35
amalgamation: *see* free product
angular efficiency: **299**, 300, 307, 310
Anick, D. J.: [Ani86]
area, combinatorial: 40, 44
area$(w)$: 44
arrow: **9**, **34**
Artin, E.: [Art47]
associated $k$-history automaton: 141
asynchronous automation: 28, 45, 135, **139**, 142
– –, characterization, 145
– –, independence on generators: 151
– – that is not synchronous: 154
– – with uniqueness: **150**, 153, 169
– automatic group, *see* asynchronous automation
– automaton: **136**, 138, 143, 249
– combing, combable group: **153**
– correspondence: 146
– factor: 141
– isoperimetric inequality: 135, **152**, 154
– lipschitz constant: 146
*automata* program: 98
automatic group: *see* automation
automatic groupoid: 246
– structure: *see* automation
automation, 27: **45**–48, 49
–, asynchronous: *see* asynchronous automation
–, characterization: 48
– for euclidean group: 90
– for factor group: 280
– for finite group: **49**, 56, 73
– for finite-index subgroup: 87–**89**, 95, 153, 166, 211, 273, 282
– on Seifert fibre space: 283
– on three-manifolds: 302
–, symmetric: 58, 68, 89, 198–199

automation *(cont'd.)*
– that is not biautomation: 59, 89, 161
– with invariance: **91**, 94–95, 177, 267, 271, 301
– with uniqueness: 28–29, **57**–58, 61, 84, 88, 108, 149
automaton: **7**
–, asynchronous: **136**
–, auxiliary: 142
–, deterministic: 9
–, non-deterministic: 10, 16
–, normalized: **9**, 12, 16, 46
–, partial: 11
axiom checking: 102–116, 135, 161

$B$, $B_n$: 182
basepoint, in Cayley graph: 34
– and automatic groupoids: **247**, 253
– designator: 305
Baumslag, G.: [BGSS]
Baumslag–Solitar group: 154
Bendix, P. B.: [KB70]; *see also* Knuth–Bendix
biautomatic group: *see* biautomation
– groupoid: **248**–249
biautomation: **58**
– and conjugacy problem: **59**, 135
–, characterization: 58
– for braid group: 199
–, for euclidean group, 95
– with uniqueness: 59
bicombable, bicombing: 84
binary numbers: 11
Bieberbach: [Bie12]
Birman, J.: [Bir75]
boolean-valued function: 22
boundary: **289**, 298–300
bounded CW-complex: **213**
– cochain: **288**, 293–295, 301, 303, 305
– length difference for accepted words: **49**, 51, 83–84, 88, 175
– number of relators in word decomposition: 43
boundedly asynchronous: **141**, 150–153
bouquet: 40
Bowditch, B. H.: 243, 267, [Bow, Bow00]
Bowen, R.: [BS79]
braid: 182
– group: 183
– – is automatic: 198–199
– – is a lattice: 200
–, closed: 201
broken path: 84
Brown, K. S.: [Bro89]
Buchberger, B.: [BL82]
Burnside problem: 61

$C(\alpha)$: 66
$C(g)$: 66
$C(\tilde{R})$: 295
Canary, R. D.: [CEG87]
Cannon, J.: [Can84]
canonical form: 116; *see also* normal form
– – in braid group: 190–198
– – in free products: 278–279
– reduction, residue: 121
Cayley graph of a group: **34**–37
– –, filled: 154
– –, partial: 109
– – of a groupoid: **246**
cell complex: 211
cellular chain: 214
– map: 212, 217
cellular map: 212
centralizer: 176–177
chain: 214
change of generators is pseudoisometry: **72**, 85, 176
– – preserves automation: **52**, 59, 81, 90, 247
– – preserves asynchronous automation: 151
characterization: *see* automation, asynchronous automation, biautomation, regular subgroup

Charlap, L.: [Cha86]
Chazelle, B.: 205
closed braid: 201
closure of hyperbolic space: 266
– under inversion: 29
cocycle: 294
combable: **84**–86, 220-221
–, asynchronous: **153**
– group and contraction: 214
– isoperimetric inequality: **86**, **154**
combinatorial area: 40, 44
combing: *see* combable
common prefix: 149–150
commuting generators: 183
compiler: 3
complement: 22, 96, 137
complete, $k$-complete set of rules: **117**–122, 124, 126, 130
complexity: 27, 205–208; *see also* quadratic, exponential
component structure graph: 304
composite generator: 301, 308
composition of pseudomaps: 73
computer languages: 3, 5, 22
concatenation: **4**, **5**, 7
cone point: 290
– type: 66
confluence: 117
conjugacy: 170, 172–173, 237
– problem: 60, 201
contraction: 219
convexity: 76, 92, 266
Coornaert, M.: [CDP90]
correspondence, asynchronous: 146
coset: 33
cutting up a three-manifold: 274 ff.
– – the base space: 285
– – the dual Dehn diagram: 42–43
CW-complex: *see* cell complex
CW-diameter: 214
CW-lipschitz: 212–213
CW-mass: 214
cycle: 215

$d_P$: 84
$D$, $D_n$: 187
D lemma: 260
de la Harpe, P.: [GdlH89]
dead state: 9
decimals: 9–10
deformation theorem: **223**, 230
degree: 227–229
Dehn, M.: [Deh12, Deh87]
Dehn diagram: 40–42
Delzant, T.: [CDP90]
departure function: 145
deterministic: 9
diameter: 221
dictionary order: 56
difference function: 250
dipping: 259, 263
direct product: 87–88, 153
directed system: 190
Dirichlet units theorem: 237
disk: 40, 211
dividing path: 148
dodecahedron: 34–35
double cosets: 276–280
dual Dehn diagram: 40
duration function: 213

effect of map on language: 17, 25, 55
Eilenberg–Maclane space: 220
endpoint map: 84
$\varepsilon$: 4, 6
Epstein, D. B. A.: [EHR91, ET88]
equality recognizer: 45, 139
estimable vertex: 270
euclidean group: 88, 97, 117
– – is automatic: 90
– – is biautomatic: 87, **95**, 97, 275
– metric: 235
– torus: 231
euler class: 287
exponential growth: 18
– isoperimetric inequality: 86, 152, 154

$F(A)$: 31
factor, asynchronous: 141
– group: 274–275, **280**
factorization of trivial word: 40–42, 106–107
failure state: 9
– type: 66
Farb, B.: 165
Federer, H.: [Fed69, FF60]
Feighn, M.: [FM91]
fibre bundle: 162–163, 234
filled Cayley graph: 154
final rules for groups: 126
– state: 7
finite group: **49**, 56, 73
finite-index subgroup: 87–**89**, 95, 153, 166, 174, 177, 211, 273, 282
finite state automaton (deterministic): 5, **7**
– – –, generalized: **12**
– – –, non-deterministic: **10**
finitely generated group: 31
– presented group: 31, 108, 110, 113
– – groupoid: 246
fitting pieces together: 297
Federer, H.: [FF60]
flip: 190
formal inverse: 29
free action: 105–107
– abelian group: 30–31, 49, 67
– group: 29, **31**, 36, 49, 68, 81, 211, 280
– groupoid: 245
– product: 8, 274–275, **277**, 279–280
– semigroup: 114
frontier: 289
fundamental group: 76 ff., 162, 166, 273 ff.
– groupoid: 244

$G(T)$: 110
$\Gamma(G, A)$, $\Gamma(H\backslash G, A)$: 34
Garside, F. A.: 194, [Gar69]
generalized finite state automaton: 12
generating function: 18
– graph: 245
generator: **28**, **31**, **243**
geodesic: **56**, **65**, 76
– automation: **56**–57, 61, 66-71, 79-81, 250, 252–253, 272, 296, 302, 311, 313
– hierarchy: 57, 113
– in Seifert complex: 295
– ray: 269
– space: 65
geometrically finite group: 243 ff.
– – – is biautomatic: 266
Geometry Center, Geometry Supercomputer Project: vii
Gersten, S. M.: 59, [BGSS, Ger91, GS00a, GS00b]
Ghys, E.: [GdlH89]
Gilman, R.H.: [Gil87, Gil84b, Gil84a, Gil79]
greedy canonical form: 190-198
Green, P.: [CEG87]
Grigorchuk, R. I.: [Gri80]
Gromov, M.: 52, 71, 79–80, [Gro87]
group, abelian: 58, 61, 80, 87–89 ff., 96, 168; *see also* euclidean
–, alternating: 35
–, asynchronous automatic: *see* asynchronous automation
–, automatic: *see* automation
–, Baumslag–Solitar: 154
–, biautomatic: *see* biautomation
–, braid: 183 ff.
–, combable: **84**–86, **153**–154, 214, 220-221
–, euclidean: 87–97, 117, 275; *see also* abelian
–, finite: **49**, 56, 73
–, finitely generated: 31
–, finitely presented: 31, 108, 110, 113
–, geometrically finite: 243 ff.
–, Heisenberg: 30, 162 ff.
–, Kleinian: 27
–, mapping class: 181, 200

group *(cont'd.)*
–, nilpotent: 30, 33, 161–162, 167 ff., 274
–, small cancellation: 52, 181
–, special linear: 230 ff.
–, symmetric: 34–35, 183–185
–, torsion: 61
–, torsionfree group: 167, 170, 173
–, vertex: 244
–s as languages: 28 ff.
groupoid: 243 ff.
growth function: 18–19

$H_n$: 237
Harpe, P. de la: [GdlH89]
hausdorff distance: 68
– closeness implies uniform closeness: **69**, 71, 251
Hayashi, C.: 101
head: 188 ff.
Heisenberg group: 30, 162 ff.
Hemion, G.: [Hem79]
Hempel, J.: [Hem76]
hexagon: 34
higher-dimensional isoperimetric inequality: 211 ff.
history automaton: 141
HNN extension: 281
Holt, D.: [EHR91, HR00]
homogeneous: 34
homomorphism of labelled graphs: 110
Hopcroft, J. E.: [HU69]
horizontal foliation: 286
horoball, horosphere: 254
horospherical automatic structure: 267
– segment: 263
Hurewicz isomorphism theorem: 216
hyperbolic plane: 38–39, 158–159
– quasigeodesic: 261
– space in the sense of Gromov: 79
– –, closure of: 266

identification: 217
*Illustrator*: vii

image of language: 17
– under pseudomap: 71
improved automatic structure: 56 ff., 308–309
inaccessible states: 9
independence on generators: 52 ff., 151
infinite order: 177-178
– torsion group: 61
inherited: 175
initial state: 7, 10
injectivity radius: 232
integers: 33, 89
intersection: 22
interval: 64
invariance under change of generators: 52 ff., 151
inverse generators: **29**, 31, 34, 58, 66, 108
inversion, closure under: 29
– image of language: 17
$\iota$: 29
irreducible: 117
isomorphism problem: 27, 51, 135
– theorem: 216
isoperimetric function, inequality: 44
– inequality, asynchronous: 135, **152**, 154
– –, exponential: 86, 152, 154
– –, higher-dimensional: 211 ff.
– –, quadratic: **51**, 83, 86, 108, 152–154, 162

Jaco, W.: [JS79]
Johannson, K.: [Joh79]
Joyce, James: 5

$k$-complete: 117, 122, 125
$k$-dimensional Jacobian: 221, 225
$k$-mass: 221
Kapovich, M.: [KP91]
Kleene closure, Kleene star: **6**, 18
–, Rabin, and Scott's theorem: **12**, 23
Kleinian groups: 27

Kneser, H.: 274
Knuth, D. E.: vi, 113, [KB70]
Knuth–Bendix algorithm: 116, **121**–124, 126
Koebe, P.: [Koe27, Koe29]

$L$: 277
$L(M)$: 8, 12
$L$-regular: 174
label: 10, 12
language: 4, 23
– over $(A_1, \ldots, A_n)$: 23
–, regular: 6, 24
LaTeX: vii
lattice: 231 ff.
left-greedy form: 195
length: 4, 12, 39, 64, 186
length order: 56
letter: 4
level: 70
lexicographic order: 56
limit set: 27, 266
lipschitz chain: 221
– constant: 45, 146
– extension: 217–221
– map and cell complexes: 211
– property: 45–**47**, 67, 83, 98, 126, 146
– pseudomap: 72
live state: 9
loop: 52, 109
Loos, R.: [BL82]
Lundell, A.T.: [LW69]

$M$: 191
$M_\epsilon$, $M_x$: 45, 139
many-variable language: 23
Macdonald, I. D.: [Mac68]
mapping class group: 181, 200
Marden, A.: vi–vii
Margulis, G.: [Mar70]
marked point: 294
mass: 221
– times diameter estimate: 154, **215**–217, 219, 230

match: 6, 170
*Mathematica*: vii
Mess, G.: [FM91]
metric, metric space: 64–65.
– properties of cycle: 240
Milnor, J.W.: [Mil65]
minimal automaton: 16
– relation: 96
Minnesota, University of: vi
mixed canonical form: 198
monoid: 4
Morse, M.: [Mor87]
Mosher, L.: [Mos, Mos87]
Mostow, G. D.: [Mos68]
multiplier automaton: 45, 49, 139
Munkres, J.: [Mun66, Mun84]
Myhill–Nerode's theorem: 15

$n$-disk, $n$-sphere: 211
$n$-level: 70
$n$-variable automaton: 24
$n$-variable language: 23
naive algorithm: 108
National Science Foundation: vii
natural language: 4
nearby basepoints: 305
negation: 200
negative curvature: 68, 77-80, 243
nilpotent group: 30, 33, 161–162, 167 ff., 274
– – has polynomial growth: 169
– – is not automatic: 169, 274
– subgroup: 177
non-commuting generators: 183
non-deterministic automaton: 10, 16
non-empty axiom: 102, 105
non-horospherical segments: 263
non-repeating: 186
non-terminating procedure: 112
normal form: 27, 31, 165; *see also* canonical form
normalized automaton: **9**, 12, 16, 46
nullstring: 4

$\omega$: 185
$\Omega$: 188
omitting redundant rule: 123, 125
one-letter automaton: 19
one-variable language: 23
open question: 51, 57–60, 85, 88–89, 119, 135, 201, 204, 209, 220
order: 56, 116, 118
– of generators, dependence on: 53, 59, 97
ordered set: 29

$P$, $P_n$: 186
$P_{(k,\epsilon)}(X)$, $P^*_{(k,\epsilon)}(X)$: 84
padded alphabet, padded extension: 24
– image, – inverse image: 25
– language, restriction, string: 24
padding symbol: 24
Papadopoulos, A.: [CDP90]
parabolic fixed point: 266
parametrization by pathlength: 85
partial automaton: 11
partial Cayley graph: 109
partially homogeneous: 109
parsing: 3
path: 64
– in Cayley graph: 37
– metric, – metric space: 64
– of arrows: 12, 141
Paterson, M.: [PR00]
permutation: 183–185, 187 ff.
$\pi$: 28, 29
picture: 40
piecewise geodesic: 74, 78, 86, 255
politicians: 145
polynomial growth: 18–22, 169
positive braid: 186
positive crossing: 183
Potyagailo, L.: [KP91]
powers and conjugates: 168, 172–173
– of strings: 20
predicate: 22
– over $(A_1, \ldots, A_n)$: 23
predicate *(cont'd.)*
– calculus: 3, 22, 25–26, 59, 129, 176, 248, 252
– –, closure under: 25, 57, 68–69, 276
– –, non-closure under: 137
prefix: **4**, 39, 45, 60, 69
prefix closure: 16, 28, 60, 73
–, common: 149–150
$p$-th prefix: 170
presentation, group: **31**, 40, 51, 97, 108, 110, 114, 162, 183
– by finite state automata: 27
–, semigroup: 114
product, direct: 87–88, 153
–, free: 8, 274–275, **277**, 279–280
properly discontinuous: 77
pseudoisometry: 63, **72**
– of Cayley graph: **75**, 79, 215, 267, 288–289
pseudomap: 71–72
pumping lemma: **17**, 61
punctured torus: 269
quadratic algorithm for word problem: 33, **50**, 105, 129, 181
– isoperimetric inequality: **51**, 83, 86, 108, 152–154, 162
quasiconvex: 174
quasigeodesic: 63, 72, 258–259,
– near geodesics: **77**, 255, 258, 261, 263, 288, 307, 310
– near horospheres: 254, 261, 263
– outside horoballs: 254, 261–263, 265, 270, 272, 307
quotient semigroup: 114

$R_\sigma$: 183
Rabin, M. O.: *see* Kleene, Rabin, Scott
Ramsay, B.: 204
rational growth: 19
– subgroup: 173
Razborov, A.: [PR00]
recognized: *see* accepted
rectifiable: 64
recursive: 32, 118

recursively enumerable: 32
reduced: 31
reduction: 117
redundant: 123
Rees, S. E.: vi, 98, [EHR91, HR00]
regular expression: 5–**6**, 23
– language: 6, 24
– predicate: 22
– structure, subgroup: 174, 177
– transversal: 275–278
$L$-regular: 174
regularly generated: **31**, 66–67
relation: 71, 114, 246; *see also* presentation, relator
relators: 31, 245; *see also* presentation, relation
repeats $N$ times: 170
represented: 28
residue: 117
restriction language: 49, 57
reversal: 15, 58, 184–185, 194, 196
Riemann hypothesis: 32
riemannian isoperimetric inequality: 230, 239
– manifold: 76 ff.; *see also* hyperbolic
right coset: 33
right-greedy form: 193–195
rule: 117
Rumsby, S.: vii, 38–39

$s(v)$: 288
$S_L$, $S_L^\$$, $S_R$, $S_R^\$$, $S^\$$: 143
Schreier generators: 90
Scott, D.: *see* Kleene, Rabin, Scott
Scott, G. P.: [Sco73, Sco83]
Scrabble: 136
Seifert fibre space: 282–283, 296
semigroup: 4
– generators: 28
– presentation: 114
SERC: vii
Series, C. M.: [BS79]
Shalen, P.: [JS79]
Shapiro, M.: [ABC+00, BGSS]
sheet: 157
Short, H.: 59, [ABC+00, BGSS, GS00a, GS00b, Sho90]
shortcut: 123, 125
shortest path: 65
– strings: 30, 56, 66
– – are ShortLex-automatic: 57, 61, 80, **96**
ShortLex automation: 56, 61, 118
– language, order: 56
shuffle: 136
$\sigma$: 234
Simon, L.: [Sim84]
simple loop: 109
simply starred: 19–20, 311
SL($n$, **Z**): 230 ff., 242
slanting cochain: 287
– foliation: 283–284, 286, 293
– section basepoints: 304
small cancellation group: 52, 181
Solitar, D.: 154
solvable group: 166
– problem: **44**, 86, 135
sophic system: v
source: 10, 12
special linear groups: 230
specialized axioms: 115, 132
speed: 65; *see also* complexity
sphere: 211, 215
Squier, C. G.: [Squ87]
standard asynchronous automata: 142, 147
– – multiplier theorem: 143, 145, 147, 148
– automata: 47–48, 109, 111
– – theorem: 47, 109, 112, 139
– slanting section: 293
star central: 170
– closure: 6
–, Kleene: **6**, 18
– of David: 93
– length: 170

start state: 7
state set: 7, 9
string: 4
strongly geodesically automatic: 56
structure component: 302, 304
– graph: 305
– plane, – torus: 303
subgroup: *see* finite-index, regular, torsion
subspace: 72, 85, 253
substitution: 17
substring: 4
success state: 7, 10
suffix: 4
Sullivan, J. M.: vii, 18, 20, 56, 98
symmetric automation: 58, 68, 89, 198–199
– group: 34–35, 183–185
– lattice: 186
– space metric: 235
synchronous: 45

$t_i$: 183
tail: 188
target: 10, 12
$\tau$: 140
$\tau_i$: 183
three-chain: 263
three-dimensional manifolds: 273 ff., 296, **314**
Thurston, N. J.: vii
Thurston, W. P.: [Thu82, Thu88]
tiling of hyperbolic plane: 38–39, 154–155
timing: *see* complexity
Todd–Coxeter algorithm: 109–112
torsion group: 61
torsion subgroup: 167–169, 177
torsionfree group: 167, 170, 173
torus theorem: 274
totally real: 236
transducer: 192
transition function: 7
$x$-transition: 10
transitivity: 104, 106
translation number: 176–177
tree: 156
2-3-7 triangle group: 38–39
triangulation: 222
trivial stabilizer: 105–108
truncated octahedron: 34–35
two-tape automaton: 136
two-variable bounded difference machine: 249

Ullman, J. D.: [HU69]
unambiguous: 109
uniform metric: 45
union: 15, 22
uniqueness property, 29
– – for asynchronous automatic groups: **150**, 153, 169
– – for automatic groups: 28–29, **57**–58, 61, 84, 88, 108, 149
– – for biautomatic groups: 58
University of Minnesota: vi
University of Warwick: vii, 26, 98
Unix: 5
upper central series: 161
upper sheet: 157

van Kampen's theorem: 274
vertex group: 244
– of groupoid: 244
vertical foliation: 283
virtually $P$: 87
visits: 13

Warwick University: vii, 26, 98
weak invariance: 186, 191
weakly geodesically automatic: 56
weakly monotonic: 146
weight: 249
Weingram, S.: [LW69]
White, B.: 211

word: **31**
– acceptor: 45, 139
– – for base: 288 ff.
– difference: 47, 126, 140 ff.
– – machine: 127, 251
– hyperbolic: 52, 58, 61, 63, **79**–80, 113
– length, – metric: 39, 81, 176, 184
– problem: 31; *see also* quadratic, exponential
wordlike: 85
$w_v$: 288

$X_{\mathcal{H}}$: 287

$\mathbf{Z}$: 33, 89
$\mathbf{Z}^2$: 126, 128
$Z_i$: 162
Zwick, Uri: vii, 101, 116

## Jones and Bartlett Books in Mathematics

Baum, R. J., *Philosophy and Mathematics*
ISBN 0-86720-514-2

Eisenbud, D., and Huneke, C., *Free Resolutions in Commutative Algebra and Algebraic Geometry*
ISBN 0-86720-285-8

Epstein, D.B.A., and Gunn, C., *Not Knot Supplement*
ISBN 0-86720-297-1

Geometry Center, University of Minnesota, *Not Knot*
ISBN 0-86720-240-8

Gleason, A., *Fundamentals of Abstract Analysis*
ISBN 0-86720-238-6

Loomis, L.H., and Sternberg, S., *Advanced Calculus*
ISBN 0-86720-122-2

Protter, M.H., and Protter, P.E., *Calculus, Fourth Edition*
ISBN 0-86720-093-6

Redheffer, R., *Differential Equations: Theory and Applications*
ISBN 0-86720-200-9

Ruskai, M.B. *et al.*, *Wavelets and Their Applications*
ISBN 0-86720-225-4

Serre, J.-P., *Topics in Galois Theory*
ISBN 0-86720-210-6

9780867202441